thelightningpress.com

C-UAS/SUAS SMARTBOOK

(C-UAS) Counter-Unmanned Aircraft System & DRONE WARFARE

Planning & Conducting C-UAS/sUAS Operations

The Lightning Press
Norman M. Wade

The Lightning Press

12144 Stonelake Ranch Blvd
Thonotosassa, FL 33592
24-hour Voicemail/Fax/Order: 1-800-997-8827
E-mail: SMARTbooks@TheLightningPress.com

www.TheLightningPress.com

C-UAS/SUAS(1): The Counter-Unmanned Aircraft System & Drone Warfare SMARTbook

Planning & Conducting C-UAS/sUAS Operations

C-UAS/SUAS(1): The Counter-Unmanned Aircraft System & Drone Warfare SMARTbook is new for 2026. At 344-pgs, chapters and topics include the strategic environment (DoD C-UAS, operational environment, multidomain operations, force structure transformation initiative); the unmanned aircraft system (UAS) threat (China, Russia, Iran, North Korea, Hybrid, Homeland); air & missile defense (SHORAD/HIMAD); planning & preparing (roles, AMD, asset protection); layered defense (passive and active defense measures); engagement operations & techniques (ADA & non-dedicated assets, AD warnings & controls); integrating sUAS into operations; fires and targeting (D3A); and equipment (ADA/C-UAS systems and capabilities).

** This is the second printing of C-UAS/SUAS(1), incorporating Addendum 1.* ***(Readers of the first printing (Nov '25) can obtain Addendum 1 for free at: www.TheLightningPress.com/smartupdates)***

ISBN: 978-1-935886-53-2

Notice of Liability

The information in this SMARTbook and quick reference guide is distributed on an "As Is" basis, without warranty. While every precaution has been taken to ensure the reliability and accuracy of all data and contents, neither the author nor The Lightning Press shall have any liability to any person or entity with respect to liability, loss, or damage caused directly or indirectly by the contents of this book. If there is a discrepancy, refer to the source document. This SMARTbook does not contain classified or sensitive information restricted from public release.

"The views presented in this publication are those of the author and do not necessarily represent the views of the Department of Defense or its components."

Photo Credits. Cover image: Army photo by Spc. Luke Sullivan. Other photos courtesy Department of Defense/Army or as credited.

Printed and bound in the United States of America.

C-UAS/SUAS(1) Notes to Reader

Planning & Conducting C-UAS/sUAS Operations

Across the globe—from Ukraine to the U.S. homeland—unmanned systems are transforming warfare by reshaping tactics, compressing innovation cycles, and challenging operational principles. Their affordability and accessibility have democratized precision strike.

Since 2022, small tactical unmanned aircraft system (UAS) and first-person-view (FPV) drones have revolutionized warfare. Unmanned aircraft systems have been fielded in every military service, ranging from handheld micro-UAS to medium-sized tactical systems to full-grown remotely piloted aircraft (RPA). Globally, nations such as China, Russia, North Korea, and Iran have accelerated drone innovation, producing systems rivaling Western models and reshaping the modern battlefield.

Air defense artillery assets provide the backbone of layered counter-UAS (C-UAS) defense, integrating sensors, weapons, and command systems to detect, track, and defeat UAS. Effective C-UAS operations blend Soldier actions, survivability, and systems from handheld EW to advanced defeat technologies. Continuous protection planning aligns capabilities to defend critical areas and maintain mission effectiveness.

Layered, defense-in-depth structures combine passive and active measures to overlap defenses and eliminate gaps. Active defenses deny enemy initiative, while passive measures reduce vulnerability through coordination of sensors, weapons, and systems. Layered defenses increase in density as UAS threats approach protected areas, requiring coordination and co-location of sensors, weapons, and systems to close dead spaces.

Ground and aerial sensors provide airspace surveillance and early warning, feeding vertically and horizontally integrated kill chains. Engagement authorities select the "best shooter" to achieve keep-out ranges and altitudes, coordinating AMD C2 to defeat threats while minimizing fratricide. ADA defends ballistic and cruise missiles, hypersonics, manned aircraft and UAS. Non-ADA units and maneuver forces must supplement defenses with organic weapons and C-sUAS; engagements, lethal or non-lethal, use the best system per ROE, enabled by detection.

sUAS are now employed by maneuver units in large numbers in a wide range of roles, fundamentally altering the conduct of military ground operations. sUAS missions include reconnaissance (SRR, MRR, LRR) with varying altitudes, ranges, and deconfliction needs, and attack/strike (launched effects) including loitering munitions and one-way attacks.

SMARTbooks - the Essentials of Warfighting!

Recognized as a "whole of government" doctrinal reference standard by military, national security and government professionals around the world, SMARTbooks comprise a comprehensive professional library designed for all levels of Service.

SMARTbooks can be used as quick reference guides during actual operations, as study guides at education and professional development courses, and as lesson plans and checklists in support of training. Visit **www.TheLightningPress.com!**

C-UAS/SUAS(1): The Counter-Unmanned Aircraft System & Drone Warfare SMARTbook

C-UAS/SUAS(1): The Counter-Unmanned Aircraft System & Drone Warfare SMARTbook is new for 2026. At 344-pgs, chapters and topics include the strategic environment (DoD C-UAS, operational environment, multidomain operations, force structure transformation initiative); the unmanned aircraft system (UAS) threat (China, Russia, Iran, North Korea, Hybrid, Homeland); air & missile defense (SHORAD/HIMAD); planning & preparing (roles, AMD, asset protection); layered defense (passive and active defense measures); engagement operations & techniques (ADA & non-dedicated assets, AD warnings & controls); integrating sUAS into operations; fires integration & targeting (D3A0; and equipment.

Chap 1: Strategic Environment

From the Middle East to Ukraine and across the globe – including in the U.S. homeland – unmanned systems are reshaping tactics, techniques, and procedures; challenging established operational principles; and condensing military innovation cycles. At the operational level, these systems are making it more difficult for forces to hide, concentrate, communicate, and maneuver. The relatively low-cost, widely available nature of these systems has, in effect, democratized precision strike. Unmanned aerial systems (UAS) have reshaped the character of conflict by enabling persistent sensing, low-risk strike, and novel methods of information and influence. Adversary employment of UAS across the five domains forces joint forces to adapt doctrine, sensors, and defenses. Understanding the physical, information, and human dimensions of the operational environment is essential to anticipate how adversaries employ capabilities—including UAS—to achieve multi-dimensional effects.

Chap 2: Unmanned Aircraft System (UAS) Threat

Since 2022, the rapid spread of small tactical unmanned aerial systems (UAS) and first-person-view (FPV) drones has revolutionized modern warfare, as evidenced in Ukraine. What began as limited experimentation with commercial drones has evolved into a vital component of ground combat operations. In recent decades, unmanned aircraft systems (UAS) have been fielded in every military service, ranging from handheld micro-UAS to medium-sized tactical systems to full-grown remotely piloted aircraft (RPA). Globally, nations such as China, Russia, North Korea, and Iran have accelerated drone innovation, producing systems rivaling Western models and reshaping the modern battlefield.

Chap 3: Air & Missile Defense (AMD)

Integrating air defense artillery (ADA) assets is essential to effective counter–unmanned aircraft system (C-UAS) operations, providing a layered defense-in-depth structure against aerial threats. ADA units, as the Army's air and missile defense (AMD) force, enable coordination of sensors, weapons, and command systems to detect, track, identify, and defeat hostile UAS. Short-range air defense (SHORAD) systems protect maneuvering and fixed forces from low-altitude threats, while high-to-medium altitude air defense (HIMAD) systems counter longer-range air and missile attacks, creating a cohesive, multidomain C-UAS network.

Chap 4: C-UAS Planning & Preparing

Countering unmanned aircraft system (UAS) threats requires an integrated approach combining Soldier actions, survivability measures, and systems ranging from handheld electromagnetic warfare (EW) tools to advanced detection and defeat technologies. Because many units lack specialized C-UAS equipment, efforts often emphasize mitigating UAS effects rather than destroying them. Effective planning synchronizes all warfighting functions within a holistic protection framework, using complementary active and passive defenses to reduce vulnerability, increase survivability, and maintain mission effectiveness against evolving aerial threats. **Protection planning** is a continuous process that must include an understanding of the threats and hazards that may affect operations from the deep area back to the strategic support area. Protection capabilities are aligned to protect critical capabilities, areas, and information and mitigate effects from threats and hazards.

Chap 5: Layered Defense (Passive/Active)

Effective counter–unmanned aircraft system (C-UAS) operations require defense in depth, built through layered capabilities that overlap area and local defenses to eliminate coverage gaps. A layered approach strengthens protection by combining individual awareness, survivability measures, and collective defensive techniques to safeguard personnel and critical assets from UAS threats. Commanders must integrate both **passive and active measures** within C-UAS planning to ensure redundancy, mutual support, and complementary coverage. Active measures deny enemy initiative, while passive measures reduce vulnerability and mitigate effects. Layered defenses increase in density as UAS threats approach protected areas, requiring coordination and co-location of sensors, weapons, and systems to close dead spaces.

Chap 6: Engagement Operations & Techniques (Execution)

Ground and aerial sensors provide airspace surveillance and early warning, feeding vertically and horizontally integrated kill chains. Engagement authorities select the "best shooter" to achieve keep-out ranges and altitudes, coordinating AMD C2 to defeat threats while minimizing fratricide. ADA defends ballistic and cruise missiles, hypersonics, manned aircraft and UAS. Non-ADA units must supplement defenses with organic weapons and C-sUAS; engagements, lethal or non-lethal, use the best system per ROE, enabled by detection. ADA defends ballistic and cruise missiles, hypersonics, manned aircraft and UAS. Non-ADA units must supplement defenses with organic weapons and C-sUAS; engagements, lethal or non-lethal, use the best system per ROE, enabled by detection.

Chap 7: Integrating sUAS into Operations

Over the last decade, OEs worldwide experienced a remarkable surge in the number, capability, and significance of sUAS supporting ground combat formations. sUAS are now employed by units and maneuver forces in huge numbers in a wide range of roles, fundamentally altering the conduct of military ground operations. The U.S. Army is buying one million new drones of all types over the next two to three years to provide real-time reconnaissance, surveillance, and target acquisition capabilities. sUAS missions are reconnaissance (SRR, MRR, LRR) with varying altitudes, ranges, and deconfliction needs, and attack/strike (launched effects) including loitering munitions and one-way attacks.

Chap 8: Fires Integration & Targeting (D3A)

Air & missile defense, counter-unmanned aircraft system (C-UAS), and small unmanned aircraft (sUAS) operations -- including strike coordination and reconnaissance (SCAR) -- are all integrated within the flres warfighting function. Targeting is the process of selecting and prioritizing targets and matching the appropriate response to them, considering operational requirements and capabilities. The D3A methodology -- decide, detect, deliver and asses -- is an integral part of the MDMP. Options may include lethal or nonlethal, organic or supporting assets at all levels, including maneuver, electronic attack, psychological operations, attack aircraft, surface-to-surface fires, air to surface fires, other information-related capabilities, or a combination of these options.

Chap 9: Equipment (ADA/C-UAS Systems & Capabilities)

Air defense artillery (ADA) systems are arrayed throughout a theater. Composite ADA systems are generically referred to as high-to medium-altitude air defense (HIMAD), and singular systems typically in the context of short-range air defense (SHORAD). Countering the threat requires the employment of equipment and systems ranging from organic weapons, to specialized handheld electromagnetic (EW) and lethal devices, to complex systems designed specifically to detect, identify, and defeat UAS.

C-UAS/SUAS(1) References

The following references were used to compile C-UAS/SUAS(1) SMARTbook. All references are considered public domain, available to the general public, and designated as "approved for public release; distribution is unlimited." C-UAS/SUAS(1) SMARTbook does not contain classified or sensitive material restricted from public release.

Field Manuals (FMs)

FM 3-0	Mar 2025	Operations
FM 3-01	Aug 2025	U.S. Army Air and Missile Defense Operations
FM 3-01.44	July 2022	Short-Range Air Defense Operations
FM 3-52	Oct 2016	Airspace Control

Army Techniques Publications (ATPs)

ATP 3-01.81	May 2025	Counter-Unmanned Aircraft System (C-UAS) Operations
ATP 3-01.8	Jul 2016	Techniques for Combined Arms for Air Defense
ATP 3-20.15	Jul 2025	Tank Platoon
ATP 3-60 *	Aug 2023	Army Targeting

Army Doctrinal Publications (ADPs)

ATP 3-19 *	Jul 2019	Fires
ADP 3-37	Jan 2024	Protection
ADP 5-0	Jul 2019	The Operations Process

CRS Reports

IF11797	Apr 2025	The Army's Multi-Domain Task Force (MDTF)
IF12397	Jan 2025	U.S. Army's Maneuver Short-Range Air Defense (M-SHORAD) System
IF12421	Jan 2025	The U.S. Army's Indirect Fire Protection Capability (IFPC) System
IF12668	Nov 2024	The U.S. Army's Small Uncrewed Aircraft Systems
IF11426	May 2022	DoD Counter-Unmanned Aircraft Systems
IN12382	Jun 2024	Proposal to Create a U.S. Army Drone Corps
IN12476	Jan 2025	Drone Encounters Prompt Calls for Restrictions and Other Protections
R47985	Feb 2025	The 2024 Army Force Structure Transformation Initiative
R48477	Mar 2025	DoD Counter Unmanned Aircraft Systems

Additional Resources

JAPCC - A Comprehensive Approach to Countering Unmanned Aircraft Systems

Center for Army Lessons Learned (CALL) #25-14, Small Unmanned Aircraft System (sUAS) Airspace Management and Control handbook.

Army Force Structure Transformation White Paper

Fact Sheet: Department of Defense Strategy for Countering Unmanned Systems

C-UAS/SUAS(1) Table of Contents

Chap 1

Strategic Environment

Chap 2

Unmanned Aircraft System (UAS) Threat

Chap 3

Air & Missile Defense (AMD)

Chap 4

Planning & Preparing (C-UAS Operations)

Chap 5

Layered Defense (Passive/Active)

Chap 6

Engagement Operations (Execution)

Chap 7

Integrating sUAS into Operations

Chap 8

Fires Integration & Targeting (D3A)

Chap 9

Equipment (Systems/Capabilities)

Chap 1

I. DoD Counter-UAS (Overview/Introduction)

Ref: CRS Report R48477, Department of Defense Counter Unmanned Aircraft Systems, Daniel M. Gettinger (Mar '25).

On January 28, 2024, a drone strike on a U.S. base in Jordan killed three U.S. servicemembers and injured more than 40. Department of Defense (DOD, or the Department) officials attributed the attack to Iran-backed militia groups in Iraq and Syria, which in the preceding months had escalated drone and rocket attacks on U.S. forces and allies in the region. In Iraq, U.S. military forces have recovered parts of Iranian-origin Shahed series of one-way attack drones—including the Shahed-101, Shahed-131, and Shahed-136—the largest of which DOD officials have said could be equipped with as much as 50 kilograms (101 pounds) of explosives. Similar drones have been used by the Houthi group in its attacks on commercial shipping in the Red Sea, and by Russia in its strikes on Ukrainian cities and energy infrastructure. Both Russia and Ukraine have reportedly produced growing numbers of drones for battlefield use and equipped front-line forces with first-person view (FPV) one-way attack drones armed with explosives to supplement artillery munitions and conventional missiles.

DOD has identified unmanned air, ground, and sea systems as an "urgent and enduring threat to U.S. personnel, facilities, and assets overseas" and said that drones "pose the most significant threat at this time and increasingly in the U.S. homeland." (Unless otherwise specified, CRS uses the terms unmanned aircraft, uncrewed aircraft, and drone interchangeably.) In response, U.S. policymakers and DOD officials have said that the Department should accelerate its investments in defensive counter-unmanned aircraft system (counter-UAS) capabilities.

I. DOD C-UAS Strategy

DOD published its Strategy for Countering Unmanned Systems in December 2024,superseding two earlier strategies for countering UAS published by the Army and the Department in 2016 and 2020, respectively. The unclassified fact sheet that accompanied the release of the 2024 strategy acknowledges that the "rapid development and proliferation of unmanned systems is changing the character of conflict."In the fact sheet, DOD describes the 2024 strategy as an effort to "unify the Department's approach" to mitigating UAS threats and identified five "strategic ways" by which DOD plans to seek to implement its vision. In the near-term, DOD has stated that it intends to focus on defensive countermeasures "with an emphasis on detection as well as active and passive defense." DOD also has stated that it intends to endeavor to make countering uncrewed systems a "key element of warfighting" and a "key element of our thinking of future force development and design" according to the unclassified fact sheet.

See following pages (pp. 1-2 to 1-3) for an overview and further discussion.

FACT SHEET: Department of Defense Strategy for Countering Unmanned Systems

The rapid development and proliferation of unmanned systems is changing the character of conflict. To meet this challenge, the Department of Defense (DoD) has developed a classified strategy to unify the Department's approach to countering adversary use of these systems that looks across domains, characteristics, missions, and timeframes.

The strategy builds on other major DoD initiatives, including the standup of the Joint Counter-Small UAS Office, the establishment of a Warfighter Senior Integration Group to meet urgent operational needs, and the launch of the Replicator 2 initiative to defend against the threats of small aerial systems at our most critical installations and force concentrations.

The Strategic Environment

Enabled by growing commercial innovation and the increasing sophistication of artificial intelligence (AI), autonomy, and networking technology, unmanned systems are fundamentally changing how militaries of all sizes, capacities, and capabilities – as well as non-state actors – achieve their objectives.

From the Middle East to Ukraine and across the globe – including in the U.S. homeland – unmanned systems are reshaping tactics, techniques, and procedures; challenging established operational principles; and condensing military innovation cycles. At the operational level, these systems are making it more difficult for forces to hide, concentrate, communicate, and maneuver. They allow adversaries to more easily surveil, disrupt, or attack our forces, assets, and installations, potentially without attribution. At the strategic level, unmanned systems provide aggressors with the ability to reduce the initial human, financial, and reputational costs of conflict. The relatively low-cost, widely available nature of these systems has, in effect, democratized precision strike.

Technological advances in the mid- to long-term will likely render unmanned systems increasingly capable, affordable, autonomous, and networked – able to loiter for longer timespans, to communicate better with other systems, move and act as swarms, and to carry larger payloads. These dynamics risk eroding deterrence and creating new and uncertain escalation dynamics.

Our Strategic Approach

The Department is mitigating the potential negative effects of unmanned systems on U.S. forces, assets, and installations – at home and abroad. A critical portion of our efforts, particularly in the near-term, comes from improving our defenses, with an emphasis on detection as well as active and passive defenses. The Department will ensure our forces and priority installations have protection. To stay ahead of advances in unmanned systems – and their growing prevalence – the Department will prepare for more advanced challenges, pacing our future capabilities to more stressing cases (e.g., larger numbers of increasingly capable and autonomous systems). Over the mid- and long-term, the Department will also develop and design our future force to reduce their vulnerability and increase their resilience to these threats. Taken together, these approaches will allow the Department to maintain our advances and our ability to fight and win our Nation's wars, if called upon.

Our Strategic Ways

To mitigate the effects of unmanned systems over the near-, mid-, and long-term, the strategy articulates five "strategic ways," each of which provides specific direction that the Department will pursue in implementation. They are to:

- **Deepen our Understanding and Awareness of Unmanned Systems Trends and Threats**. The Department will "sense and make sense" of threats that unmanned systems pose, including by gaining a greater understanding of unmanned systems threats and by improving the ability of our operational forces to detect, track, and characterize these threats.
- **Disrupt & Degrade Unmanned Systems Threat Networks**. The Department will address the threat networks that drive the development and proliferation of unmanned systems whenever possible, including by launching and executing deliberate campaigns to counter these networks, in partnership with other U.S. departments and agencies.
- **Defend Against Unmanned Systems Threats to U.S. Interests**. The Department will adapt fully to defending against unmanned systems as a core element of warfighting, including by: improving our active and passive defenses; clarifying, streamlining, and delegating authorities, as needed; and institutionalizing approaches across doctrine, organization, training, materiel, leadership, personnel, facilities, and policy.
- D**eliver Solutions with Greater Speed, Adaptability, and Scale**. The Department will deliver robust counter-unmanned systems at speed and scale, including by leveraging rapid acquisition approaches; prioritizing integrated, open, modular solutions; employing systems engineering and predictive analytics; reducing the cost imbalance between unmanned systems and countermeasures; expanded budget agility; increasing experimentation; creating conditions for rapid and realistic testing; and maximizing exportability, codevelopment, and co-production of capabilities with our closest allies and partners.
- **Develop & Design the Future Joint Force for Unmanned Systems-Driven Ways of War**. The Department will make countering unmanned systems a key element of our thinking about future force development and design, including by pursuing changes to our force structure, employing our forces differently, and seeking technologies that could enable us to offset adversary advantages.

Implementation

The Department's ability to successfully execute this strategy requires a campaign mindset – denying adversaries and competitors the benefits of using unmanned systems. We will work in concert with U.S. interagency partners and Congress to align our authorities, resources, and approaches. We will fully incorporate allies and partners in our efforts – leveraging our collective strengths, maximizing interoperability and integration, and promoting responsible norms of behavior surrounding unmanned systems and related technologies. We will work with the defense industry and broader defense innovation ecosystem to address advances in unmanned systems rapidly and at scale. Most of all, in implementing this approach, the Department will draw upon our enduring advantages, including the U.S. military's unrivaled initiative and creativity in adapting to new challenges and the strength of our diverse, dynamic society.

This strategy marks a critical next step in the Department of Defense's efforts to counter unmanned systems, but much work lies ahead. The Department will establish clear metrics and measures of effectiveness to track progress in achieving outcomes. Although the rapidly evolving nature of the threats posed by adversary use of unmanned systems means that the Department will need to continually reassess our efforts, this strategy is a foundation for action to meet this challenge.

II. Overview of Service's Counter-Unmanned Aircraft Systems

Ref: CRS Report IF11426, Department of Defense Counter Unmanned Aircraft Systems, John R. Hoehn and Kelley M. Sayler, (May '22).

Unmanned aircraft systems (UAS), commonly called drones, have proliferated rapidly and are available to nation states and to nonstate actors and individuals. These systems could provide U.S. adversaries with a low-cost means of conducting intelligence, surveillance, and reconnaissance missions against—or attacking—U.S. forces. Furthermore, many smaller UASs cannot be detected by traditional air defense systems due to their size, construction material, and flight altitude. As a result, in FY2023, the Department of Defense (DOD) plans to spend at least $668 million on counter-UAS (C-UAS) research and development and at least $78 million on C-UAS procurement.

C-UAS Technology

C-UAS can employ a number of methods to detect the presence of hostile or unauthorized UAS. The first is using electro-optical, infrared, or acoustic sensors to detect a target by its visual, heat, or sound signatures, respectively. A second method is to use radar systems. However, these methods are not always capable of detecting small UAS due to the limited signatures and size of such UAS. A third method is identifying the wireless signals used to control the UAS, commonly using radio frequency sensors. These methods can be—and often are—combined to provide a more effective, layered detection capability.

Once detected, the UAS may be engaged or disabled. Electronic warfare "jamming" can interfere with a UAS's communications link to its operator. Jamming devices can be as light as 5 to 10 pounds and therefore man-portable, or as heavy as several hundred pounds and in fixed locations or mounted on vehicles. UAS can also be neutralized or destroyed using guns, nets, directed energy, traditional air defense systems, or even trained animals such as eagles. DOD is developing and procuring a number of different C-UAS technologies to try to ensure a robust defensive capability.

Air Force

The Air Force is testing high-powered microwaves and lasers—both forms of directed energy—for C-UAS missions. For example, in October 2019, the Air Force received delivery of a vehicle-mounted C-UAS prototype—the High-Energy Laser Weapon System (HELWS)—that will undergo a year-long overseas field test. HELWS is intended to identify and neutralize hostile or unauthorized UAS in seconds and, when connected to a generator, to provide "a nearly infinite number of shots." As stated in its 2016 Small UAS Flight Plan, the Air Force may additionally pursue airborne C-UAS options, although the status of such efforts is unclear.

Navy

In 2014, the Navy fielded the first—and, to date, only—operational directed-energy weapon, the Laser Weapon System (LaWS), aboard the USS Ponce (LPD-15). LaWS is a 30-kilowatt laser prototype capable of performing a C-UAS mission. The Navy also plans to deploy ODIN, an optical dazzler that interferes with UAS sensors, and HELIOS, a 60-kilowatt laser, aboard the USS Preble (DDG-88) in 2021. Both systems are intended to protect U.S. assets from UAS attacks. In addition, in a March 28, 2019, memorandum, the Department of the Navy announced that it would be partnering with the Defense Digital Service to "rapidly develop new [cyber-enabled] C-UAS products to address the evolving UAS threats."

Marine Corps

The Marine Corps funds a number of C-UAS systems through its Ground Based Air Defense (GBAD) program office. For example, in 2019, the Corps completed overseas tests of the Marine Air Defense Integrated System (MADIS), which employs jamming and guns. The system can be mounted on MRZR all-terrain vehicles, Joint Light Tactical Vehicles, and other platforms (see Figure 2). In July 2019, Marines aboard the USS Boxer (LHD-4) used MADIS to neutralize an Iranian UAS that was deemed to be within "threatening range" of the ship. As part of GBAD, the Marine Corps is also procuring the Compact Laser Weapons System (CLaWS), the first DOD-approved ground-based laser. This system—which reportedly comes in variants of 2-, 5-, and 10-kilowatts—is also in use by the Army.

Army

In July 2016, the Army published a C-UAS strategy to guide the development of its C-UAS capabilities (to date the only service to do so publicly). This was followed in April 2017 by Army Techniques Publication 3-01.81, Counter-Unmanned Aircraft System Techniques, which outlined "planning considerations for defending against low, slow, small [LSS] unmanned air threats during operations," as well as "how to plan for, and incorporate, C-UAS soldier tasks into unit training events." C-UAS is also part of the U.S. Army Combat Capabilities Development Command's six-layer air and missile defense concept, composed of (1) Ballistic, Low-Altitude Drone Engagement (BLADE), (2) Multi-Mission High-Energy Laser (MMHEL), (3) Next-Generation Fires Radar, (4) Maneuver Air Defense Technology (MADT), (5) High-Energy Laser Tactical Vehicle Demonstrator (HEL-TVD), and (6) Low-Cost Extended Range Air Defense (LOWER AD). Although these systems are still in development, the Army has fielded some man-portable, vehicle-mounted, and airborne C-UAS systems. In addition, like the Navy, it has partnered with the Defense Digital Service to develop computer-enabled C-UAS products.

DOD-Wide Developments

DOD is researching and developing a number of C-UAS technologies. For example, the Joint Staff and other DOD agencies have participated in C-UAS efforts such as Black Dart, an exercise intended to "[assess and validate] existing and emerging air and missile defense capabilities and concepts specific to the C-UAS mission set" and "[advocate] for soldiers' desired C-UAS capabilities." In addition, the Defense Advanced Research Projects Agency (DARPA) funds technology development programs for C-UAS such as CounterSwarmAI, which is to "develop systems for anticipating and defeating autonomous systems of the future," and the Multi-Azimuth Defense Fast Intercept Round Engagement System for ship-based point defense.

In December 2019, DOD streamlined the Department's various counter-small UAS (C-sUAS) programs, naming the Army as the executive agent tasked with overseeing all DOD C-sUAS development efforts. On January 6, 2020, the Secretary of Defense approved the implementation plan of the new office, known as the **Joint C-sUAS Office (JCO)**. Working in consultation with the combatant commands and the Office of the Under Secretary of Defense for Acquisition and Sustainment, JCO assessed over 40 fielded C-sUAS systems. To date, it has selected 10 C-sUAS defensive systems and one standardized command and control system for further development.

JCO has produced a Joint Capability Development Document outlining operational requirements for future systems and, in January 2021, released a DOD C-sUAS strategy. JCO is to additionally produce a DOD Directive on C-sUAS and a threat assessment of C-sUAS capabilities. DOD also plans to establish a **Joint C-sUAS academy at Fort Sill, Oklahoma** by FY2024. The academy is to synchronize training on counter-drone tactics across the military services.

See p. 1-6 for discussion of the Joint Counter-Small UAS Office (JCO) and pp. 1-9 to 1-11 for discussion of the Joint Counter-Small UAS University (JCU).

III. Joint Counter-Small UAS Office (JCO)

The Joint Counter-Small UAS Office (JCO) is the lead Army organization for counter-UAS and is tasked with directing DOD's counter-UAS "doctrine, requirements, material, and training" efforts and establishing "joint solutions within a common architecture" to address current and future UAS threats. The Secretary of Defense established the JCO in 2020 under DOD Directive 3800.01E Counter-Small Unmanned Aircraft Systems (C-sUAS) for Unmanned Aircraft Groups 1, 2, and 3, to implement its responsibilities as DOD's executive agent for counter-small UAS, including the evaluation and oversight of material requirements for the services, the coordination of policy and strategy, and the development of training curricula and courses. One of the JCO's immediate tasks following its establishment was to evaluate the suitability of the interim counter-UAS products fielded by military services in the preceding years. In 2020, the JCO evaluated 40 interim counter-UAS systems, of which it selected and endorsed eight for potential acquisition by the military services.

Since then, the JCO, in collaboration with the Army Rapid Capabilities and Critical Technologies Office (RCCTO), has hosted semi-annual, themed exercises to demonstrate and validate material solutions for countering UAS. Based on these demonstrations, the JCO may determine that a given product is not suitable or that it may require more investment to mature, or the military services may select a product for further development or for immediate purchase and fielding. While the JCO's operating statute does not extend to acquiring equipment for the services, using the series of themed exercises, the JCO has expanded its lists of equipment that it recommends to the military services. Additionally, in partnership with the Army Fires Center of Excellence at Fort Sill, the JCO has developed training materials and courses that are made available to members of the joint force. Congress, in Section 353 of the FY2025 NDAA, further directed the Army, through the JCO, to "establish and maintain a threat library, or expand and maintain an existing threat library" of UAS threats, to include suspected threats, proposed counter-UAS solutions, and a "comprehensive list of global incursions" by UAS at DOD installations.

Besides the Army, each of the military services appears to be participating in the JCO in some capacity. Then-Commandant of the Marine Corps David H. Berger said in congressional testimony in 2023 that the Marine Corps is "actively engaged with the JCO" and that it "takes lessons learned after actions from the JCO to inform decision about air defense." The Navy, said then-Chief of Naval Operations Admiral Michael Gilday in 2023, views the JCO's role as imperative "to reduce wasted resourcing and time in pursuit of material solutions and procedural changes" and considers the organization an opportunity to "eventually establish a truly Joint solution that can benefit all Services." Meanwhile, in congressional testimony in 2023, then-Secretary of the Air Force Frank Kendall said that the Air Force works with the JCO, including on the development of high-energy laser and high-powered microwave technology and by leading the JCO's "low collateral effects kinetic interceptor evaluations." None of the military services other than the Army appear to have allocated personnel to fill positions at the JCO.

IV. DOD Establishes Joint Interagency Task Force to Deliver C-sUAS

Ref: https://www.war.gov/News/Releases/Release/Article/4289621/dod-establishes-joint-interagency-task-force-to-deliver-affordable-c-suas-capab/ (accessed 19 Oct 25.)

Secretary Hegseth announced today, actions the Department is taking to address the growing threat of hostile drones and to enhance its small unmanned systems to protect personnel, equipment, and facilities at home and abroad.

To ensure that American airspace remains secure, Secretary Hegseth directed Secretary of the Army to formally establish the Joint Interagency Task Force 401 (JIATF 401) — a new, unified team that will bring together our best talent from across all agencies to counter these threats and restore control of our skies. The Task Force will report to the Deputy Secretary of Defense, to better align authorities and resources to rapidly deliver Joint C-sUAS capabilities to America's warfighters, defeat adversary threats, and promote sovereignty over national airspace.

"We're moving fast — cutting through bureaucracy, consolidating resources, and empowering this task force with the utmost authority to outpace our adversaries," said Secretary Hegseth. "Make no mistake — under this administration and President Trump's leadership, we will innovate, we will lead, and we will win."

The Department of Defense is committed to enhancing its C-sUAS capabilities to protect personnel, equipment, and facilities at home and abroad.

SECRETARY OF DEFENSE
1000 DEFENSE PENTAGON
WASHINGTON, DC 20301-1000

AUG 2 7 2025

MEMORANDUM FOR SENIOR PENTAGON LEADERSHIP
COMMANDERS OF THE COMBATANT COMMANDS
DEFENSE AGENCY AND DOD FIELD ACTIVITY DIRECTORS

SUBJECT: Establishment of Joint Interagency Task Force 401

References: (a) DoD Directive 3800.01E, "DoD Executive Agent for Counter Small Unmanned Aircraft Systems for Unmanned Aircraft Groups 1, 2, and 3," February 21, 2020
(b) Executive Order 14305, "Restoring American Airspace Sovereignty," June 6, 2025
(c) Title 10, United States Code, section 130i, "Protection of certain facilities and assets from unmanned aircraft"
(d) DoD Strategy for Countering Unmanned Systems, December 1, 2024
(e) Secretary of Defense Memorandum, January 17, 2025, "Implementation Plan for the Department of Defense Strategy for Countering Unmanned Systems"

Purpose. To direct the Secretary of the Army (SecArmy) to disestablish the Joint Counter-small Unmanned Aircraft Systems (C-sUAS) Office (JCO) and to establish a Joint Interagency Task Force 401 (JIATF 401) that reports to the Deputy Secretary of Defense, to better align authorities and resources to rapidly deliver Joint C-sUAS capabilities to America's warfighters, defeat adversary threats, and promote sovereignty over national airspace. The JCO disestablishment will take effect upon the establishment of JIATF 401. The JIATF 401 will have a sunset review at 36 months.

Background. DoD Directive (DoDD) 3800.01E designated the SecArmy as the Department of Defense (DoD or Department) Executive Agent (EA) for C-sUAS for Unmanned Aircraft Groups 1, 2, and 3, under the oversight of the Under Secretary of Defense for Acquisition and Sustainment (USD(A&S)), who serves as the principal staff assistant for the DoD EA. The SecArmy established the JCO to lead and synchronize DoD efforts to counter the small unmanned systems threat while minimizing unnecessary duplication and redundancy. The Department has maintained pace with its adversaries in conventional warfighting capabilities. However, the small UAS threat continues to grow exponentially and is becoming increasingly sophisticated. The number of DoD organizations involved in C-sUAS since DoDD 3800.01E was issued in 2020 has increased, and those organizations often are unconnected to each another. DoD needs a single focal point to centralize, coordinate, and lead these efforts.

My priorities for transformation and acquisition reform include improving C-sUAS mobility and affordability and integrating capabilities into warfighter formations. DoD must focus on speed over process by disestablishing the JCO and establishing JIATF 401 with expanded authorities to execute capability development and delivery timelines that outpace the threat. This change is fully aligned with the President's direction to reestablish air

The memo is posted at https://media.defense.gov/2025/Aug/28/2003790021/-1/-1/0/ESTABLISHMENT-OF-JOINT-INTERAGENCY-TASK-FORCE-401.PDF.

V. DOD Counter-UAS Personnel and Force Structure

Career air defense and electronic warfare personnel in the **Army and Marine Corps** are responsible for operating certain counter-UAS systems and short-range air defense systems that conduct counter-UAS.

The Army's Air Defense Artillery (ADA) *(See chap. 3.)*

The Army's Air Defense Artillery (ADA) branch operates the Army's air defense systems, including short-range air defense (SHORAD) and high to medium air defense (HIMAD) systems and counter-rocket, artillery, and mortars (C-RAM) systems. ADA personnel are further responsible for operating certain counter-UAS systems, such as the Low, Slow, Small Integrated Defense System (LIDS). Likewise, the Marine Corps personnel responsible for operating the Marine Air Defense Integrated System (MADIS) counter-UAS platform belong to the Low Altitude Air Defense occupational specialty.

The Army has focused on acquiring systems that are specific to a military occupational specialty (MOS), such as air defense, and systems that are MOS-agnostic, as in systems that allow a broader range of soldiers to employ counter-UAS capabilities regardless of their specialized career field.

A. Army "Force Structure Transformation" *(See p. 1-33.)*

Under the "Force Structure Transformation" initiative, announced in February 2024, the Army intends to create new units dedicated to countering short-range air threats, including UAS. These units include four additional battalions dedicated to operating the Integrated Fire Protection Capability (IFPC) system, the successor to the Army's counter-RAM system, as well as four additional SHORAD battalions. The Army also intends to create nine counter-UAS batteries located within IFPC and divisional air defense battalions. The Army expects to activate the first of the nine planned counter-UAS batteries in FY2029, according to congressional testimony in 2024 by then-Secretary of the Army Christine Wormuth.

Air Defense Artillery (ADA) Force Structure

The challenges posed by UAS proliferation have exposed long-running historical debates over the size, composition, and mission of the Army's air and missile defense forces, specifically those forces for short-range air defense. The Army is in the process of reversing the downsizing of its SHORAD capabilities it undertook in the early 2000s and plans on reintroducing SHORAD battalions to Army divisions and adding units specialized in conducting counter-UAS. As a result, according to Colonel Glenn Henke, commandant of the Air Defense Artillery (ADA) school at Fort Sill, "the ADA branch is undergoing our most significant modernization and growth in arguably the last 40 years."

B. Drone Branch or MOS? *(See p. 1-44.)*

None of the military services have a designated an MOS for counter-UAS. Instead, Army and Marine Corps officials have expressed the view that the responsibility for countering UAS, especially small drones in Groups 1 and 2, is a task shared by all military personnel. In the introduction to Army Techniques Publication (ATP) 3-01.81 Counter-Unmanned Aircraft System (C-UAS), the Army's published doctrine on counter-UAS, the Army describes counter-UAS as "not a stand-alone effort or the sole responsibility of any warfighting function or branch"; rather, counter-UAS is "part of local security and counter reconnaissance missions that is the responsibility of every Soldier and unit." Both the Army and Marine Corps are acquiring counter-

VI. DOD Counter-UAS Training: Joint Counter-Small UAS University (JCU)

Ref: CRS Report R48477, Department of Defense Counter Unmanned Aircraft Systems, Daniel M. Gettinger (Mar '25), pp. 9 to 11.

The Joint Counter-Small UAS University (JCU), part of the Army's Fires Center of Excellence and located at Fort Sill, OK, offers training in counter-UAS equipment and operations to Army, Navy, and Air Force personnel, among others. The Army established the JCU to centralize and institutionalize the Department's training in an effort to ensure a "shared understanding of current and evolving threats among all military services, federal agencies, allied, and partner nations," according to JCU Director Lt. Col. Moseph Sauda. The JCU also offers courses on protecting installations from UAS and contingency planning, all of which appear to be available to military personnel from any career field.

See following pages (pp. 1-10 to 1-11) for an overview and further discussion.

Several efforts are underway to introduce counter-UAS tactics, techniques, and procedures to a wider audience. For instance, the JCU has integrated counter-UAS training into the Basic Combat Training course for new soldiers at Fort Sill, one of the Army's four basic training sites. In addition to the training at Fort Sill, the JCU offers virtual courses through the Joint Knowledge Online (JKO) platform and manages counter-small UAS Mobile Training Teams. In the Marine Corps, the 1st Marine Division offers a course on integrating small UAS and counter-UAS tactics, techniques, and procedures at and below the company level.

Military personnel also gain proficiency with counter-UAS systems and tactics during military exercises, including those that focus specifically on counter-UAS. For example, in an effort to incorporate lessons learned from counter-UAS operations in the Middle East, the Army established in 2023 the Green Sands training initiative at the Army Central Command (ARCENT) Training and Readiness Center. In August 2024, the Army held the Operation Hard Kill to demonstrate various counter-UAS weapons in a live-fire environment. Meanwhile, in October 2024, NORTHCOM conducted the Falcon Peak exercise that, for the first time, sought to evaluate counter-UAS solutions for DOD installations within the United States. Counter-UAS has also been the focus of joint training exercises conducted by the United States and its allies and partners, such as the U.S.-Saudi Arabia Red Sands exercise.

See following pages (1-10 to 1-11) for an overview and further discussion.

Drones 101: Welcome to DOW's Joint Counter-Small UAS University

Ref: Excerpted DoD News article by Katie Lange (Sept. 17, 2025). https://www.war.gov/News/Feature-Stories/Story/Article/4306645/drones-101-welcome-to-dows-joint-counter-small-uas-university/ (accessed 19 Oct 25.)

In the subtropical landscape of Fort Sill, Oklahoma, those who are up early can often hear a constant buzzing coming from part of the installation. But it's not their tinnitus acting up, or a swarm of killer bees. That low droning sound is likely coming from the Joint Counter-Small Unmanned Aircraft Systems University.

The school is the War Department's only joint training center aimed at countering the growing number of aerial threats against U.S. troops and coalition forces. It teaches soldiers, sailors, airmen and Marines how to operate drone equipment, plan counter-small UAS strategies and train others to defend their comrades from harm.

"Coming out of a long period of counterterrorism ... we've really seen the proliferation of drones [that] we call small-UAS ... at low altitudes that have speed and accuracy that we've never seen before," explained Army Lt. Col. John Peterson, JCU director. "That alone has changed the landscape."

The urgent need to understand how to counter these threats is obvious from the conflicts unfolding in Ukraine, the Middle East and the Indo-Pacific region. Enemy drone tactics, techniques and procedures change almost daily, JCU officials said, leaving experts to constantly search for ways to update their defenses.

"Just understanding how easily these drones can be converted for malicious intent — that was something that was very surprising [to] me," said Marine Corps 1st Lt. Jacob Ceglenski, JCU student.

However, air defense is nothing new to the U.S. military. American short-range air defense has been dominant for decades, so despite the new threats that are persistently cropping up, it's nothing the U.S. can't handle.

"For our [counter-small] UAS battle drills, they're really not any different than anything our air defenders have already been doing," said Army Sgt. 1st Class Alan Buhl, JCU instructor. "Everyone else is just now learning entry-level air defense doctrine."

"[We are] making sure that we're sustaining the training for joint partners," Peterson said. "That they have the understanding and they're comfortable using the equipment that they're being provided."

Courses Cover Operation, Planning and Protection

The JCU couples lessons learned in the field with intelligence assessments and other capabilities to give deploying service members the knowledge they need to mitigate ever-changing threats. About 30 experts run the school, which offers three courses that can last from one to three weeks.

The operator's course goes in-depth to teach service members, most of whom are enlisted, how to use counter-small UAS equipment and understand the technicalities. The planner's course teaches officers and staff how to place and employ those assets. Those students gain an understanding of U.S. systems and enemy tactics, then develop defensive plans to protect critical capabilities and assets. The installation protection course teaches post leaders how to sharpen their defenses, including how to tell the difference between potential threats and benign hobbyists' drones.

Each course starts in a joint classroom environment where students from all career fields learn the basics: the types of aircraft they may see downrange, where they come from, how to identify them, and how to employ passive air defense measures, such as aircraft shelters, camouflage to hide assets and dispersion to spread them out.

"Looking up in the air — the third dimension — is not a requirement for most jobs in the military. But now everyone — the infantrymen, the artillerymen, the logistician — all have to look up," explained Army Maj. Dennis "Chip" Stanford, JCU executive officer and a career air defender. "People [have to] adopt the mindset that everything is being observed from multiple angles, and it's realistically a transparent battlespace."

After a few days, the classes are split into service-specific tracks. At the JCU's training range, they're taught how to operate and maintain various systems and pieces of equipment, depending on what their service uses downrange.

Students also gain an understanding of 24 types of commercial off-the-shelf UASs. Many are known as "blue air" drones, which are used by DOW during operations. Others, known as "red air" aircraft, are used by countries perceived as our enemies.

Air Force Capt. Sarah Dahlke, assigned to Air Forces Central Command Force Protection at Al Udeid Air Base in Qatar, took the planner's course in August. Approaching the UAS problem from the intel side, Dahlke often monitors aerial threats from a desk, so it's good for her to see all the various aircraft up close.

"It looks a lot different when you're out here on the ground, especially when you're talking about visually identifying things," she said. "Working in intel, you see all the red [adversary threats], but you don't really get into the intricacies of the blue side. So, for me, that's been the most valuable thing."

All of the students know how important the curriculum is to the current global environment.

"We need to learn, and we need to be able to train people, as well as have knowledgeable planners," said Army Chief Warrant Officer 2 Jilliene Huffman, air and missile defense systems tactician assigned to the 82nd Airborne Division. "[The 82nd] did just stand up the 3rd Battalion, 4th Air Defense Regiment. They're going to be one of the first air battalions under a division that has actual counter-UAS systems ... which is why it's important that I came here."

The Threat Detection Systems

When it comes to students in the operator's course, there are 15 fixed-site systems they may learn, each of which identifies, tracks and mitigates drone threats using a combination of components.

One of the main fixed-site systems students train on is the Fixed-Site Low, Slow, Small Unmanned Aircraft Integrated Defeat System. It includes components to counter electromagnetic attacks, as well as ground-based radar and an electro-optical/infrared camera that can identify potential threats. The FS-LIDS can detect and even shoot down missiles, school officials said — a scenario students practice during simulations.

The simulations operate from trailers on the compound that mimic forward operating base command and control centers. Students learn the roles of each player in the room so they can understand the capabilities and limitations of every portion of the process, from the camera systems to the radar and the Forward Area Air Defense System.

"[FAADS] controls everything that's supposed to happen ... from shooting a rocket to electronic warfare," explained Fred Hill, manager of the JCU operator's course. "Everything the service member does — how he would do it downrange — he's going to do it that exact way right here."

While Army students heavily focus on FS-LIDS, the Navy and Air Force both use a fixed-site system known as a Counter-Remote Control Model Aircraft Integrated Air Defense Network. Other systems taught include the Negation of Improvised Non-State Joint Aerial system, used to counter small UASs; the Drone Restricted Access Using Known Electromagnetic Warfare system; and the Tactical Intelligence Targeting Access Node, a next-generation intelligence ground system.

The culmination of the classes is a joint exercise, during which all the students work together to counter a simulated threat.

UAS equipment suitable to personnel independent their occupational specialization, and have encouraged through doctrine and training the development of passive air defense tactics such as dispersing forces, hardening equipment and shelters, and applying camouflage.

Some commentators have argued that the military services could go further to integrate units specialized in counter-UAS equipment and tactics, perhaps by creating new organizational constructs and communities for countering UAS threats. One commentator has argued that the absence of a dedicated counter-UAS community in the Navy has resulted in an insufficient number of trained personnel to protect domestic military installations and the lack of training opportunities for operationally relevant personnel. Another commentator has argued that "entry-level training for all Marines should incorporate basic C-UAS considerations," in the same way that Marines were trained to counter improvised explosive devices (IEDs) during the Iraq and Afghanistan wars. Yet another commentator has argued that the Army should explore decentralizing counter-UAS batteries so that they are embedded within a greater number of lower-echelon units.

These concerns appear to be reflected to some extent in recent congressional activity and proposed legislation. The House Armed Services Committee, in a report accompany its version of the FY2025 NDAA (H.Rept. 118-529; H.R. 8070), expressed concern that the Army had not developed integrated counter-UAS capabilities for echelons at or below that of Brigade Combat Teams with the same urgency as it had for those at the division-level and above. Meanwhile, a provision of the House-passed version of H.R. 8070 would have established the "Drone Corps" as a basic branch of the U.S. Army, one responsible for small and medium UAS and defensive counter-UAS. Army officials and the Biden Administration opposed the creation of the Drone Corps, arguing that doing so would create additional bureaucracy and that sufficient specialization already existed within the Army. The provision was not adopted in the enacted version of the FY2025 NDAA (P.L. 118-159).

VII. DOD Systems for Countering UAS

Following its establishment in 2020, the JCO endorsed eight interim counter-UAS systems for potential acquisition and use by the military services, some of which have since progressed to programs of record. The selected systems included kinetic and nonkinetic effectors, command-and-control systems, and sensors. In the years since 2020, the JCO has expanded its list of systems it recommends to the military services based on product evaluations it has conducted with the Rapid Capabilities and Critical Technologies Office and the Army's Program Executive Office (PEO) Missiles and Space. Army officials have said that the service expects to continually evaluate and field new counter-UAS systems in order to maintain pace with technological developments. Collectively, with these systems and others, DOD seeks to provide what Army doctrine describes as a "layered" defensive capability against UAS threats. Army officials and some commentators have identified layered defenses against UAS as an imperative for military forces.

See pp. 9-13 to 9-32.

Chap 1

II. (C-UAS) Operational Environment (OE)

Ref: ATP 3-01.81, Counter-Unmanned Aircraft System (C-UAS) Operations (Mar '25), chap. 1.

The exponential growth of UAS creates new risks for the Army. Technology trends are dramatically transforming legitimate applications of UAS while simultaneously making them increasingly capable weapons in the hands of state actors, non-state actors, and criminals. Small UAS may also pose hazards to Army operations across all domains when controlled by negligent or reckless operators. The Army must protect and defend personnel, facilities, and assets in an environment where increasing numbers of UAS will share the skies with standard aircraft, operate in the airspace over Army installations, and be employed by our Nation's adversaries.

Operational Environment (OE)

An operational environment is the aggregate of the conditions, circumstances, and influences that affect the employment of capabilities and bear on the decisions of the commander (JP 3-0). For Army forces, an operational environment includes portions of the land, maritime, air, space, and cyberspace domains understood through three dimensions (human, physical, and information). The land, maritime, air, and space domains are defined by their physical characteristics. Cyberspace, a man-made network of networks, transits and connects the other domains as represented by the dots shown in fig. 1-4.

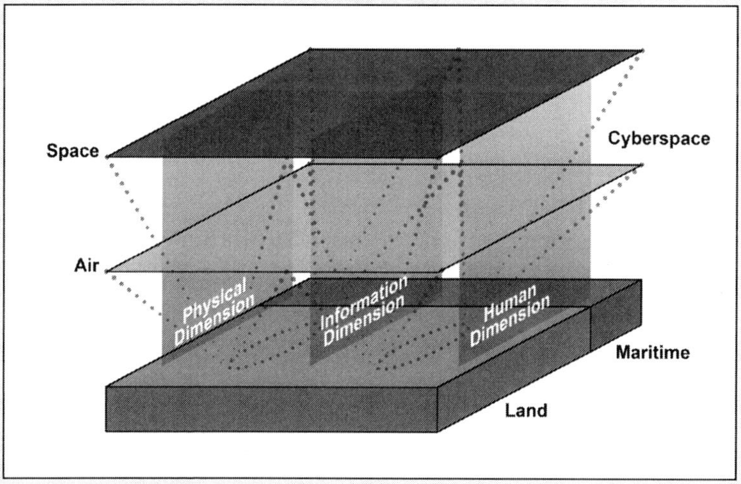

Ref: FM 3-0 (Mar '25), fig. 1-4. Domains and dimensions of an operational environment. See following pages (pp. 1-14 to 1-15) for further discussion.

This section describes the operational environment in which ground units at all levels must plan, coordinate, and synchronize actions to UAS threats. The UAS threat continues to advance technologically and have proliferated exponentially over the past decade. As technology has progressed, both reconnaissance and attack capabilities have matured to the point where enemy UAS represent a significant threat to Army, joint, and multinational partner operations from both state and non-state actors.

Impact of UAS on the Operational Environment

Across the (Five) Domains...

Unmanned Aerial Systems (UAS) have reshaped the character of conflict by enabling persistent sensing, low-risk strike, and novel methods of information and influence. Adversary employment of UAS across the five domains forces joint forces to adapt doctrine, sensors, and defenses. Threat UAS increase the tempo and dispersion of operations in the land and maritime domains, broaden the character of air-domain contestation, create acute dependencies on space services, and expose new cyber vulnerabilities.

Land Domain. On land, adversary UAS complicate close combat by providing near-real-time observation, target acquisition, and small-warfare strike options that support maneuver, ambush, and counter-insurgency operations. Threat actors use both commercially available quadcopters for tactical reconnaissance and weaponized loitering munitions to attack troops and fixed positions.

Threat example (Russian): In the Ukraine conflict, Russian forces have employed swarms of small unmanned systems and loitering munitions to locate and strike Ukrainian maneuver units, interdict rear areas, and saturate local air defenses—forcing ground units to disperse, operate under constant ISR, and prioritize short-range air defense and electronic protection measures.

Maritime Domain. UAS employed by adversaries extend maritime domain awareness, enable remote targeting of ships and offshore infrastructure, and support asymmetric attacks in littoral waters. Small maritime or ship-launched UAS increase the reach of coastal surveillance and enable precision or swarm raids on surface vessels.

Threat example (Chinese): In the South China Sea and Taiwan Strait, Chinese military and paramilitary forces increasingly use maritime ISR UAS to monitor foreign naval activity, track merchant shipping, and support coast-guard and gray-zone harassment operations—compelling regional navies to invest in integrated air-sea sensing and counter-UAS layers to preserve sea control and operational access.

Air Domain. Adversary UAS operate within the air domain as persistent ISR platforms, armed UCAVs (Unmanned Combat Aerial Vehicles), and electronic warfare nodes. Their speed, endurance, and low-risk profile change how air superiority is contested and how commanders plan air operations.

Threat example (Iranian): Iran and Iranian proxies have deployed long-endurance and attack UAS for overland targeting and maritime strikes. Use of such systems to conduct strikes or provide battle damage assessment forces coalition air planners to account for dispersed, low-signature aerial threats and to adapt air defense and counter-UAS employment to deny adversary freedom of maneuver.

Space Domain. While UAS do not operate in space, their functionality depends critically on the space domain—GNSS for navigation, SATCOM for beyond-line-of-sight control, and space-based ISR for cueing. Adversary denial or disruption of space services thus degrades UAS effectiveness and complicates joint operations.

Threat example (North Korean): A conflict in which North Korea attempts GNSS jamming or satellite link interference would limit the precision navigation and remote control of UAS across the peninsula, reducing the range and accuracy of both reconnaissance and weaponized drones and forcing reliance on line-of-sight control or inertial navigation backups.

Cyberspace Domain

UAS are inherently cyber-dependent: command links, telemetry, and payload sensors traverse networks that can be degraded, intercepted, or commandeered. Cyberspace operations therefore can neutralize or repurpose UAS and create cascading effects across other domains.

Threat example (Terrorist groups): Terrorist organizations have exploited commercial UAS and unsecured data links to conduct surveillance, mark targets, and drop explosives in urban environments. Cyber exploitation of poorly secured UAS control systems—through simple spoofing, capture, or data interception—has allowed non-state actors to conduct high-impact attacks while evading attribution and conventional air defenses.

Across the (Three) Dimensions...

Understanding the physical, information, and human dimensions of the operational environment is essential to anticipate how adversaries employ capabilities—including Unmanned Aerial Systems (UAS)—to achieve multi-dimensional effects. Modern threat actors such as Russia, China, Iran, North Korea, and terrorist organizations integrate unmanned systems across these dimensions to gain initiative, shape perception, and erode the will of opponents.

Effective countermeasures require recognizing these interdependencies—protecting electromagnetic signatures, contesting hostile information flows, and sustaining the morale and adaptability of friendly forces.

UAS are thus not merely tools of tactical advantage but instruments that bridge physical action, informational influence, and human will in contemporary warfare.

Physical Dimension

In the physical dimension, UAS provide threat actors with a material means to generate combat power, hold key terrain, and exploit electromagnetic and geographic advantages. Physical dominance often translates into freedom of movement, targeting flexibility, and the ability to impose material losses on the enemy.

Example (Iran): In the Gulf region, Iran's Islamic Revolutionary Guard Corps (IRGC) uses small UAS to surveil and harass naval vessels, supporting asymmetric tactics within the littorals. These drones enable real-time targeting and propaganda footage, magnifying limited physical capabilities into perceived regional dominance.

Information Dimension

The information dimension encompasses the content, flow, and manipulation of information that connects people and systems. Threat actors use UAS not only as reconnaissance or strike platforms but as instruments for information warfare—collecting imagery, shaping narratives, and conducting psychological operations.

Example (Terrorist groups): ISIS and other extremist organizations weaponized commercial drones to film attacks and disseminate videos online, turning small-scale strikes into global propaganda events. By fusing physical attacks with viral information effects, they magnify perceived power, recruit followers, and erode adversary confidence—all while operating on minimal budgets.

Human Dimension

The human dimension reflects decision making, morale, and the will to fight. Adversary UAS usage directly targets human perception, leadership confidence, and societal resilience. By intimidating troops and populations or eroding trust in protection measures, drones generate disproportionate psychological effects.

Example (Russia): Persistent drone surveillance and loitering attacks over Ukrainian trenches contribute to combat fatigue and morale degradation. The psychological burden of constant exposure to unseen aerial threats affects decision cycles, risk tolerance, and unit cohesion—illustrating how the physical employment of drones creates decisive human effects.

II. Unmanned Aircraft Systems (Overview)

Ref: ATP 3-01.81, Counter-Unmanned Aircraft System (C-UAS) Operations (Mar '25), pp. 2-4.

An **unmanned aircraft system (UAS)** is a system whose components include the necessary equipment, network, and personnel to control an unmanned aircraft (JP 3-30). A UAS is controlled remotely through a specific datalink. An unmanned aircraft (UA) is an aircraft that can fly autonomously or remotely. UAS is the totality of everything that makes up a UA platform including its position, navigation, and timing (PNT) module, ground control module, transmission systems, camera, all software, and the operator on the ground controlling the UAS.

UAS Groups

As depicted in the table, UAS are categorized into Groups 1 through Group 5, and this designation is typically based on weight, operating altitude, and speed. The bigger the platform the more robust its suite of capabilities.

Group	*Weight (lbs)*	*Speed (kts)*	*Normal Operating Altitudes*	*Notes*	*Threat & COTS Examples*	*Friendly Examples*
Group 1: micro/mini UAS	0 - 20	<100	< 1,200 AGL	Generally, hand launched, possibly commercial-off-the-shelf, radio-controlled platforms. They have limited ranges and small payload capabilities. They offer real time video. Operated within line of sight of the user.	DJI MAVIC, Enterprise Dual	RQ-11 Raven
Group 2: small tactical UAS	21 - 55	101 - 250	< 3,500 AGL	Small airframes with low radar cross sections provide medium range and endurance. Launched from unimproved areas with a small number of people involved. Typically line of sight to the ground control station.	SKY-09Ps	Scan Eagle
Group 3: Tactical UAS	56 – 1,320		< 18,000 AGL	Similar to Group 1 and 2 UAS, requires a larger logistical footprint. Range and endurance vary significantly among platforms.	Shahed 136	RQ-7B Shadow
Group 4: Strategic or theater UAS	> 1,320	Any speed	< AGL 18,000	Relatively large systems operated at medium to high altitudes. This group has extended range and endurance capabilities. Normally requires a runway for launch and recovery.	Forpost	MQ-1C Gray Eagle MQ-1A/B Predator
Group 5: Strategic UAS	> 1,320	Any speed	> AGL 18,000	Operates at medium to high altitudes having the greatest range, endurance, and airspeed. Requires large logistical footprint like that of manned aircraft and has a suite of optics for targeting and weaponry for engagements.	Wing Loong II	RQ-4 Global Hawk MQ-9 Reaper

Ref: ATP 3-01.81 (Mar '25), table 1-1. Unmanned aircraft system groups.

ATP 3-01.81 focuses on Groups 1 through 3. While the group designations can clarify UAS and their capabilities, it is important that commanders and staff understand that lines of differentiation between the groups operationally are not rigid. Both passive and active defense efforts should include capabilities that provide overlapping, redundant protection against all UAS threats.

Group 1 UAS consist mostly of hand-launched radio-controlled platforms. Due to construction, range, cost, and operation, these UAS may be operated using launch and forget tactics. These tactics allow one operator to conduct limited ISR gathering operations.

Adding weapons to these platforms is possible and can provide significant effects for the adversary. Additional weight, however, will change the range of the aircraft, significantly reducing its effectiveness. These Group 1 platforms are most capable of ISR gathering tasks.

Group 2 and 3 UAS are larger than micro and mini UAS. They are relatively small airframes constructed of lightweight materials. The increased size, range, and load bearing capacity of these larger UAS enable a wide range of mission and payload capabilities. This increases threat capability well beyond ISR operations.

Loitering munition is a type of UAS designed to engage beyond line-of-sight ground targets with an explosive warhead. Loitering munitions are often portable and meant to provide ground units such as infantry with a guided precision munition. They are equipped with high-resolution electro-optical and infrared cameras that enable the operator to locate, surveil, and guide the vehicle to the target. A defining characteristic of loitering munitions is the ability to "loiter" in an area of airspace for an extended period before striking, giving the operator time to decide when and what to strike.

Group 1 thru 3 UAS have a lower radar cross-section larger than Groups 4 and 5 UAS. This low radar cross-section is harder to detect by friendly forces' early warning and detection capabilities. Launch and recovery requirements allow UAS launching capability from unimproved areas by a single person or small team, making them more difficult to find. Advances in artificial intelligence capabilities enable threat actors to employ UAS with little to no operator interaction. Artificial intelligence capability is also increasing the threat of swarm attacks.

UAS Types

There are three types of UAS: rotary-wing/vertical takeoff and landing, fixed-wing, and hybrid. Rotary-wings UAS operate like helicopters, so their greatest advantages are their vertical flight capability, which facilitates takeoffs and landings, and their hovering capability, which allows them to fly in narrow spaces and around obstacles. A fixed-wing UAS can fly at a higher altitude than other UAS. It will typically have greater ranges and is more efficient than its rotary-wing counterparts are and is thus often better suited for surveying larger plots of ground or geography on a single flight. A hybrid UAS is a combination of a fixed and rotary-wing UAS that can hover like a rotary-wing and fly like a fixed-wing. The advantages are they can conduct vertical take-off in addition to flying like a traditional fixed-wing aircraft.

Type	*Capabilities*	*Limitations*
Fixed wing	• Increased flight time. • Larger payload capacity. • Greater speed.	• May have increased operational and sustainment requirements. • Needs room for takeoff and landing. • Must constantly move forward, not able to hover.
Rotary-wing	• Vertical take-off and landing capable, which means they can make off from almost anywhere. • Ability to hover and stare capability. • Requires little training. • Low cost easy to acquire and conceal.	• Low load carrying capacity. • Limited speed.
Hybrid	• Includes advantages of fixed wing and VTOL unmanned aircraft.	• Includes disadvantages of fixed wing and VTOL unmanned aircraft.

Ref: ATP 3-01.81 (Mar '25), table 1-2. Capabilities and limitations by unmanned aircraft ***type****.*

III. (C-UAS) Operational Environment (OE)

Ref: ATP 3-01.81, Counter-Unmanned Aircraft System (C-UAS) Operations (Mar '25), pp. 1 to 2.

The **operational environment** is the aggregate of the conditions, circumstances, and influences that affect the employment of capabilities and bear on the decisions of the commander (JP 3-0). The operational environment (OE) includes enemy, friendly, and neutral elements. This environment must be understood across the entire competition continuum, from peace to war, and across the entire range of military operations, from military engagement, security cooperation or deterrence mission to multi-domain operations *(refer to FM 3-0 for further discussion on OE).*

Commanders and planners must understand the physical environment, governance, technology, local resources, and culture of elements within the area of operations (AO). They must then translate this understanding into actionable mission orders to their units. Defining and understanding the OE enables the staff to focus their planning efforts on defending against low-level air threats and assists the unit in creating an effective C-UAS strategy.

Commanders and staffs require information that allows them to understand and influence the environment where their forces operate. They use this information as a basis to fill information requirements, complete thorough assessments, develop and refine plans, and conduct operations. **Specific to C-UAS, commanders and staffs must apply methods that identify the effects UAS have within the OE on friendly and enemy courses of action to include:**

- Determining likely avenues of approach.
- Identifying potential target areas.
- Estimating threat UAS standoff ranges.
- Employing layered C-UAS capabilities based on OE factors.

Rapidly advancing technology and ready availability of UAS by state and non-state actors has resulted in a significant increase in UAS threats against U.S. forces and our allies. Threat UAS information is critical and likely differs between adversaries. Commanders at all levels face the challenge of being able to counter the air threat and adversarial reconnaissance within their OE. Countering air threats and UAS threats are shared joint and combined arms responsibilities. Commanders and staffs must be prepared to address these across the entire competition continuum.

Leaders and Soldiers must understand how UAS operate and where they are vulnerable in order to plan a defense. A UAS has three main components: the sensor, the controller, and the communication links. Each component has its own weakness. The sensor is vulnerable to Global Positioning System [US system], GLONASS [Russian system], Galileo [EU system], and BeiDou [Chinese system] signals. The controller to link protocols and the communications link are vulnerable to video and telemetry downlink. See figure 1-2, for visual representation of the communication path for a UAS.

- The **sensor** (UAS) relies on a commercial position, navigation, and timing PNT (L1 channel) signal for stabilization. The PNT down link is needed to establish which way is up or down and its altitude. It is required if the UAS needs to fly to a specific point if a C2 signal is not being utilized.
- The **controller** is not always required to operate the UAS. The C2 uplink signal can be used to geo-locate the actual operator of the UAS.
- The **video and telemetry downlink signals** from the UAS to the controller are one directional and these signals can be used to geo-locate the actual UAS. The signal can be intercepted using cyber effects to locate the UAS reporting location and possibly its video feed.

- **Pilot in the loop operations** maintains C2 via line of sight (LoS) and beyond line of sight (BLoS) methods. **Autonomous operations** do not need a data link, can be programmed from launch to recovery

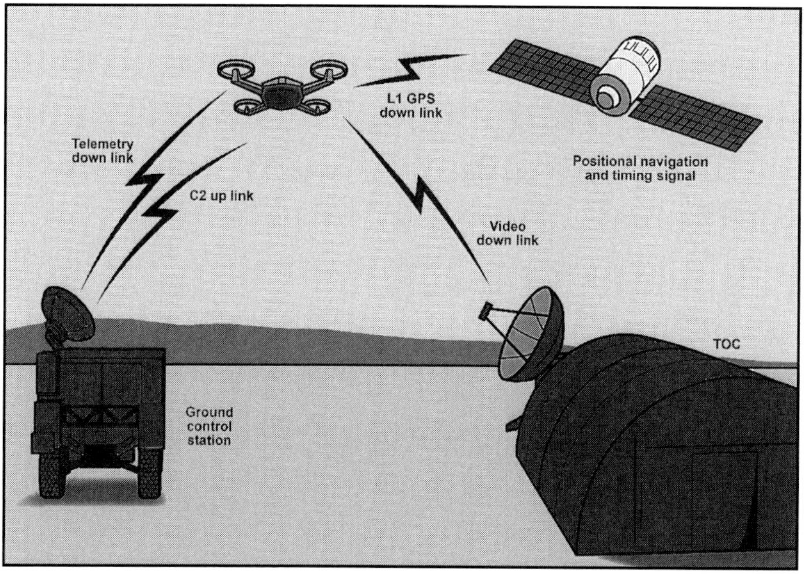

Ref: ATP 3-01.81 (Mar '25), fig. 1-2. Unmanned Aircraft System communication link.

Multiple devices we use operate on various waveform frequencies. When jamming, you can jam across one or more frequencies. Operators and leaders must understand what frequencies are being jammed and what other equipment may be impacted as a second or third order effect. A UAS relies on PNT signal for stabilization, lost link protocols, and video downlink/telemetry; UAS video downlink presents a weakness that may be exploited by friendly forces. Leaders must understand the electromagnetic frequency and the factors that influence it for planning and how to mitigate the risk of potentially jamming friendly equipment.

Factors that may influence the range of signals are-

- Antenna
- Receiver noise
- Output power
- Radio pollution
- Environment
  - Line of Sight
  - Obstructions
  - Reflections
  - Atmospheric effects

Refer to AODS7-1: The Army Operations & Doctrine SMARTbook, w/SMARTupdate 1 (Multidomain Operations), pp. 1-17 to 1-22. An ***operational environment*** *is the totality of factors that affect what occurs in an assigned area. The complex and dynamic nature of an operational environment makes determining the relationship between cause and effect challenging, and it contributes to the uncertain nature of war and human competition.*

IV. (C-UAS) Operational View

Ref: ATP 3-01.81, Counter-Unmanned Aircraft System (C-UAS) Operations (Mar '25), pp. 1 to 2.

Figure 1-1 depicts a notional C-UAS operational view.

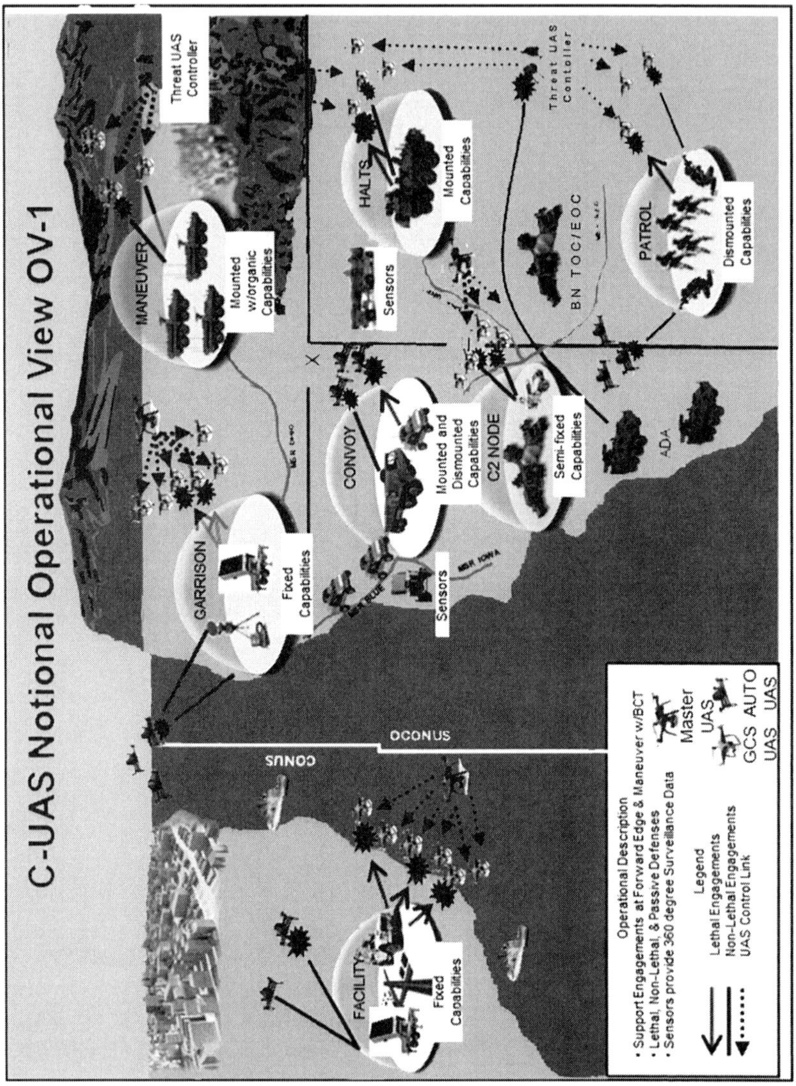

Ref: ATP 3-01.81 (Mar '25), fig. 1-1. Notional C-UAS operational view

See pp. 6-2 to 6-3 for an expanded discussion of a C-UAS "operational concept."

III. Multidomain Operations

Ref: FM 3-0, Operations (Mar '25).

The Army's primary mission is to organize, train, and equip its forces to conduct prompt and sustained land combat to defeat enemy ground forces and seize, occupy, and defend land areas. It supports four strategic roles for the joint force. Army forces shape operational environments, counter aggression on land during crisis, prevail during large-scale ground combat, and consolidate gains.

Army forces meet a diverse array of challenges and contribute to national objectives across a wide range of operational categories, including large-scale combat operations, limited contingency operations, crisis response, and support to security cooperation.

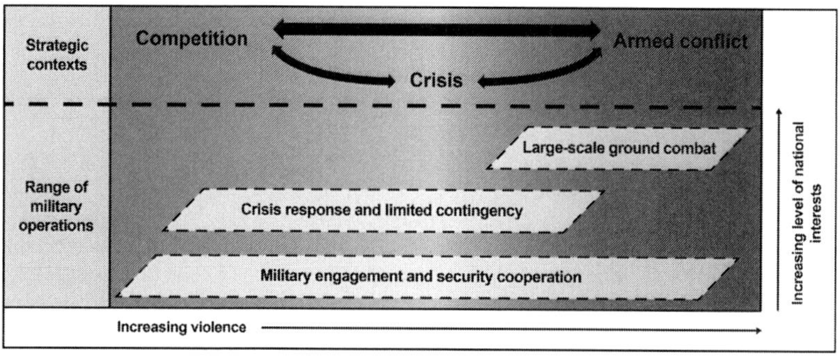

Ref: FM 3-0 (Mar '25), fig. 1-3. Army strategic contexts and operational categories.

I. Multidomain Operations

Multidomain operations are the combined arms employment of joint and Army capabilities to create and exploit relative advantages that achieve objectives, defeat enemy forces, and consolidate gains on behalf of joint force commanders. Employing Army and joint capabilities makes use of all available combat power from each domain to accomplish missions at least cost. Multidomain operations are the Army's contribution to joint campaigns, spanning the competition continuum.

Below the threshold of armed conflict, multidomain operations are how Army forces accrue advantages and demonstrate readiness for conflict, deterring adversaries while assuring allies and partners. During conflict, they are how Army forces close with and destroy the enemy, defeat enemy formations, seize critical terrain, and control populations and resources to deliver sustainable political outcomes.

Refer to AODS7-1: The Army Operations & Doctrine SMARTbook, w/SMARTupdate 1 (Multidomain Operations). Completely updated with the 2025 edition of FM 3-0, AODS7 focuses on Multidomain Operations and features rescoped chapters on generating and applying combat power: command & control (ADP 6-0), movement and maneuver (ADPs 3-90, 3-07, 3-28, 3-05), intelligence (ADP 2-0), fires (ADP 3-19), sustainment (ADP 4-0), & protection (ADP 3-37).

Multidomain operations are the combined arms employment of joint and Army capabilities to create and exploit relative advantages that achieve objectives, defeat enemy forces, and consolidate gains on behalf of joint force commanders. Employing Army and joint capabilities makes use of all available combat power from each domain to accomplish missions at least cost. Multidomain operations are the Army's contribution to joint campaigns, spanning the competition continuum.

Army forces conduct operations in support of joint campaigns which for the most part occur as part of a larger coalition operation. Leaders must understand the interdependencies between their own assigned forces and the forces or capabilities provided by others to generate the complementary and reinforcing effects of combined arms approaches. Army forces employ joint and other unified action partner capabilities to the degree they are available. However, because peer threats can contest the force in all domains, Army forces must be prepared to conduct operations when some or all joint capabilities are unavailable to support mission accomplishment.

All operations are multidomain operations. Army forces employ organic capabilities in multiple domains, and they continuously benefit from air and maritime strategic transportation and space and cyberspace capabilities that they do not control, including global positioning, satellite communications, and intelligence, surveillance, and reconnaissance (ISR). Lower echelons may not always notice the opportunities created by higher echelons or other forces that operate primarily in other domains; however, leaders must understand how the absence of those opportunities affects their concepts of operations, decision making, and risk assessment.

During operations, small advantages can have significant impacts on the outcome of the mission, particularly when they accrue over time. Creating and exploiting relative advantages are therefore necessary for all operations, and they become even more critical when opposing sides are evenly matched. A relative advantage is a location or condition, in any domain, relative to an adversary or enemy that provides an opportunity to progress towards or achieve an objective. Commanders seek and create relative advantages to exploit through action, and they continually assess the situation to identify ways to expand opportunities.

Army leaders are accustomed to creating and exploiting relative advantages through the combined-arms approach that traditionally focuses on capabilities from the land, air, and maritime domains. The proliferation of space and cyberspace capabilities further requires leaders who understand the advantages those capabilities create in their operational environment. The ability to integrate and synchronize space and cyberspace capabilities at the most effective tactical echelon expands options for creating advantages to exploit.

Multidomain operations fracture the coherence of threat operational approaches by destroying, dislocating, isolating, and disintegrating their interdependent systems and formations, and exploiting the opportunities these disruptions provide to defeat enemy forces in detail. Army forces therefore require timely, accurate, relevant, and predictive intelligence to understand threat characteristics, capabilities, objectives, and courses of action. Intelligence initially drives what combinations of defeat mechanisms commanders pursue as they employ the capabilities of their forces in space and time against enemy forces. Army forces combine maneuver and targeting methods to defeat enemy formations and systems. Army forces employ maneuver to close with and destroy enemy formations in close operations. Targeting generally sets priorities for information collection, fires, and other key capabilities to disintegrate enemy networks and systems. Leaders execute the targeting process to create advantages that enable freedom of maneuver and exploit the positional advantages created by maneuver. Targeting is a key way for leaders to integrate the joint capabilities required to create depth in the battlefield and protect friendly formations.

II. Tenets & Imperatives (of Operations)

Ref: FM 3-0, Operations (Mar '25).

Uncrewed Aerial Systems (UAS) have a direct impact on how the Army applies the tenets and imperatives of operations in modern warfare. Their agility and rapid deployability enable commanders to see themselves and the enemy more clearly, maintaining awareness in an environment of constant observation and contact. UAS extend operational depth by projecting reconnaissance, precision fires, and logistical support far beyond traditional lines, while their persistence underpins endurance in prolonged operations. They create convergence across domains—linking sensors, shooters, and decision-makers—to impose multiple dilemmas on the enemy and achieve decision dominance. By allowing forces to make initial contact with the smallest elements, UAS reduce risk, preserve combat power, and enhance adaptability.

Tenets

The tenets of operations are desirable attributes that should be built into all plans and operations, and they are directly related to how the Army's operational concept should be employed. Commanders use the tenets of operations to inform and assess courses of action throughout the operations process. The tenets of operations are—

- Agility
- Convergence
- Endurance
- Depth

The Army provides forces capable of transitioning to combat operations, fighting for information, producing intelligence, adapting to unforeseen circumstances, and defeating enemy forces. Army forces employ capabilities from multiple domains in a **combined arms approach** that creates **complementary and reinforcing effects** through **multiple domains**, while preserving **combat power** to maintain options for the joint force commander (JFC). **Creating and exploiting relative advantages** require Army forces to operate with endurance and in depth. Endurance enables the ability to absorb the enemy's attacks and press the fight over the time and space necessary to accomplish the mission. Depth applies combat power throughout the enemy's formations and the operational environment, securing successive operational objectives and consolidating gains for the joint force.

Imperatives *(See p. 2-12.)*

Imperatives are actions Army forces must take to defeat enemy forces and achieve objectives at acceptable cost. They are informed by the operational environment and the characteristics of the most capable threats Army forces can encounter. Imperatives include—

- See yourself, see the enemy, and understand the operational environment
- Protect against constant observation and all forms of enemy contact
- Own the electromagnetic spectrum
- Create and exploit relative physical, information, and human advantages
- Make initial contact with sensors, unmanned systems, or the smallest element possible
- Impose multiple dilemmas on the enemy
- Anticipate, plan, and execute transitions
- Designate, weight, and sustain the main effort
- Consolidate gains continuously
- Understand and manage the effects of operations on units and Soldiers

III. Combat Power

Ref: FM 3-0, Operations (Oct. '22), chap. 2.

Combat power is the total means of destructive and disruptive force that a military unit/ formation can apply against an enemy at a given time (JP 3-0). It is the ability to fight. The complementary and reinforcing effects that result from synchronized operations yield a powerful blow that overwhelms enemy forces and creates friendly momentum.

The complementary and reinforcing effects that result from synchronized operations yield a powerful blow that overwhelms enemy forces and creates friendly momentum. **Army forces deliver that blow through a combination of five dynamics: leadership, information, survivability, firepower, and mobility.**

Dynamics of Combat Power

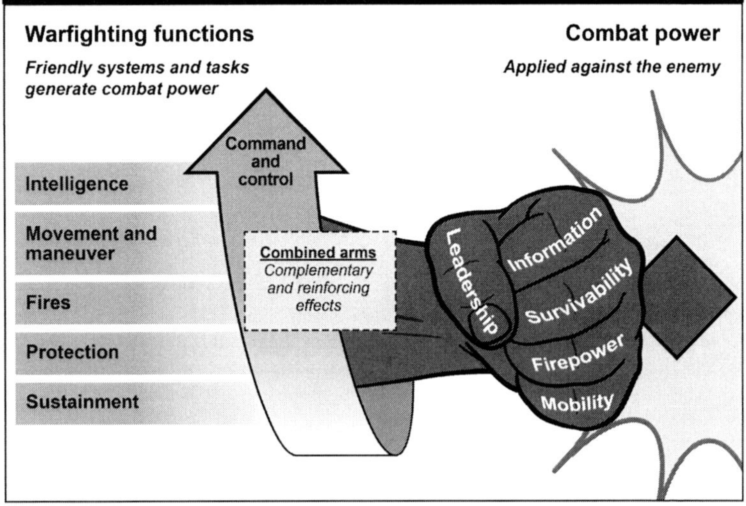

All **warfighting functions** contribute to generating and applying combat power. Well sustained units able to move and maneuver bring combat power to bear against the opponent. Joint and Army indirect fires complement and reinforce organic firepower in maneuver units. Survivability is a function of protection tasks, the protection inherent to Army platforms, and schemes of maneuver that focus friendly strengths against enemy weaknesses. Intelligence determines how and where to best apply combat power against enemy weaknesses. C2 enables leadership, the most important qualitative aspect of combat power.

See following pages for an overview and discussion of these dynamics which may offer insight into planning and conducting C-UAS / sUAS operations: information (pp. 1-26 to 1-27), mobility (pp. 1-28 to 1-29), and survivability (pp. 1-30 to 1-31).

Refer to AODS7-1: The Army Operations & Doctrine SMARTbook, w/SMARTupdate 1 (Multidomain Operations). Completely updated with the 2025 edition of FM 3-0, AODS7 focuses on Multidomain Operations and features rescoped chapters on generating and applying combat power: command & control (ADP 6-0), movement and maneuver (ADPs 3-90, 3-07, 3-28, 3-05), intelligence (ADP 2-0), fires (ADP 3-19), sustainment (ADP 4-0), & protection (ADP 3-37).

The Six Warfighting Functions *(See p. 5-7.)*

1. Command and Control

The command and control warfighting function is the related tasks and a system that enable commanders to synchronize and converge all elements of combat power. The primary purpose of the command and control warfighting function is to assist commanders in integrating the other elements of combat power (leadership, information, movement and maneuver, intelligence, fires, sustainment, and protection) to achieve objectives and accomplish missions.

Refer to AODS7-1, chap. 3 for further discussion.

2. Movement and Maneuver

The movement and maneuver warfighting function is the related tasks and systems that move and employ forces to achieve a position of relative advantage over the enemy and other threats. Direct fire and close combat are inherent in maneuver. The movement and maneuver warfighting function includes tasks associated with force projection. Movement is necessary to disperse and displace the force as a whole or in part when maneuvering. Maneuver directly gains or exploits positions of relative advantage. Commanders use maneuver for massing effects to achieve surprise, shock, and momentum

Refer to AODS7-1, chap. 4 for further discussion.

3. Intelligence

The intelligence warfighting function is the related tasks and systems that facilitate understanding the enemy, terrain, weather, civil considerations, and other significant aspects of the operational environment. Other significant aspects of an operational environment include threats, adversaries, and operational variables, which vary with the nature of operations. The intelligence warfighting function synchronizes information collection with primary tactical tasks of reconnaissance, surveillance, security, and intelligence operations. Intelligence is driven by commanders, and it involves analyzing information from all sources and conducting operations to develop the situation.

Refer to AODS7-1, chap. 5 for further discussion.

4. Fires

The fires warfighting function is the related tasks and systems that create and converge effects in all domains against the adversary or enemy to enable operations across the range of military operations (ADP 3-0). These tasks and systems create lethal and nonlethal effects delivered from both Army and joint forces, as well as other unified action partners.

Refer to AODS7-1, chap. 6 for further discussion.

5. Sustainment

The sustainment warfighting function is the related tasks and systems that provide support and services to ensure freedom of action, extended operational reach, and prolong endurance. Sustainment determines the depth and duration of Army operations. Successful sustainment enables freedom of action by increasing the number of options available to the commander. Sustainment is essential for retaining and exploiting the initiative.

Refer to AODS7-1, chap. 7 for further discussion.

6. Protection

The protection warfighting function is the related tasks and systems that preserve the force so the commander can apply maximum combat power to accomplish the mission. Commanders incorporate protection when they understand and visualize threats and hazards in an operational environment. This allows them to synchronize and integrate all protection capabilities to safeguard bases, secure routes, and protect forces. Preserving the force includes protecting personnel (combatants and noncombatants) and physical assets of the United States, unified action partners, and host nations.

Refer to AODS7-1, chap. 8 for further discussion.

A. Information (as a Dynamic of Combat Power)

All military activities produce information. Informational aspects are the features and details of military activities observers interpret and use to assign meaning and gain understanding. Those aspects affect the perceptions and attitudes that drive behavior and decision making. The JFC leverages informational aspects of military activities to gain an advantage; failing to leverage those aspects may cede this advantage to others. Leveraging the informational aspects of military activities ultimately affects strategic outcomes.

Information Advantage*

An information advantage is a condition when a force holds the initiative in terms of situational understanding, decision making, and relevant actor behavior. There are several forms of information advantage.

An information advantage can result from and exploit human and physical advantages or enable those advantages. Like human and physical advantages, information advantages are often temporary and change over time relative to the threat and changes in an OE. While friendly forces are seeking information advantages, threat forces are doing the same. As such, an information advantage is something to gain, protect, and exploit across as many domains as possible.

Threat UAS have significantly altered the information environment by challenging the Army's ability to gain and maintain information advantage across all domains. Their persistent surveillance, real-time data transmission, and psychological impact allow adversaries to "see" friendly forces continuously, eroding operational security and exposing friendly activities to observation, targeting, and exploitation. Their presence forces friendly forces to operate under near-constant observation, complicating concealment, maneuver, and command and control. In the modern information fight, threat UAS have become key tools that compress decision cycles, contest informational space, and undermine the Army's ability to understand, decide, and act faster than the adversary—making counter-UAS and information resilience essential to sustaining advantage.

Information Advantage (Activities)

I **Enable** *(INFO2, pp. 1-21 to 1-28.)*

II **Protect** *(INFO2, pp. 1-29. to 1-34.)*

III **Inform** *(INFO2, pp. 1-35 to 1-44.)*

IV **Influence** *(INFO2, pp. 1-45 to 1-48.)*

V **Attack** *(INFO2, pp. 1-49 to 1-56.)*

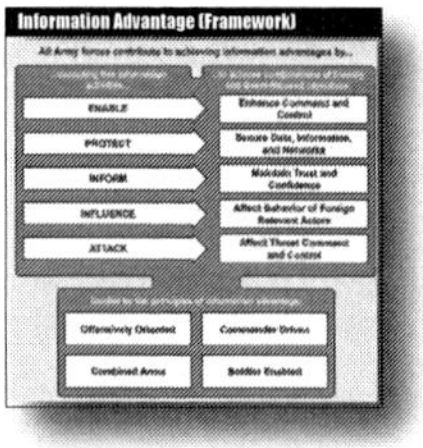

The **information advantage framework** presents a framework for creating and exploiting information advantages. Within this framework, Army forces integrate all relevant military capabilities through the execution of five information activities -- enable, protect, inform, influence, and attack.

An information advantage can result from and exploit human and physical advantages or enable those advantages. Like human and physical advantages, information advantages are often temporary and change over time relative to the threat and changes in an OE. While friendly forces are seeking information advantages, threat forces are doing the same. As such, an information advantage is something to gain, protect, and exploit across as many domains as possible.

Advancements in information technologies and increased global connectivity continue to shape how we interact with each other and how forces fight. These advancements accelerate and expand the ability of joint and Army forces to collect, process, analyze, store, and communicate information at a scale previously unimaginable.

ADP 3-13, Information (Nov '23), is the Army's first publication dedicated to information. It provides a framework for creating and exploiting information advantages during the conduct of operations and at home station. It represents an evolution in how Army forces think about the military uses of data and information, emphasizing that everything Army forces do, to include the information and images it creates, generates effects that contribute to or hinder achieving objectives. **As such, creating and exploiting information advantages is the business of all commanders, leaders, and Soldiers.**

ADP 3-13 operationalizes the two big ideas inherent in multidomain operations—combined arms and positions of relative human, information, and physical advantage. **We no longer regard information as a separate consideration or the sole purview of technical specialists. Instead, we view information as a resource that is integrated into operations with all available capabilities in a combined arms approach** to enable command and control; protect data, information, and networks; inform audiences; influence threats and foreign relevant actors; and attack the threat's ability to exercise command and control.

*** Editor's Note on Changes in Information Terminology**

Based on changes to joint information doctrine, **Army forces will no longer use the terms** *information operations, information-related capabilities, or information superiority*.

A significant joint doctrinal change is the transition from *joint information operations (IO)* to *operations in the information environment (OIE)*. Joint doctrine, however, retains the term *information environment*. The Army's new model of an operational environment established in FM 3-0 no longer includes an information environment. The term *informational considerations* aligns with the joint term *information environment*.

The Army is currently revising all its doctrine, to include FM 3-13, to account for these changes and the Army's new information advantage framework.

- ADP 3-13 (Nov '23)

Refer to INFO2 SMARTbook: Information Advantage, 2nd Ed. (Activities, Tasks & Capabilities). Completely updated for 2024, INFO2 chapters and topics include information advantage (enable, protect, inform, influence, attack); information in joint operations (information joint function & OIE: operations in the information environment); information capabilities (PA, CA, MILDEC, MISO, OPSEC, CO, EW, Space, STO), information planning (information environment analysis, IPB, MDMP, JPP); information preparation, execution, fires & targeting, and assessment.

B. Mobility (as a Dynamic of Combat Power)

Ref: Adapted from ADRP 3-90, Offense and Defense (Aug '12), pp. 3-12 to 3-13.

Assured mobility is a framework of -- processes, actions, and capabilities -- that assure the ability of a force to deploy, move, and maneuver where and when desired, without interruption or delay, to achieve the mission.

Threat UAS have profoundly impacted assured mobility by transforming the predict, detect, prevent, avoid, neutralize, and protect aspects of maneuver operations. Their persistent surveillance and real-time targeting capabilities allow adversaries to monitor routes, detect movements, and rapidly strike or cue fires, effectively turning once-secure corridors into contested terrain. This constant aerial observation undermines the ability of commanders to conceal or reposition forces, forcing more dispersed, time-consuming, and resource-intensive movements. UAS also enable the enemy to emplace dynamic obstacles—such as loitering munitions or improvised aerial-delivered explosives—at critical choke points, further degrading freedom of movement. As a result, friendly forces must now integrate counter-UAS measures into mobility planning, rely on deception and electronic protection to mask movements, and continuously reassess routes and timing to maintain tempo. In essence, threat UAS compress decision windows, increase risk across maneuver corridors, and challenge the Army's ability to move where and when desired without interruption—fundamentally redefining assured mobility.

Assured Mobility

- **A** Predict
- **B** Detect
- **C** Prevent
- **D** Avoid
- **E** Neutralize
- **F** Protect

A. Predict

Commanders and staffs must accurately predict potential obstacles to force mobility by analyzing the enemy's capabilities and tactics, techniques, and procedures. This involves understanding how the enemy will evolve in reaction to friendly force countermeasures. It also involves understanding how the effects of terrain and the effects of the population, such as vehicular traffic and dislocated civilians, will impact force mobility. This helps build the mobility portion of the common operational picture and facilitates decisionmaking.

B. Detect

Commanders and staffs use intelligence products and information collection assets to identify the location of natural and manmade obstacles and potential means the enemy can use to create obstacles. Commanders employ available information collection assets to detect enemy obstacle preparations and also identify areas where there are no or only limited obstacles to ground movement and maneuver. This knowledge can be obtained through sustained surveillance of an area. Commanders identify both actual and potential obstacles and propose solutions and alternate COAs to minimize or eliminate their potential impact.

C. Prevent

Commanders and staffs apply this fundamental by preventing civilian interference with operations and denying the enemy's ability to influence friendly mobility. This is accomplished by forces acting proactively to elicit local populace support, or at least non-interference, and to eliminate enemy countermobility capabilities before those capabilities can emplace or activate obstacles, and by mitigating the factors that result in natural obstacles to friendly force movement and maneuver. This may include the employment of information-related capabilities to decrease uncertainty among the population to build support for or acceptance of operations.

Prevention may also consist of aggressive action to destroy enemy assets and capabilities before they can be used to create obstacles. In recent operations this included disrupting terrorist bomb-making cells by all available means, such as cutting off their funding, eliminating safe houses where bombs can be constructed, jamming frequencies to prevent remote detonators from being triggered, and either capturing or killing members of these cells. Forces also apply this fundamental by conducting countermobility operations to shape enemy movement and maneuver that may affect friendly movement and maneuver.

D. Avoid

If prevention fails, the commander will move or maneuver forces to avoid impediments to mobility, if this is viable within the scheme of maneuver. If detection efforts can tell the commander where the enemy has not been, this frees up the unit to maneuver rapidly through those areas, even if they are not the most favorable movement routes.

E. Neutralize

Commanders and staffs plan to neutralize, reduce, or overcome obstacles and impediments as soon as possible to allow unrestricted movement of forces. The specific tactics, techniques, and procedures employed will depend on the mission variables of METT-TC, the rules of engagement, and where along the range of military operations the unit finds itself. For example, a small unit involved in major operations encountering surface-laid mines on a road in an urban area might attempt to destroy the mines in place using organic methods, such as aimed rifle or machinegun fire, after only minimal checks to reduce the danger to local civilians and accepting collateral damage to civilian buildings before proceeding on with its mission. That same unit encountering the same situation during the conduct of a peace-keeping operation would more likely secure the site, evacuate civilians from the area, and call for an explosive ordnance disposal team to disarm the mines in place to preclude any collateral damage.

F. Protect

Commanders and staffs plan and implement survivability and other protection measures that will prevent observation of the maneuvering force and thereby reduce the enemy's ability to engage or otherwise interfere with that force. This includes the use of combat formations and movement techniques. It may involve the use of electronic warfare systems—such as counter-radio controlled improvised explosive device electronic warfare systems, mine plows and rollers, and modifications to the rules of engagement. This may also include the conduct of countermobility missions to deny the enemy the capability to maneuver in certain directions and thereby provide additional protection to friendly maneuvering forces. It can also be as simple as altering patrol routes.

While engineers are the principal staff integrators for assured mobility, other staff sections play critical roles in ensuring the effective application and integration of mobility, countermobility, and protection tasks. In the case of amphibious operations, this would include naval forces that are responsible for assured mobility from amphibious shipping to beach and landing zone exits. These critical roles include providing information on threats to the routes. The senior engineer staff officer's role within assured mobility is similar to the role of the assistant chief of staff, intelligence (G-2) or the intelligence staff officer's (S-2s) integrating role with intelligence preparation of the battlefield. Ultimately, assured mobility is the commander's responsibility.

C. Survivability (One of the Five Dynamics of Combat Power)

Ref: FM 3-0, Operations (Mar '25), pp. 2-3 to 2-6, ADP 3-37, Protection (Jan '24), pp. A-4 to A-5 and ADP 3-90, Offense and Defense (Jan '24), p. 4-16.

Threat UAS have reshaped the concept of survivability on the modern battlefield by eroding traditional assumptions about concealment, protection, and force preservation. Their persistent aerial surveillance and precision-strike capabilities reduce the effectiveness of camouflage, cover, and concealment, making even well-dispersed or fortified positions vulnerable to detection and engagement. The ubiquity of small, low-cost drones allows adversaries to maintain continuous observation, identify movement patterns, and cue fires with unprecedented speed, forcing friendly forces to adopt constant mobility, deception, and emission control measures to survive. Survivability operations must now integrate counter-UAS systems, electromagnetic discipline, and active obscuration to deny enemy drones targeting opportunities.

Survivability *(See pp. 5-17 to 5-20.)*

Combat power is the means for conducting warfare and consists of five dynamics (leadership, firepower, information, mobility, and survivability). Each dynamic is interdependent and enabled by the protection warfighting function. Survivability is both a primary task of the protection warfighting function and a dynamic of combat power. In both uses, survivability is a quality or capability of military forces that permits them to avoid or withstand hostile actions or environmental conditions while retaining the ability to fulfill their primary mission. As a primary task of the protection warfighting function, survivability includes supporting tasks that enhance a formation's ability to avoid or withstand an enemy's actions by providing or improving camouflage, cover, and concealment; controlling electromagnetic emissions; and increasing personal protective posture (adding small arms protective insert plates, adding mission-oriented protective posture, or hardening command facilities).

Protection considerations are not the same at every echelon or in every area of operations. While some multidomain protection considerations, such as cybersecurity and defense, are executed at a strategic level across the operational framework, protection priorities diverge based on the time and proximity to different threat capabilities and capacities. Protection prioritization changes are anticipated and assets are reassessed as transitions occur throughout operations, changes to the commander's priorities occur, or decision criteria is met.

Conduct Survivability

Leaders assess survivability as the ability of a friendly force to withstand enemy effects while remaining mission-capable. It represents the degree to which a formation is hard to destroy. Survivability is a quality or capability of military forces which permits them to avoid or withstand hostile actions or environmental conditions while retaining the ability to fulfill their primary mission (ATP 3-37.34). Personnel and physical assets have inherent survivability qualities or capabilities that are enhanced through various means and methods. For example, the body armor or mission-oriented protective posture gear enhances survivability when worn correctly in the corresponding threat environments. These qualities are especially important when elements are targeted by threats and protection capabilities are limited.

Survivability operations are those military activities that alter the physical environment to provide or improve cover, camouflage, and concealment (ATP 3-37.34). Survivability operations enhance the ability to avoid or withstand hostile actions by altering the physical environment. Units accomplish this by providing or improving camouflage, cover, and concealment by—

- Constructing fighting positions.
- Constructing protective positions.
- Hardening infrastructure.
- Employing camouflage and concealment.

Survivability and survivability operations are not interchangeable. Survivability refers to a quality or capability, while survivability operations refer to tasks that enhance survivability. Survivability and survivability operations employ active and passive measures to enable freedom of action when synchronized and integrated with the other protection capabilities. Survivability operations are the responsibility of every unit at all echelons. Some units, such as engineer and CBRN, possess capabilities that enable survivability operations. Survivability focuses on providing camouflage, cover, and concealment. Some of the means and methods that enhance survivability become a part of the protected personnel or physical assets, which safeguard assets even while mobile. In addition to activities that alter the physical environment, other capabilities and activities are used to conduct or support survivability by—

- Establishing local security.
- Conducting military deception.
- Employing protective obstacles.
- Mitigating the emission of electromagnetic radiation.
- Countering unmanned aircraft systems.
- Maintaining dispersion.
- Using obscuration.
- Remaining mobile.
- Modifying the tempo.
- Taking evasive actions.
- Maintaining noise and light discipline.
- Implementing CBRN defensive measures.

Survivability operations are ongoing activities throughout the Army's strategic context and are not limited to crisis and armed conflict. Commanders include survivability and survivability operations into training during competition below armed conflict to validate their unit's individual and collective proficiency. Proficiency is an inherent form of passive protection. Organizations may also employ hardening measures, such as emplacing bollards to provide standoff and prevent threat effects from damaging critical deployment-related infrastructure or command and control facilities. Army forces, especially those stationed in forward locations or tasked with protecting critical capabilities, areas, and information, never stop improving the survivability of their positions. Plans and orders should specify dispersion and survivability requirements.

During crisis and armed conflict, units improve survivability within the limits of their capabilities. Competing demands and limited resources for conducting survivability operations are predominant factors throughout an operational environment and are considered during planning. Commanders identify those critical capabilities, areas, and information requiring additional protection because their loss or degradation would have a significant and/or debilitating effect on operations. When existing terrain features offer insufficient cover, concealment, and dispersion, altering the physical environment provides or improves cover and concealment, which enhances survivability. Similarly, using natural or artificial materials such as camouflage may confuse or mislead the enemy.

Refer to ATP 3-37.34 for additional information on survivability.

IV. Methods of Warfare

Although the nature and principles of war reflect the continuity of war, the conduct of warfare, like dynamic operational environments, reflects wide variation. Therefore, depending on the situation, strategic actors pursue their objectives in war through different methods of warfare. There are many different methods, but they generally fall into two broad categories: conventional and irregular. Each method of warfare serves the same strategic purpose—to defeat an enemy—but they take fundamentally different approaches to achieving their purpose. Both methods share one characteristic, which is that they involve the use of lethal force to achieve a political end.

A. Conventional Warfare

Conventional warfare is a violent struggle for domination between nation-states or coalitions of nation-states. Conventional warfare is generally carried out by two or more military forces through armed conflict. It is commonly known as conventional warfare because it means to fight enemy forces directly, with comparable military systems and organizations. A nation-state's strategic purpose for conducting conventional warfare is to impose its will on an enemy government and avoid imposition of the enemy government's will on it and its citizens. Joint doctrine refers to conventional warfare as "traditional" because it has been understood that way in the West since the Peace of Westphalia (1648), which reserved, for the nation-state alone, a monopoly on the legitimate use of force. However, irregular warfare has a longer history, and it has been just as common as the "traditional" method of warfare in some societies.

B. Irregular Warfare *(See pp. 2-45 to 2-48.)*

Irregular warfare is the overt, clandestine, and covert employment of military and non-military capabilities across multiple domains by state and non-state actors through methods other than military domination of an adversary, either as the primary approach or in concert with conventional warfare. Irregular warfare may include the use of indirect military activities to enable partners, proxies, or surrogates to achieve shared or complementary objectives.

V. Large-Scale Combat Operations *(See pp. 3-13 to 3-32.)*

The focus of Army readiness is on large-scale combat operations. Large-scale combat operations (LSCO) are extensive joint combat operations in terms of scope and size of forces committed, conducted as a campaign aimed at achieving operational and strategic objectives (ADP 3-0). During ground combat, they typically involve operations by multiple corps and divisions, and they typically include substantial forces from the joint and multinational team.

Conflicts encompassing large-scale combat operations are more intense and destructive than limited contingencies, often rapidly amassing heavy casualties. Peer threats employ networks of sensors and long-range massed fires that exploit electromagnetic signatures and other detection methods to create high risk for ground forces, particularly when they are static. Army forces must account for constant enemy observation, including the **threat from unmanned systems that saturate the operational environment.**

Refer to AODS7-1: The Army Operations & Doctrine SMARTbook, w/SMARTupdate 1 (Multidomain Operations). Completely updated with the 2025 edition of FM 3-0, AODS7 focuses on Multidomain Operations and features rescoped chapters on generating and applying combat power: command & control (ADP 6-0), movement and maneuver (ADPs 3-90, 3-07, 3-28, 3-05), intelligence (ADP 2-0), fires (ADP 3-19), sustainment (ADP 4-0), & protection (ADP 3-37).

Chap 1

IV. Army Force Structure Transformation Initiative

Ref: CRS Report R47985, 2024 Army Force Structure Transformation Initiative, Andrew Feickert (Feb '25).

On February 27, 2024, the Army publicly announced "changes to its force structure that will modernize and continue to transform the service to better face future threats. Under the plan, the Army will reorganize over the next decade to ensure it can deliver trained, cohesive and lethal forces to meet future challenges in increasingly complex operational environments." In conjunction with this announcement, the Army published *Army White Paper: Army Force Structure Transformation.* In early 2025, it was reported the Army is further examining force structure and reconfigurations, with recommendations due to Army leadership in March 2024.

> Over the past year, the Army conducted a rigorous assessment of its force structure using a process known as Total Army Analysis. Following extensive consultations with Congress, Army leaders are moving forward with a significant force structure transformation that will build on the modernization and organizational shifts of recent years.
>
> This transformation will enable the Army to bring in new capabilities to meet requirements under the National Defense Strategy. It will also allow the Army to narrow the gap between force structure, which was designed to accommodate 494,000 soldiers, and current Active Duty end strength, which is set by law at 445,000.
>
> TAA decisions will bring down "authorized" troop levels to approximately 470,000 soldiers by FY2029. Guided by the results of the TAA process, the Army will shrink excess, largely unmanned "hollow" force structure and build new formations equipped with new capabilities needed for large scale combat operations. By bringing force structure and end strength into closer alignment, the Army will ensure its formations are filled at the appropriate level to maintain a high state of readiness. At the same time, the Army will continue to transform its recruiting efforts so that it can build back its end strength, which is needed to provide strategic flexibility, reduce strain on frequently deploying soldiers, and add new capabilities to the force.
>
> *- Army White Paper, Army Force Structure Transformation (2024)*

As part of this transformation, the Army plans to develop new

- Multi-Domain Task Forces (MDTFs),
- Indirect Fire Protection Capability (IFPC) battalions,
- Counter-Small Unmanned Aerial Systems (C-sUAS) batteries, and
- Maneuver Short-Range Air Defense (M-SHORAD) battalions.

See p. 1-35 for an overview and further discussion.

To transform its force structure, the Army plans to initiate a number of force structure reductions and, at the same time, undertake a number of actions to address recruiting challenges that directly affect the Army's ability to populate both existing and planned force structures. In addition to forces covered in its February 2024 white paper, the Army reportedly also plans to make force structure changes to its Combat Aviation Brigades (CABs).

I. Army Force Structure Reductions

To identify force structure reductions, the Army employed a multifaceted approach to realign and resize the force. The Army "looked carefully at each military occupational specialty, and examined each skill set and functional area for efficiencies." In one example, the Army reallocated BCT-level counterinsurgency-focused engineer force structure to the division level, thereby "allowing the Army to reduce the overall number of engineer positions while giving division and corps commanders flexibility to concentrate assets as necessary during large scale combat operations." The Army notes that by identifying efficiencies in other types of units, it was able to reduce its authorizations by almost 10,000 spaces.

The Army further noted it reduced 2,700 authorizations

[b]ased on modeling that included factors such as demand over time, capacity to meet National Defense Strategy requirements and past deployment stress. Units that deploy infrequently or not at all pointed to areas where the Army could reduce manning authorizations within a formation at minimal risk. The Army will also decrease the number of transients, trainees, holdees and students (TTHS) by approximately 6,300 authorizations as it resizes for a 470,000 soldier Regular Army.

The majority of additional reductions "resulted from adjustments to the close combat forces" and targeted legacy formations that had "previously been sized and structured for soldier-intensive COIN operations." These reductions included

- inactivation of cavalry squadrons in continental U.S.-based Stryker brigade combat teams and infantry brigade combat teams;
- conversion of infantry brigade combat team weapons companies to platoons, and,
- elimination of some positions across Regular [Active Component] Army Security Force Assistance Brigades (SFABs).

These force structure reductions accounted for roughly 10,000 additional reductions in authorizations.

The Army also intends to reduce Army Special Operations Forces (ARSOF). The white paper states

Working closely with the Assistant Secretary of Defense for Special Operations and Low-Intensity Conflict and U.S. Special Operations Command, the Secretary of Defense determined the Army could reduce Army special operations forces by approximately 3,000 authorizations.... Positions and headquarters elements that are historically vacant or hard to fill will be prioritized for reduction.

> The U.S. Army is cutting approximately 6,500 active-duty aviation jobs over fiscal years 2026 and 2027 as part of a major transformation initiative to shift from manned aircraft to unmanned systems. (There are about 30,000 aviation branch soldiers in the ranks, including flight crews, maintainers and pilots.)This reduction is driven by the changing character of warfare, the need for modernized capabilities, and lessons from conflicts that show manned aircraft can be vulnerable to modern air defenses. The Army is also eliminating all Army Reserve helicopter units to focus on future conflicts and technological advancements.

While not specifically addressed in the Army's February 2024 white paper, the Army reportedly plans to convert Combat Aviation Brigades (CABs) to a previously used customized force structure as opposed to the current modular CAB force structure construct adopted in the early 2000s.

II. New Units (with C-UAS Capabilities) (Army Force Structure Transformation)

The *Army Force Structure Transformation White Paper (2024)* outlines the U.S. Army's ongoing effort to reshape its organization and capabilities to meet emerging threats across multiple domains of warfare—land, air, maritime, space, and cyberspace. Central to this transformation are four new types of units that represent the Army's drive to integrate advanced technologies, increase mobility and survivability, and enhance joint interoperability.

Collectively, these new formations reflect a pivot toward multi-domain operations (pp. 1-21 to 1-32) and layered defense integration (see chap. 5.)

A. Multi-Domain Task Forces (MDTFs) *(See pp. 1-36 to 1-37.)*

At the core of transformation are the Multi-Domain Task Forces, which embody the Army's shift toward integrated operations across domains. The Army is building five MDTFs—tailored, theater-level units capable of combining intelligence, cyber, space, and long-range fires effects. Each MDTF includes specialized elements: a headquarters, a multi-domain effects battalion, a long-range fires battalion, an Indirect Fire Protection Capability (IFPC) battalion, and a brigade support battalion. These task forces are designed to sense, target, and strike across great distances, protecting joint and coalition forces while shaping operational environments through synchronized non-kinetic and kinetic means. Full fielding is expected in the near term, establishing a scalable blueprint for future formations.

B. Indirect Fire Protection Capability (IFPC) Battalions *(See pp. 1-38 to 1-39.)*

The Army's plan to add four new IFPC battalions reflects the critical need for integrated air and missile defense against evolving aerial threats. IFPC units bridge the gap between short- and long-range air defense systems, providing a mobile, layered defense against rockets, artillery, mortars, cruise missiles, and unmanned aerial systems (UAS). These formations will likely complement the Army's forthcoming Long-Range Hypersonic Weapon (LRHW) batteries, offering protection for maneuver and support elements in high-threat theaters.

C. Counter-Small Unmanned Aerial Systems (C-sUAS) Batteries *(See pp. 1-40 to 1-41.)*

Recognizing the proliferation of small drones on modern battlefields, the Army is standing up nine C-sUAS batteries to bolster defense against this rapidly growing threat. These units will integrate into both IFPC and divisional air defense battalions, employing a variety of detection, jamming, directed-energy, and kinetic systems to defeat Group 1–3 UAS. The creation of dedicated C-sUAS units marks a doctrinal acknowledgment that drone defense is now a core requirement for every echelon of combat operations.

D. Maneuver Short-Range Air Defense (M-SHORAD) Battalions *(See pp. 1-42 to 1-43.)*

To restore organic air defense capabilities lost after decades of counterinsurgency focus, the Army plans to form four additional M-SHORAD battalions. Mounted on Stryker platforms and future variants, these mobile formations provide protection against low-altitude threats such as rotary-wing, fixed-wing, and unmanned aircraft. M-SHORAD enhances survivability and freedom of maneuver for maneuver brigades, particularly Infantry and Armored Brigade Combat Teams (IBCTs and ABCTs).

A. Multi-Domain Task Force (MDTF)

Ref: CRS Report IF11797, The Army's Multi-Domain Task Force (MDTF) (Apr '25), Andrew Feickert.

Congress has expressed concern about the threat to U.S. national security posed by Russia and China. The Army believes to address this threat, it must be able to operate in a multi-domain (air, land, water, space, cyber, information) environment, requiring new operational concepts, technologies, weapons, and units. The Multi-Domain Task Force (MDTF) is the Army's self-described "organizational centerpiece" of this effort.

What Is a Multi-Domain Task Force?

The Army's Chief of Staff Paper #1: Army Multi-Domain Transformation Ready to Win in Competition and Conflict dated March 16, 2021, describes the Multi-Domain Task Force (MDTF) as "theater-level maneuver elements designed to synchronize precision effects and precision fires in all domains against adversary anti-access/area denial (A2/AD) networks in all domains, enabling joint forces to execute their operational plan (OPLAN)-directed roles."

Anti-Access/Area Denial (A2/AD)

Anti-Access is any action, activity, or capability designed to prevent an advancing military force from entering an operational area. Area Denial is any action, activity, or capability designed to limit an adversarial force's freedom of action within an operational area. Threat A2/AD defenses are composed of layered and integrated long-range precision-strike systems, littoral anti-ship capabilities, air defenses, and long-range artillery and rocket systems.

The diagram at Figure 1 depicts a generic MDTF.

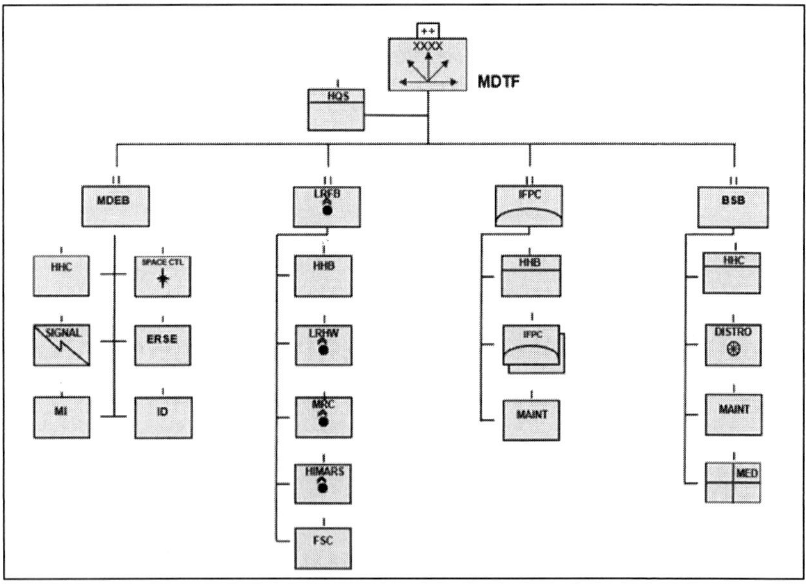

Figure 1. Generic MDTF. Army Information Paper provided to CRS, April 10, 2025.

Notes: MDEB = Multi-Domain Effects Battalion; HHC = Headquarters and Headquarters Company; MI = Military Intelligence; ERSE = Extended Range Sensing Element; ID = Information Dominance; LRFB = Long-Range Fires Battalion; HHB = Headquarters and Headquarters Battery; LRHW = Long-Range Hypersonic Weapon; MRC =Mid-Range Capability (also referred to by the Army as Strategic Mid-Range Fires (SMRF)); HIMARS = High Mobility Artillery Rocket Systems; FSC = Forward Sustainment Company; IFPC = Integrated Fire Protection Capability; BSB = Brigade Support Battalion.

The Army notes each MDTF is to be tailored to Combatant Commander requirements, so it is possible the generic MDTF in Figure 1 may contain more, fewer, or other types of units depending on the requirements of its assigned theater of operations.

April 2024 Update on MDTF Alignment and Stationing

According to an April 2024 Defense News article, the Army has updated MDTF pre-decisional plans. The Army reportedly plans to consolidate Mid-Range Capability and Long-Range Hypersonic batteries under a Long-Range Fires Battalion (LRFB) and complete programming of the remaining Indirect Fire Protection Capability (IFPC) battalions over the next five years. The Army also plans to convert all brigade support companies to battalions. In terms of specific MDTFs

- The 1st MDTF is to be fully established in FY2024 with a Multi-Domain Effects Battalion (MDEB), an IFPC battalion, a BSB, and a LRFB.
- The 2nd MDTF is to be headquartered in Germany with an MDEB and other elements stationed at Fort Drum, NY, with an IFPC battalion, a BSB and an LRFB. The 2nd MDTF is to be fully operational in FY2025 with the addition of the LRFB in FY2026. The 2nd MDTF is to focus on supporting the European theater.
- The 3rd MDTF, headquartered in Hawaii with an MDEB, will have an IFPC battalion, a BSB and an LRFB stationed at Joint Base Lewis-McChord, WA, and is to be fully operational by FY2026.
- The 4th MDTF is to be stationed at Fort Carson, CO, but focused on the Indo-Pacific theater, and is to be fully operational by FY2027.
- The 5th MDTF is to be stationed at Fort Bragg, NC, and is to concentrate on regions as determined. The 5th MDTF is to be fully operational by FY2028.

2nd MDTF Strategic Fires Battalion to be Stationed in Germany

On July 10, 2024, the White House issued a joint statement with Germany "following discussions ahead of the NATO Summit"

The United States will begin episodic deployments of the long-range fires capabilities of its Multi-Domain Task Force in Germany in 2026, as part of planning for enduring stationing of these capabilities in the future. When fully developed, these conventional long-range fires units will include SM-6, Tomahawk, and developmental hypersonic weapons, which have significantly longer range than current land-based fires in Europe. Exercising these advanced capabilities will demonstrate the United States' commitment to NATO and its contributions to European integrated deterrence.

3rd MDTF to Stand Up Long-Range Fires Battalion

Reportedly, in March 2025, the 3rd MDTF's commander stated that "the Army's 3rd MDTF is to stand up its long-range fires battalion over the next 12 to 18 months, as part of a "sizeable amount of growth" for the task force." It was further noted that "during the same timeframe the MDTF is standing up the fires battalion, it will also establish a sustainment battalion at Joint Base Lewis-McChord," and that the Army plans to add additional personnel to the unit's Hawaii-based Effects Battalion and MDTF Headquarters.

B. Indirect Fire Protection Capability (IFPC)

Ref: CRS Report IF12421, The U.S. Army's Indirect Fire Protection Capability (IFPC) System (Jan '25), Andrew Feickert.

Protecting high-value military sites against enemy cruise missile (CM), unmanned aerial system (UAS), and rocket, artillery, and mortar (RAM) attacks has long been an important consideration in protecting military forces. According to the U.S. Army, the Indirect Fire Protection Capability (IFPC) System "is a mobile, ground-based weapon system designed to defeat cruise missiles (CM), unmanned aircraft systems (UAS), and rocket, artillery, and mortars (RAM)." IFPC is to consist of a launcher and interceptors and is intended to "protect critical fixed-or semi-fixed assets," and "bridge the gap between short-range air defense (SHORAD) systems, the Patriot air and missile defense system, and the Terminal High Altitude Area Defense (THAAD) system."

Ref: U.S. Army Acquisition Program Portfolio 2023-2024.

Origins of the IFPC Program

The Army initiated the IFPC program, known as IFPC Increment 1, in 2004. Concerned with the pace and direction of the Army's counter RAM (C-RAM) development, some in Congress expressed an interest in acquiring Israel's Iron Dome C-RAM system as an interim solution. In both the FY2019 National Defense Authorization Act (NDAA) (P.L. 115-232) and the FY2019 Department of Defense Appropriations Act (P.L. 115-245), Congress directed the Army to deploy four batteries—two in 2020 and two in 2023—of a medium-range air defense system to counter cruise missiles and other threats. In response to this mandate, the Army selected Iron Dome as providing "the best value to the Army based on its schedule, cost per kill, magazine depth, and capability against specified threats." Because of difficulties integrating the first two Iron Dome batteries into existing and planned Army air and missile defense command and control architecture, the FY2021 NDAA (P.L. 116-283) waived the requirement for the final two Iron Dome batteries. The decision not to adopt Iron Dome reportedly served as the basis for the Army initiating the IFPC Increment 2 program.

IFPC Increment 2

On September 24, 2021, the Army announced "the award of a three-year prototype Other Transaction Authority Agreement (OTA) to Dynetics (a subsidiary of Leidos) for the development and delivery of 16 launcher prototypes, 60 interceptors, and associated all-up-round magazines for the Indirect Fire Protection Capability Increment 2" (Figure 1). The OTA was valued at approximately $237.38 million over two and a half years. Plans called for prototype development in Huntsville, AL, and Tucson, AZ, with deliveries to support testing beginning in the fourth quarter of FY2022. The first IFPC Increment 2 combat-capable battery was to be available to the Army in the fourth quarter of FY2023.

Other IFPC Variants

The Army is presently developing three IFPC variants, the Increment 2 Interceptor variant, the High Energy Laser (HEL) variant (Figure 2), and the High-Power Microwave (HPM) variant (Figure 3). Each variant is in a different stage of development, and variants are intended to operate in a complementary manner.

IFPC Increment 2 Interceptor Variant

The interceptor variant is to utilize an open architecture design to enable future missile integration. Reportedly, Increment 2 can employ the AIM-9X Sidewinder missile and the AGM-114L Longbow variant of the Hellfire missile. The Army is also testing the Israeli Tamir missile used by Israel's Iron Dome system.

Reportedly, the Army is considering developing a second interceptor for IFPC Increment 2, with one official noting, "It is now clearer that the service needs a second interceptor that is more capable against lower flying, supersonic cruise missiles." While it had been suggested that the Norwegian Advanced Surface to Air Missile System (NASAMS) could help fill the aforementioned capability gap with the AIM-120D Advanced Medium Range Air-to-Air Missile, the Army believes it is not an option, as NASAMS has a magazine depth of six rounds. Instead, the Army reportedly envisions "an AIM-120D-like capability, but with smaller rounds that fit inside an IFPC Increment 2 launcher holding 18 interceptors." If the Army pursues a second interceptor, development would continue through 2029 to 2030 before low-rate initial production.

IFPC High Energy Laser (HEL)

IFPC HEL is being designed by Dynetics to protect critical fixed-or semi-fixed assets against CMs, UASs, and RAMs using a laser as opposed to interceptor missiles. The Army planned to mount four operational 300 kilowatt (kW)-class IFPC HEL prototypes onto tactical vehicles by FY2024. If testing proved successful, the Army would transition the IFPC HEL to a Program of Record in FY2025.

FY2024 Army budget documents and reports note the Army plans to cut approximately $4.8 billion from planned IFPC HEL future spending, which the Army attributed to "changing priorities." The FY2025 IFPC HEL budget request is a $327 million reduction compared with the Army's forecast in the FY2024 budget. Future funding is eliminated starting in FY2026, suggesting almost $4.5 billion is to be redirected to higher-priority needs. While the Army says it remains committed to HELs, it is reportedly adopting a "buy-try-decide strategy."

IFPC High Power Microwave (HPM)

According to the Army, the IFPC HPM is intended to provide short-range protection for fixed and semi-fixed sites against small UAS (weighing about 55 pounds or less) swarm attacks. In January 2023, the Army reportedly awarded a $66.1 million contract to Epirus to deliver four high-powered microwave prototypes to the Army in FY2024 for testing. Reportedly on May 15, 2024, Epirus delivered the last two IFPC-HPM systems to the Army, with Epirus further noting the systems had completed new equipment training and engineering developmental testing.

C. Counter-Small Unmanned Aerial Systems (C-sUAS) Batteries

Ref: CRS Report R48477, Department of Defense Counter Unmanned Aircraft Systems (Mar '25), Daniel M. Gettinger.

The U.S. Army plans to establish nine new Counter-Small Unmanned Aerial Systems (C-sUAS) batteries as part of its Force Structure Transformation initiative. These batteries are designed to counter the growing threat of small, low-altitude drones and other aerial systems.

The nine new C-sUAS batteries will be assigned to Indirect Fire Protection Capability (IFPC) and division-level air defense battalions.

The first battery is scheduled to be fielded in the third quarter of Fiscal Year 2025, starting with the 1st Cavalry Division and followed by the 82nd Airborne Division.

The C-sUAS batteries will utilize a layered approach to detect, track, and neutralize small drone threats, employing a combination of kinetic and non-kinetic effects.

Potential equipment includes:

Kinetic Interceptors

This includes the Coyote anti-drone interceptor, for which the Army plans to significantly expand its inventory.

Non-Kinetic Systems

These can include radio frequency jammers and other systems that disrupt the command-and-control links of enemy drones.

Mobile-Low, Slow, Small Unmanned Aerial Vehicle Integrated Defeat System (M-LIDS)

Some units, such as the 1st Cavalry Division, are already receiving M-LIDS, a vehicle-mounted system of sensors and weapon systems.

Ref: Mobile Low, Slow, Small Integrated Defeat System (M-LIDS). (Sgt. Brandon Hernandez, U.S. Army, Defense Visual Information Distribution Service.) Notes: This November 2024 photo shows the original configuration of M-LIDS consisting of two M-ATV vehicles.

The Army's Low, Slow, Small Unmanned Aircraft System Integrated Defeat System, or LIDS, is a family of systems composed of the Fixed Site LIDS (FS-LIDS) and Mobile LIDS (M-LIDS), both of which are designed to defend against drones in Groups 1-3, as well as other low-flying threats. The core elements of the fixed and mobile versions of LIDS are similar and include a radar system and electro-optical/infrared (EO/IR) cameras, radar, a command-and-control system, and kinetic and electronic means of engaging and debilitating or destroying the perceived aerial threat. The Army reportedly developed the requirements for the M-LIDS in 2017-2018.115 Both the Army and Air Force have deployed the LIDS systems.

In M-LIDS Increment 2.1, initiated by the Army in 2022, the Army consolidated the M-LIDS sensors and weapons on a single Stryker vehicle, instead of the two Mine Resistant Armored Platform All-Terrain Vehicles (M-ATVs)—one vehicle for electronic warfare and another for kinetic defeat capabilities—on which the M-LIDS relied previously. The M-LIDS has incorporated products from multiple companies, including the Moog, Inc. Reconfigurable Integrated weapons Platform (RIwP) turret; Raytheon's Coyote interceptors and eXpeditionary Battlefield active electronically-scanned array External Unit (XBAEU) radar; and the Counter-small Unmanned Aircraft Electronic Warfare System (CUAEWS) from SRC. However, the individual products on the M-LIDS could change as the Army continues to evaluate and test new capabilities.

The FS-LIDS, an immobile system that provides a counter-UAS capability to a fixed military installation, features many of the same core elements as its mobile counterpart, with a few differences. The FS-LIDS lacks the weapons turret on the M-LIDS; instead, Coyote interceptors (see "Coyote" below for more details) feature prominently in the FS-LIDS. The detect and track capability on the FS-LIDS is provided by Raytheon Ku Band Radio Frequency System (KuRFS) radars and electro-optical and infrared cameras. In addition to Coyote interceptors, the FS-LIDS is equipped with an electronic warfare system for mitigating drones. Both the FS-LIDS and M-LIDS use the Army's Forward Area Air Defense Command and Control (FAADC2) system.

Forward Area Air Defense (FAAD)

The FAAD system is a battle-hardened command and control system that is used to integrate sensors and effectors.

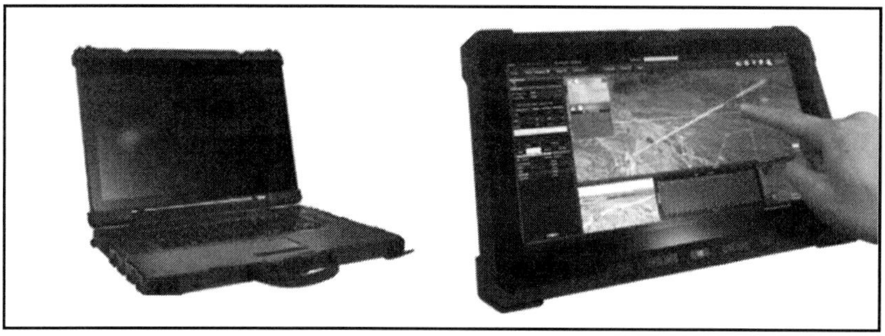

Ref: U.S. Army Acquisition Program Portfolio 2023-2024.

Forward Area Air Defense Command and Control (FAAD C2) provides critical C2, situational awareness, and automated air track information by integrating engagement operations software to support multiple missions. FAAD C2 supports air defense and Counter-Rocket, Artillery, Mortar (C-RAM) weapon systems engagement operations. The system also performs C2 engagement operations.

FAAD C2 provides a real-time correlated air picture of the battlespace with positive aircraft identification for early detection, warning, and response to forward deployed forces.

D. Maneuver Short-Range Air Defense (M-SHORAD) System

Ref: CRS Report IF12397.17 - Maneuver Short-Range Air Defense (MSHORAD) System (Jan '25), Andrew Feickert.

The Army is developing a new maneuver short-range air defense system, or M-SHORAD, to perform short-range air defense (SHORAD). The Army defines SHORAD as capabilities that provide air defense against low-altitude air threats. In the early 2000s, the Army divested SHORAD units from Army force structure to free up personnel to create other types of units deemed more mission-critical at the time. After 2005, SHORAD force structure was reduced to two battalions of Active Component Avenger systems and Counter-Rocket, Artillery and Mortar (C-RAM) batteries and seven Army National Guard Avenger battalions.

Since 2005, there has been a dramatic increase in air and missile platforms that could threaten U.S. ground forces. The use of unmanned aerial systems (UASs) has increased exponentially and UASs have been used successfully in a variety of conflicts, including the current Ukrainian conflict. Given the increase in threat and limited air defense assets available to Army divisions, the Army decided to improve the air defense posture of its maneuver forces.

Ref: U.S. Army Acquisition Program Portfolio 2023-2024.

M-SHORAD Requirement

In response to the growing aerial threat, the Army originally planned to field 144 M-SHORAD Increment 1 systems to four battalions and could field additional battalions in the future. Each M-SHORAD battalion would consist of 40 M-SHORAD systems, support vehicles and equipment, and about 550 soldiers. In April 2021, the 5th Battalion, 4th Air Defense Artillery Regiment received the first four of its M-SHORAD systems, becoming fully equipped in late 2022. In addition to the 144 systems designated for operational units, the Army planned to procure 18 additional systems for training, operational spares, and testing, for a total of 162 systems. Reportedly, as of June 2024, the Army envisioned building about 312 M-SHORAD systems but that number could increase to as many as 361 "based on future Army decisions," but, at present, the Army only has approval for 162 systems.

Initially, there were three M-SHORAD variants, or "Increments," planned by the Army.

M-SHORAD Increment 1

M-SHORAD Increment 1 (Figure 1) was developed under the Other Transaction Authority contracting process. M-SHORAD uses the M-1126 Stryker combat vehicle as its chassis. The weapons and radar packages are configured by Leonardo DRS and then installed on the Stryker by General Dynamics Land Systems (GDLS), the vehicle's original manufacturer. Leonardo DRS reports the multipurpose unmanned turret includes

- two AGM-114L Longbow Hellfire missiles capable of hitting ground targets;
- four FIM-92 Stinger missiles for aerial targets in a launcher (configured by Raytheon);
- an XM914 30 mm automatic cannon;
- an M-240 7.62 mm machine gun; and
- a multi-mission radar capable of tracking both ground and air targets.

Reportedly, the Army is planning to replace the Longbow Hellfire missile launcher on M-SHORAD Increment 1 vehicles because the Hellfire mounted on the side of the vehicle "created wear and tear on the missile, leading to potential safety concerns." The Army reportedly plans to retrofit M-SHORAD vehicles "by swapping out the Hellfire launcher and for a second Stinger pod," which would give the modified vehicles eight total FIM-92 Stinger rounds.

M-SHORAD Increment 2

M-SHORAD Increment 2 is also referred to as DE (Directed Energy) M-SHORAD and incorporates a 50 kilowatt (kW) laser as its primary armament to defend against a variety of air and artillery threats. Efforts to develop the 50 kW laser began in 2019 and, in 2021, Raytheon was awarded a $123 million developmental contract after a competitive shoot-off against Northrop Grumman. The Army reports additional testing of the 50 kW laser has enjoyed success against a variety of drones, but according to Army program officials, "challenges remain" in terms of defending against rockets, artillery, and mortars. Army plans called for M-SHORAD Increment 2 to start a user assessment beginning in the fourth quarter of FY2023, running through the first quarter of FY2024. Reportedly, the Army now plans to choose a few designs from an existing group of vendors to compete to develop what it calls its Enduring High Energy Laser (HEL) system for short-range air defense in early 2025 The Army plans to fund teams through design and development and then select a winner to build an enduring HEL in the first quarter of FY2026.

M-SHORAD Increment 3

The Army reportedly plans for M-SHORAD Increment 3 to incorporate the FIM-92 Stinger replacement missile—the Next Generation Short Range Interceptor—into the Increment 1 system. In addition, these plans call for the Increment 1 30 mm automatic cannons to receive the XM 1223 Multi-Mode Proximity Airburst munition (MMPA), which features a multipurpose munition that can be employed against air, ground, and personnel targets. In March 2023, the Army reportedly selected Lockheed Martin and Raytheon Technologies to develop competing prototypes of a Next-Generation Short-Range Interceptor.

Army Issues a Request for Information for M-SHORAD Increment 4

On May 8, 2024, the Army issued a request for information (RFI) for M-SHORAD Increment 4, stating that this capability is to focus on delivering "air defense capability to support dismounted maneuver." M-SHORAD Increment 4 is to include capabilities that can be transported on a C-130 aircraft and that are air-droppable and sling load capable. It is also to be capable of integration onto platforms such as the Joint Light Tactical Vehicle (JLTV) and/or a robotic vehicle.

III. Proposal to Create a U.S. Army Drone Corps?

Ref: CRS Report R47985, 2024 Army Force Structure Transformation Initiative, Andrew Feickert (Feb '25), pp. 10 to 11.

A provision of the House-passed H.R. 8070, the Servicemember Quality of Life Improvement and National Defense Authorization Act for Fiscal Year 2025, would establish the "Drone Corps" as a basic branch of the U.S. Army, one responsible for small and medium uncrewed aircraft systems (UAS) and defensive counter-UAS. If the House provision is adopted in law, Congress would be responsible for the oversight and funding of this new branch of the Army. This product provides a background on the Army's basic branches, as well as a discussion of the proposed legislation and potential considerations for Congress.

A military occupational specialty (MOS) signifies the career path of Army enlisted and warrant officer personnel. In most cases, an MOS corresponds to a basic branch. Personnel responsible for operating and maintaining medium and large UAS belong to the MOS 15 series, the Aviation Branch. Army personnel responsible for counter-UAS systems typically belong to the MOS 14 series, the Air Defense Artillery Branch. To date, the Army has not established an MOS for counter-UAS, and Army doctrine states that countering drones is "not a stand-alone effort or the sole responsibility" of any branch. Neither small drones nor handheld counter-UAS systems are branch- or MOS-specific; instead, the Army treats these systems as individual or crew-served "weapons" much like machine guns.

Both the Army's small and medium UAS and counter-UAS capabilities are in states of transition. The Army's UAS Project Office is developing requirements for new classes of small drones in Groups 1-2 (see CRS In Focus IF12668, The U.S. Army's Small Uncrewed Aircraft Systems) and for the Future Tactical UAS, a Group 3 system. Meanwhile, under the Force Structure Transformation Initiative, the Army plans to establish new counter-UAS units.

Advocates for the proposed Drone Corps argue that a new branch would acknowledge the growing prevalence of drones on the modern battlefield and foster specialized training and skills development. Opponents of the proposal, including Under Secretary of the Army Gabe Camarillo, counter that a new branch is premature, given that the Army is still considering requirements for new small and medium drones and experimenting with how these systems might be effectively employed. In a statement on June 11, 2024, addressing a draft of H.R. 8070, the Biden Administration said that it "strongly opposes" the establishment of a Drone Corps, arguing that the new branch would create "an unwarranted degree of specialization."

I. UAS Threat (Introduction)

Ref: A Comprehensive Approach to Countering Unmanned Aircraft Systems, Joint Air Power Competence Centre (Jan '21).

I. Background

In recent decades, Unmanned Aircraft Systems (UAS) have been fielded in every military service, ranging from handheld micro-UAS to medium-sized tactical systems to full-grown Remotely Piloted Aircraft (RPA).

At the same time, the civilian market has witnessed an exponential growth of predominantly smaller systems intended for public and recreational use. However, the latter use case has gained the attention of law enforcement agencies and military force protection communities due to the increased misuse of Commercial-Off-The-Shelf (COTS) 'drones' in the vicinity of airports, public events and military installations.

The Strategic Environment *(See pp. 1-2 to 1-3.)*

Enabled by growing commercial innovation and the increasing sophistication of artificial intelligence (AI), autonomy, and networking technology, unmanned systems are fundamentally changing how militaries of all sizes, capacities, and capabilities – as well as non-state actors – achieve their objectives.

From the Middle East to Ukraine and across the globe – including in the U.S. homeland – unmanned systems are reshaping tactics, techniques, and procedures; challenging established operational principles; and condensing military innovation cycles. At the operational level, these systems are making it more difficult for forces to hide, concentrate, communicate, and maneuver. They allow adversaries to more easily surveil, disrupt, or attack our forces, assets, and installations, potentially without attribution. At the strategic level, unmanned systems provide aggressors with the ability to reduce the initial human, financial, and reputational costs of conflict. The relatively low-cost, widely available nature of these systems has, in effect, democratized precision strike.

Technological advances in the mid- to long-term will likely render unmanned systems increasingly capable, affordable, autonomous, and networked – able to loiter for longer timespans, to communicate better with other systems, move and act as swarms, and to carry larger payloads. These dynamics risk eroding deterrence and creating new and uncertain escalation dynamics.

Unmanned Aircraft Systems (UAS) have become an integral part of operations and have advanced into an invaluable asset for Intelligence, Surveillance, and Reconnaissance, as well as for combat missions.

Chinese, Russian, & Iranian UAS Threat *(See pp. 2-13 to 2-48.)*

This has not gone unnoticed by both state and non-state actors, which has led to an enormous effort by these players to catch up with or at least mimic the Western level of technology. Over the last decade, China, Russia, and to a certain extent Iran, have considerably advanced the development of UAS, and their latest models seem to have the same performance characteristics as Western models. Russian and Chinese inventories comprise the full range from small and tactical UAS, through Medium- and High-Altitude Long-Endurance (MALE/HALE) systems, to replicas of US and European stealth prototypes.

UAS Threat

II. The State of the Art of War

A Comprehensive Approach to Countering Unmanned Aircraft Systems, Joint Air Power Competence Centre (Jan '21), "Future Threats" by Dr James Rogers, pp. 498 to 505.

The Second Drone Age

The world has entered the 'second drone age'. Defined by the global proliferation of military Unmanned Aircraft Systems (UAS) and weaponized commercial drones, this new era of drone warfare has seen, and will continue to see, both state and non-state actors competing for power in the skies above (and beyond) designated zones of conflict. Hostile actors, with threatening remotely operated air power components, now vie for command of the air against NATO and allied forces. Civilian populations are at increased risk in this adjusted state of war. Ethical controversies from the first drone age have been exacerbated by the widespread use of distant lethal robotics, making it difficult to distinguish between the perpetrators of drone atrocities and attacks or accidents. This 'deniability' has important political, legal, and strategic implications. Holding actors to account, or retaliating against belligerents, is difficult in this deniable, multi-user context, where similar, if not identical systems, are deployed by myriad disparate actors. The second drone age also poses broader implications for international security, stability, and Great Power politics. Decisions about who joins the 'global drone club' are not made by accident, especially where the transfer of military UAS is concerned. The unrestricted supply of armed UAS to surrogate, partner, and proxy actors by state suppliers – of which China is one of the most prolific – will influence the fate of nations. As recent 'State versus State' drone wars in the Caucasus and Libya show, the politically motivated supply of military UAS has contributed to international instability and conflict escalation. The supply of both commercial and military-grade remote technologies to non-state actors, allegedly by countries like Iran or through commercial shell companies, exacerbate the manifest threats present in this altered security environment. Put simply, new 'Drone Powers', and the 'new drone world', present fundamentally different challenges to those faced during the first drone age.

The First Drone Age

The American monopoly over the use of military UAS defined the first drone age, with Western powers deploying remote aerial technologies in largely uncontested airspace. Operated by the United States (US) military, the Central Intelligence Agency (CIA), and the militaries of select allies, these systems were used to hunt and kill those defined as terrorists and insurgents inside and outside designated regions of active conflict. They were (and are) also used to provide close air support and over-watch for friendly forces deployed as part of the Global War on Terror (GWOT). UAS like the Predator, Reaper and unarmed Global Hawk became symbolic of a post-9/11 period where military robotics surged forward to become the spearhead of American and allied force deployment. There can be little doubt that these systems were deemed to be both politically and militarily effective. By the time President Obama came to power in January 2009, the US had lost at least 625 soldiers in Afghanistan and 4,221 personnel in Iraq. In 2010, the Improvised Explosive Device (IED) was responsible for 60 per cent of American casualties in Afghanistan alone. In this high-risk and dangerous context of Asymmetric Warfare, UAS provided the ability to transcend the threats on the battlefield, whilst simultaneously extending the operational reach of American military power to regions previously deemed too hostile or too difficult to operate in with large deployments of American personnel. As part of a broader 'Remote Warfare', or as Thomas Waldman has termed it a 'Vicarious Warfare' modus operandi, UAS were used in tandem with small detachments of Special Operations Forces (SOF), military training units, manned air power assets, and larger groups of local forces who bore the brunt of the risk. In this context, UAS played a central role as a 'panacea weapon' for President Obama, who was faced with a mounting fatality count and growing public unease about the wars America was committed to.

The rise of UAS, went hand in hand with – or as Clausewitz may have put it, 'was a continuation of' – the casualty-fatigued politics of the 2010s. Yet, whilst these unmanned systems remained popular with the American people and many political leaders, concerns were raised by human rights groups who argued that the increasing use of UAS made killing too easy. This 'easiness' was mostly due to concerns over lowering the threshold in the use of force, as raised by organizations like PAX. Alongside this, NGOs concerned with civilian protection claimed that UAS had broader gendered, traumatic, and economic effects on the populations they operated high above; well beyond those of the initial kinetic precision strike. Claims of statistical inaccuracy were also made. When it came to the counting of the dead, Non-Governmental Organizations (NGOs), such as AirWars and The Bureau of Investigative Journalism (TBIJ), asserted that American and allied UAS strikes were responsible for killing thousands more civilians than officials would admit. Even senior administration staff worried about the growing reliance on UAS. Former US Secretary of Defence Robert Gates noted that US defence leaders had the tendency to view the use of force by UAS as 'bloodless, painless, and odourless', meaning deadly strikes may be used more frequently and as a first resort. Despite these concerns, however, during the first drone age, the prevailing wisdom was that UAS fulfilled an effective military and political role for the US. This role continued into the Trump administration, where UAS were consolidated in their position as the go-to military assets when facing America's foes. Most notably, during the Trump era, new regions of active armed conflict were defined and there were re-interpretations of those designated as a terrorist. The Trump Administration also loosened the restrictions on the use of drone strikes, decentralizing command over strike decisions. One consequence of these changes was an increase in UAS strikes in places like Somalia and the first targeted killing of a state actor – General Qassem Soleimani, the head of the Iranian Revolutionary Guard Corps (IRGC).In the second drone age, such precedents will lead to new norms in the deployment of UAS by other state and non-state actors, ones which are likely to travel full circle and challenge the security of NATO states, state representatives, and civilian populations. These are the first of many worrying trends that will emerge with the global spread of UAS.

Power, Proliferation, and Escalation

It will come as no surprise that the perceived successes and sheer scale of the US and allied UAS deployments have combined to make UAS systems an attractive military investment. If the first drone age was defined by the US and its allies setting broad standards and norms for UAS use, then the second drone age will be defined by a diverse mix of new UAS actors – some friendly and some hostile – copying established forms of deployment and pushing those very same standards and norms to their limits. Whole new practices – some troubling and some improved – will also be formed, altering the character of conflict. As this section explains, the second drone age will see nation-states capitalizing on the global proliferation of military UAS, providing state militaries with a viable, and often far too easy capacity to deliver death from above. This will challenge traditional notions of Asymmetric Warfare and the effectiveness of a NATO-allied deterrence posture.

To be specific, 102 nation-states now operate a various assortment of military UAS. In 2010, just 60 nations possessed these unmanned aerial technologies. To put these figures into perspective, this equates to a 70 percent increase in the number of state militaries operating armed or unarmed UAS capabilities. When we drill down a little deeper into these statistics, provided by the Centre for the Study of the Drone, it can be seen that out of these 102 states, around 40 possess, or are in the process of purchasing, large UAS with the capacity to launch deadly attacks. These include, but are not limited to, Israel, Iraq, Iran, the United Kingdom, the US, Turkey, France, the United Arab Emirates, Saudi Arabia, Egypt, Nigeria, and Pakistan, all of whom have, according to the United Nations (UN), operated UAS for targeted killings or close air support.

III. Transformation of Battlefield Tactics (with the Introduction of sUAS and FPV Drones)

The rapid proliferation of small tactical unmanned aerial systems (UAS) and first-person-view (FPV) drones since 2022 has profoundly altered modern warfare, especially in Ukraine. What began as limited experimentation with commercial drones has evolved into a fully integrated aerial layer of ground combat operations. By 2025, small drones had become indispensable for reconnaissance, artillery direction, and precision strike—fundamentally reshaping tactics, techniques, and procedures (TTPs) across the front.

Type and Severity of UAS Threats

Army officials have acknowledged that the focus of the service's counter-UAS efforts has shifted over the years from commercially available drones (e.g., the Group 1 and 2 drones used by the self-proclaimed Islamic State in around 2016-2017) to larger—and potentially more lethal—UAS, including those sourced from state actors like Iran and used in attacks on commercial shipping in the Red Sea and against Ukraine and Israel. In its 2024 Strategy for Countering Unmanned Systems, DOD identified several technological trends that could influence the types of UAS threats U.S. military forces may face in the future, including drones that are "able to loiter for longer timespans, to communicate better with other systems, move and act as swarms, and to carry larger payloads."

At the tactical level, small UAS now provide front-line units with organic intelligence, surveillance, and reconnaissance (ISR) capabilities once reserved for higher headquarters. Infantry squads and platoons routinely deploy quadcopters and FPVs for target spotting, route reconnaissance, and damage assessment. Drone teams equipped with handheld controllers and digital video links can detect enemy positions in real time, transmit coordinates to supporting artillery, or strike targets directly using explosive payloads. This has shortened the sensor-to-shooter cycle to mere seconds, dramatically increasing accuracy and lethality.

FPV drones, adapted from commercial racing platforms, have become precision-guided munitions in their own right. Cheap, maneuverable, and expendable, they allow even light infantry units to destroy armored vehicles, artillery, and fortified positions without heavy weapons. Drone swarms are now used to suppress defenders, target logistics convoys, and harass command posts. The result is a form of continuous aerial harassment that disrupts mobility and erodes morale.

Defensive tactics have adapted accordingly. Static positions are no longer survivable for long periods; units disperse, conceal, and frequently relocate to avoid detection. Overhead cover and camouflage nets designed to obscure heat and optical signatures have become standard, and deception measures—such as decoy vehicles and false heat sources—are widely employed. Electromagnetic warfare (EW) has also become central to defense, as both sides seek to jam enemy control links, blind sensors, or spoof drone navigation systems.

The widespread use of small drones has driven organizational and doctrinal change. Dedicated drone units, repair workshops, and electronic-warfare teams are now organic to battalion-level formations. Training programs emphasize not only marksmanship and fieldcraft but also coding, radio operation, and digital literacy. Commanders at all levels have had to incorporate airspace management, deconfliction, and EW planning into routine operations.

Ultimately, small tactical UAS and FPV drones have transformed the character of combat. They enable decentralized decision-making, real-time reconnaissance, and cost-effective precision strikes, while simultaneously exposing forces to new vulnerabilities in the electromagnetic and information domains.

Line-of-Sight (LOS) Radio/Data-Link Control versus Wire-Guided Connections

During recent conflicts, particularly in Ukraine, there has been a decisive shift from system-guided unmanned aerial systems (UAS) to small, wire-guided and first-person-view (FPV) tactical drones. This transformation reflects the changing realities of modern warfare, where electronic warfare, cost pressures, and the need for precision and adaptability have reshaped how both sides employ unmanned systems. Traditional drones relying on radio data links and satellite navigation once offered long-range reconnaissance and strike capabilities, but as electronic countermeasures have grown more sophisticated, these platforms have become increasingly vulnerable to jamming, interception, and spoofing. In response, militaries and irregular forces alike have turned to FPV and wire-guided drones, which are less susceptible to these disruptions and can be built and operated at a fraction of the cost.

The primary catalyst behind this evolution is the growing prevalence of electronic warfare. In contested environments such as eastern Ukraine, dense jamming fields routinely disable or confuse system-guided drones dependent on GPS or radio telemetry. Fiber-optic or wire-guided drones, however, bypass this vulnerability by maintaining a physical connection between the operator and the aircraft, rendering them effectively immune to radio-frequency interference. Similarly, FPV drones, which rely on direct, short-range analog video feeds, provide a resilient and low-latency control method that remains effective even in heavily jammed zones. These systems allow operators to maintain precise control during terminal guidance and deliver small explosive payloads with remarkable accuracy, especially against mobile or fortified ground targets.

Another major factor driving the shift is economic. High-end drones with autonomous navigation and encrypted communications are expensive to produce and replace, while FPV drones can be assembled from commercially available components at a fraction of the cost. This affordability allows forces to field drones in large numbers and absorb attrition rates that would be unsustainable for more advanced systems. The mass employment of inexpensive FPV craft has enabled saturation tactics—sending multiple drones simultaneously to overwhelm defenses or ensure that at least one reaches the target. This economic asymmetry, where a $400 FPV drone can destroy a million-dollar vehicle, has fundamentally altered the cost calculus of warfare.

Tactically, small FPV and wire-guided drones also offer a high degree of flexibility and adaptability in complex terrain. Operators can navigate them through buildings, trenches, or wooded areas where GPS signals are unreliable or nonexistent. Their small size and agility allow for unconventional flight paths and low-altitude approaches, exploiting blind spots in radar and visual detection systems. Furthermore, the immediacy of real-time video control enhances human decision-making on the battlefield, allowing operators to adjust dynamically to target movement or defensive fire.

However, these advantages come with notable limitations. Wire-guided drones are constrained by the physical fragility and finite length of their cables, while FPV drones remain dependent on the operator's line of sight and skill. Environmental factors—wind, rain, and terrain obstacles—can easily disrupt operations. Yet, despite these drawbacks, the tactical and operational advantages of unjammable, low-cost, and precise drones continue to outweigh their limitations.

In sum, the battlefield shift toward FPV and wire-guided drones represents the convergence of necessity and innovation. Under intense electronic warfare pressure, militaries have adapted by prioritizing simplicity, resilience, and affordability over technological sophistication. The result is a new era of drone warfare—defined less by autonomous, high-end systems and more by human-guided, expendable platforms that emphasize flexibility, mass employment, and tactical precision.

IV. (Tactical- Level) UAS Threat Techniques

Ref: ATP 3-01.81, Counter-Unmanned Aircraft System (C-UAS) Operations (Mar '25), chap. 3.

Units operating in and around combat areas should assume the enemy is observing or attempting to observe them. Modern, friendly, and allied sensors, which include a host of robust long-range and short-range radars, optical devices, and audible alert systems, face challenges detecting the UAS Groups 1, 2, and 3 at sufficient ranges to mitigate effects. Threat UAS may go undetected while operating at a standoff range. ADA units cannot engage all hostile air threats, due to the lack of availability of ADA weapons and sensors, and/or equipment capability limitations. Such being the case, all units should be trained and ready to detect, identify and be prepared to engage UAS Groups 1 through 3.

Contact with unidentified UAS may be a precursor to an imminent attack. However, not all unidentified UAS encounters pose a risk. All units must react quickly and appropriately respond and report when recognizing signs of possible enemy observation or attack. Whether a counter-response capability is available or not, units must implement passive or active defense measures to nullify or reduce the effectiveness of enemy UAS operations.

Tactical level UAS operate at various altitudes. The ability for groups 1 thru 3 UAS to operate at low altitudes decreases the likelihood of friendly forces detecting the threat in a timely manner. UAS provide cost effective, high payoff means of surveillance and reconnaissance for an adversary. Integrated AMD capabilities can effectively counter larger UAS groups, but have difficulty tracking, identifying, and defeating lower class UAS. The challenge is at the brigade and below level with planning for and defending against small UAS threats.

UAS may be employed to conduct a wide variety of missions to include ISR, direct, indirect, or mass attack, fire support, harassment/nuisance and EW:

Intelligence, Surveillance, & Reconnaissance (ISR) *(See fig. 3-1.)*

Mounted cameras; UAS can provide an adversary with a contemporary ISR capability in real time. Patterns of life, such as when there is a shift change and feeding hours, can be used in enemy target acquisition and use in the information operation campaign.

Command and Control (C2)

UAS can provide an aerial view of the OE to provide commands to forces to accomplish assigned missions.

Direct Attack *(See fig. 3-2, following page.)*

UAS can be employed as a fire-and-forget weapon in a "suicide mode" with either just the UAS, or with an explosive chemical biological and radiological payload attached to cause injury.

Indirect Attack *(See fig. 3-3, following page.)*

The UAS deploys a payload while maintaining overhead coverage and video feed. Payloads may include explosives, chemical or biological agents, radiological isotopes or other payloads to disrupt, attrit, and destroy friendly forces.

Forward Observer

UAS can provide forward observer functionality that can enable adjustment of indirect fire or provide direct line of sight to the target. Forward observation concerns the employment of UAS for target acquisition and battle damage assessment.

Harassment/Nuisance

UAS can be used in manner that congests airspace or interdicts movement, which may be intentional or unintentional.

Swarm/Mass Attacks

A swarm refers to a method of operations in which multiple, sometimes-autonomous systems act as a cohesive unit by actively coordinating their actions. A swarm of UAS is a cluster, fleet, or set of aerial UAS that work together to achieve a specific goal and are controlled either manually or autonomously. What differentiates a swarm from a group of individual UAS is the capability of the UAS within the swarm to communicate with one another and work to accomplish prescribed objectives, whereas a group is any number of UAS, each of which is being controlled by a single source. *(See figure 3-4.)*

Electromagnetic Warfare (EW)

Near peer adversaries are increasingly investing in EW capabilities to include UAS based EW platforms. Uses include electromagnetic support (radio direction finding, threat recognition) and electromagnetic attack (EA) (jamming, spoofing). Jamming communications can add to the battlefield confusion resulting in reduced friendly command and control while the threat conducts offensive operations.

See following pages (pp. 2-8 to 2-9) for an overview and examples.

Transformation of Battle: UAS Employment in the Russo-Ukrainian War *(See pp. 2-24 to 2-25.)*

The widespread employment of small unmanned aircraft systems (UAS) in the Russo-Ukrainian war has fundamentally transformed the character of battle and operational art. Both Russia and Ukraine have leveraged UAS for intelligence, surveillance, reconnaissance (ISR), and precision strike, compressing decision cycles and reshaping the tempo of combat.

Russia's extensive use of Iranian-designed Shahed loitering munitions beginning in late 2022 introduced a new cost asymmetry. Low-cost one-way attack drones enable mass strikes on infrastructure and logistics nodes once reserved for expensive cruise or ballistic missiles. This has forced Ukraine and its partners to expend high-value air-defense interceptors and redirect radar coverage to defend rear areas.

At the tactical level, the fusion of reconnaissance and strike capabilities through first-person-view (FPV) and loitering munitions has blurred traditional combat roles. Small units now employ Group 1 and 2 systems to conduct ISR, designate targets, and immediately transition to attack using kamikaze drones. This has decentralized firepower, expanded engagement ranges, and placed lethal precision within reach of platoon and company-sized formations.

Russian forces have coupled massed UAS employment with aggressive electronic warfare (EW), jamming Ukrainian control links and sensors to overwhelm localized defenses. In turn, Ukraine has adapted with distributed short-range air defenses, camouflage, deception, and rapid innovation in counter-UAS tactics. The result is a dynamic contest for control of the lower airspace, where attrition, adaptation, and tempo determine local advantage.

Operationally, UAS have transformed warfare from maneuver-centric to attrition-oriented engagements characterized by persistent surveillance, rapid targeting, and continuous pressure across depth. Both sides now view small UAS as indispensable force multipliers that extend reach, enhance precision, and redefine the balance between cost, risk, and effect on the modern battlefield.

- Open-source military assessments, Ukrainian General Staff briefings, and NATO defense analysis reports, 2023–2025.

V. Threat Techniques (Examples)

Ref: ATP 3-01.81, Counter-Unmanned Aircraft System (C-UAS) Operations (Mar '25), pp. 15 to 19.

Mass UAS employment by threat forces can produce overwhelming effects on maneuver forces. Understanding threat capabilities as well as current and emerging UAS employment tactics is critical in the effort to counter UAS threats.

UAS may be employed to conduct a wide variety of missions to include ISR, direct, indirect, or mass attack, fire support, harassment/nuisance and EW:

See previous page for an overview and further discussion.

UAS Intelligence Surveillance Reconnaissance

ISR: Surveillance. UAS could parallel dismounted and mounted elements, and in most cases undetected and operating at proximity. This allows for detailed surveillance from multiple vantage points, within the commander's time and space. This may be the most probable COA as it allows for continued use of the system without destruction.

Ref: ATP 3-01.81 (Mar '25), fig. 3-1. Unmanned aircraft system intelligence surveillance reconnaissance role.

UAS Direct Attack

Direct Attack: UAS strikes target and causes damage or injury through force of impact or payload carried.

Ref: ATP 3-01.81 (Mar '25), fig. 3-2. Unmanned aircraft system direct attack.

UAS Indirect Attack

Indirect Attack: In the role of the indirect attack, the UAS can guide indirect fires, observe/spot fires, adjust fires and or serve as the indirect attack weapon or munitions. As an indirect attack platform, the UAS could carry the improvised explosive device or become the improvised explosive device.

Ref: ATP 3-01.81 (Mar '25), fig. -3. Unmanned aircraft system indirect attack.

UAS Swarm Attack

Swarm Attack: Threat uses a swarm of UAS to cause significant damage and overwhelm friendly systems and reactions.

Ref: ATP 3-01.81 (Mar '25), fig. 3-4. Swarm attack.

The advancements in unmanned technologies allow threat actors asymmetrical means to conduct attacks, collect information, or trigger other threatening events. This problem can escalate as UAS technologies become less expensive and more capable, accessible, and adaptable.

VI. Detecting and Neutralizing Mini-Drones

Ref: Detecting and Neutralizing Mini-Drones, The Journal of the JAPCC (ed. 25), Joint Air Power Competence Centre (Mar '18), Daniela Pistoia.

Small (15–150 kg), mini (<15 kg), micro (<66 J energy state) Unmanned Aerial Systems (UAS)1 will drastically proliferate in the near future with rapidly advancing performance and functionalities. Progress in power storage, avionics miniaturization, materials and design methodologies, together with the availability of commercial or open source software applications, will enable increasingly smaller and cheaper platforms for a broadening range of possible uses. While such advanced technology means a huge opportunity for the military and industry, its alternative, sinister use for criminal and terrorist purposes is also no longer a fictitious risk. As many recent examples show, small UAS have become a real threat to both civil and military targets.

The detection, identification, and neutralization of such UAS flying near key infrastructure or sensitive areas (e.g. government buildings, high-profile event locations, prisons, military compounds) has therefore become a critical capability. So far, traditional countermeasures have demonstrated their weakness in this regard. Unconventional threats require more advanced solutions, and many industrial and government initiatives are rising to meet this new threat.

New developments such as advanced passive and active multispectral technologies seem most feasible to deal with the counter-UAS challenge. Multiple domain (electromagnetic, acoustic, electro-optic), multiple sensor (active radar, passive electromagnetic interceptors, acoustic sensors, infrared cameras), multiple jamming/deception system of systems, integrated via a dedicated command & control (C2) capability, are key elements within this approach. Last but not least, exploiting cyber capabilities is an important vector to counter the mini drone threat, though respective solutions are still immature. Superior knowledge, skills, and tools in the cyber domain will probably be the most decisive factor for a successful defence.

Too Small and Simple to be a Threat?

Drones are rapidly becoming 'tools of the trade' in many industries and could be categorized into segments of the market: Government (including Military), Enterprise (Corporations/Businesses) and Consumers (Personal/Hobbyist).

Since low-altitude drones fly only hundreds of feet above ground, they mostly operate outside traditional radar coverage used to track commercial aircraft. Also, military air defence radar systems are usually not designed to detect aircraft with such a small radar cross-section. In other words, there is an airspace segment neither under control of civil authorities nor military air power.

According to the 2016 Field Guide 'Drones Operating in Syria and Iraq' published by the Center for the Study of the Drones at Bard College (Annadale-On-Hudson, New York), 'there are more drones, made in more countries, and flown by more groups, than in any other previous conflict'. Among the 38 different types of UAS, the institute counted at least eight recreational hobby drones and possibly six unidentified home-made models, asserting 'the conflict marks the first time that hobby drones have been modified with explosives and turned into flying improvised explosive devices'.

However, such use of mini-drones may not be limited to zones of war or conflict. The probability of threat proliferation to domestic areas combined with the inability of traditional airspace control and defence to effectively deal with such small and low-flying objects underline even more the pressing need for appropriate counter-UAS technology. Many companies have therefore created intense research & development programmes to provide effective solutions.

Small UAS Detection and Identification

Defending against small UAV threats is a complex issue since it is not only about eliminating the drone to prevent it from completing its mission. Successful defence must ensure the immediate detection and identification of the object prior to neutralizing it in a secure framework for the safety of the people on the scene as well as minimizing collateral damage. Given the physical and kinematic characteristics of the drone and the typical modes of use, a multiple sensor approach is necessary to improve the detection capability. Several options, each with their own strengths and weaknesses are currently being tested:

EM Sensors

Defence systems could exploit the sudden presence of radio signal used to send the commands from the pilot to the drone (uplink) and to send data and images from the drone to the command post (downlink). Those radio signals are transmitted on well-known and standardized frequencies, relatively easy to be intercepted with electronic surveillance in automatic mode, even though complex wave modulation is often superimposed to the carrier signal. Furthermore, passive geo-location techniques can be put in place to locate both the drone and the control station.

Active/Passive Radar

A sensor particularly devoted to the detection of aerial tracks is the radar. However, mini-drones are hard to detect and identify due to very low radar signatures (with a radar cross-section of the order of 0.01 m2). Furthermore, it is a tremendous challenge to distinguish the target from other objects particularly in an urban environment, with a high probability of false alerts. The challenges increase when trying to use passive (bi-static) radar.

Infrared Sensors

Together with electromagnetic sensors, other promising devices could be thermal cameras, usable under low visibility conditions and at night. Infrared sensors could reveal drones even in the presence of strong lighting due to the ability to locate thermal hotspots generated by motors. Such hotspots, located in fixed positions in relation to the structure of the drone, also contribute to automatic object identification by making use of IR image reference libraries.

Acoustic Sensors

During flight, drones generate noise both in the audible frequencies and in the ultra-sounds. Acoustic sensors reveal the presence of mini-drones as well as helping to classify the target based on noise characteristics specific to the drone model. However, the operational range of acoustic sensors is limited to a few hundred metres. At longer distances, drones are lost in background noise. On the other hand, radars have a blind spot at shorter distances. This means the acoustic sensor, made up of an array of microphones, is the ideal complement to radar systems to cover both long and short ranges. Being relatively cheap, acoustic sensors are efficient tools for the continuous surveillance of particularly sensitive areas.

VII. Adapting to the UAS Threat: Implications for the Imperatives of Operations

Imperatives are actions Army forces must take to defeat enemy forces and achieve objectives at acceptable cost. They are informed by the operational environment and the characteristics of the most capable threats Army forces can encounter. *(See p. 1-23.)*

Protect Against Constant Observation and All Forms of Enemy Contact

Small tactical UAS (Groups 1–3) flying low reduce detection time and increase the chance of surprise. Persistent ISR, mounted cameras, and C2 feeds create almost continuous observation if not countered, threatening force protection and movement discipline. To satisfy the imperative, commanders must harden signatures, enforce dispersed and deceptive routines, use camouflage/cover and active counter small-UAS measures at brigade and below, and treat denial of aerial observation as a persistent operational requirement rather than an occasional task.

Create and Exploit Positions of Physical, Information, and Human Advantage

UAS provide adversaries inexpensive physical advantage (overwatch, target acquisition), information advantage (real-time patterns of life, battle damage assessment), and leverage over human decision making (shock, confusion). Friendly forces must therefore deny enemy ISR while building their own multi-domain sensing and information operations to preserve situational understanding. Exploitation also requires turning enemy UAS detections into actionable targeting and deception opportunities — feeding false patterns, manipulating what the enemy observes, and using UAS-derived imagery to shape human and informational effects.

Make Initial Contact with Sensors, Unmanned Systems, or The Smallest Element Possible

Adversary UAS frequently become the initiator of contact — acting as forward observers, C2 relays, or harassment assets. The imperative to make initial contact at the smallest possible scale means leveraging organic sensors and small unmanned systems to detect, classify, and, where appropriate, engage enemy UAS early. That minimizes escalation, preserves larger assets, and allows commanders to fight the first tactical decision with accurate localized information.

Impose Multiple Dilemmas on the Enemy

Swarm and mass-attack tactics, combined with EW (jamming/spoofing) and mixed payloads (direct, indirect, CB/RA threats), force defenders to choose among competing countermeasures under time pressure. To impose dilemmas back on the enemy, friendly forces should integrate layered defenses (hard-kill, soft-kill, decoys), multi-domain deception, and offensive use of UAS/EW to force the enemy to expend resources, lose tempo, or reveal C2 nodes.

Understand and Manage the Effects of Operations on Units and Soldiers

Persistent UAS threats produce cognitive fatigue, movement friction, and logistic/munitions strain from constant counter-UAS actions. Commanders must manage ROE, sustainment of countermeasures, training for UAS encounters, and the psychological effects of harassment/nuisance operations.

Chap 2

II. Threat UAS Overview (Systems & Capabilities)

Editors note: This chapter is intended to give the reader an broad overview of threat UAS systems and capabilities given the current security environment as outlined and addressed by the 2025 DNI National Threat Assessment and the 2025 Homeland Threat Assessment. It is not intended to be exhaustive or all-inclusive. (See following page for a listing of useful UAS and C-UAS resources.)

I. The Security Environment

Russia, China, Iran and North Korea—individually and collectively—are challenging U.S. interests in the world by attacking or threatening others in their regions, with both asymmetric and conventional hard power tactics, and promoting alternative systems to compete with the United States, primarily in trade, finance, and security. They seek to challenge the United States and other countries through deliberate campaigns to gain an advantage, while also trying to avoid direct war. Growing cooperation between and among these adversaries is increasing their fortitude against the United States, the potential for hostilities with any one of them to draw in another, and pressure on other global actors to choose sides.

- Annual Threat Assessment (2025)

The central challenge to U.S. security is the reemergence of long-term, great power competition with China and Russia as individual actors and as actors working together to achieve common goals.

China and Russia *(See pp. 2-15 to 2-30.)*

China uses its rapidly modernizing military, information warfare, and predatory economics to coerce neighboring countries to reorder the Indo-Pacific region to its advantage. Concurrently, Russia seeks veto authority over nations on its periphery in terms of its governmental, economic, and diplomatic decisions, to subvert the North Atlantic Treaty Organization (NATO), and to change European and Middle East security and economic structures to its favor.

North Korea and Iran *(See pp. 2-31 to 2-44.)*

In addition to China and Russia, several other states threaten U.S. security. North Korea seeks to guarantee survival of its regime and increase its leverage. It is pursuing a mixture of CBRN, conventional, and unconventional weapons and a growing ballistic missile capability to gain coercive influence over South Korea, Japan, and the United States. Similarly, Iran seeks dominance over its neighbors by asserting an arc of influence and instability while vying for regional hegemony. Iran uses state-sponsored terrorist activities, a network of proxies, and its missile capabilities to achieve its objectives.

Non-state Actors (Irregular Warfare) *(See pp. 2-45 to 2-48.)*

While states are the principal actors on the global stage, non-state actors also threaten the strategic environment with increasingly sophisticated capabilities. Terrorists, transnational criminal organizations, threat cyber actors, and other malicious non-state actors have transformed global affairs with increased capabilities of mass disruption. Terrorism remains a persistent tactic driven by ideology and enabled by political and economic structures.

II. Useful UAS/C-UAS Resources/Databases (Unclassified / Open Source)

Online Resources & Publications

ODIN Worldwide Equipment Guide (https://odin.tradoc.army.mil/WEG). The OE Data Integration Network (ODIN) Worldwide Equipment Guide (WEG) was developed to support the TC 7-100 series and all OPFOR portrayal in training simulations (constructive, virtual, live, and gaming). The equipment portrayed in this WEG represents military systems, variants, and upgrades that U.S. forces may encounter now and in the foreseeable future. The authors continually analyze real-world developments, capabilities, and trends to guarantee that the OPFOR remains relevant.

CNAS Proliferated Drones "Drone Database" (https://drones.cnas.org/drones/) Proliferated Drones, a project by the Center for a New American Security (CNAS), is designed to address the challenges associated with the expanding use of drones.

Center for the Study of the Drone at Bard College – "The Drone Databook" (https://dronecenter.bard.edu/projects/drone-proliferation/ "The Drone Databook" is an in-depth survey of the military drone capabilities of over 100 countries around the globe that are known to possess and operate unmanned aircraft. This study includes information about each country's active drone inventories, drone units, training programs, active acquisition and development efforts, infrastructure, drone exports, and operational experiences.

Center for the Study of the Drone at Bard College – "Counter-Drone Systems" (https://dronecenter.bard.edu/projects/counter-drone-systems-project/) Now in its second edition, "Counter-Drone Systems," is a research report covering the fundamentals of counter-drone technology. The report includes a database of more than 530 counter-drone products available on the market.

JAPCC: A Comprehensive Approach to Countering Unmanned Aircraft Systems (https://www.japcc.org/books/a-comprehensive-approach-to-countering-unmanned-aircraft-systems/) The JAPCC is sponsored by 14 NATO nations who provide a variety of experienced Subject Matter Experts (SME) that come from all three services.

Wikipedia – "List of unmanned aerial vehicles" and associated country categories (https://en.wikipedia.org/wiki/List_of_unmanned_aerial_vehicles)

Warpower Russia "Russian Drone Force" (https://www.warpowerrussia.com/droneforce.php)

The Iran Primer "Roster of Iranian Drones" (https://iranprimer.usip.org/blog/2024/feb/02/roster-iran%E2%80%99s-drones

38 North "Current Status of North Korea's Drone Program" (https://www.38north.org/2025/09/current-status-of-north-koreas-drone-program/)

C-UAS Training Resources

See p. 4-41 for a listing of C-UAS training resources.

UAS Threat Overview (Systems & Capabilities)

II(a). China

I. Background

Over the past two decades, the People's Republic of China (PRC) has rapidly evolved into one of the world's leading producers and operators of unmanned aerial vehicles (UAVs). What began as a limited reconnaissance effort in the late 1990s has expanded into a comprehensive, multi-domain UAV program encompassing reconnaissance, strike, electronic warfare, and swarm operations. This expansion reflects the People's Liberation Army's (PLA) broader strategy of "intelligentized warfare," which integrates artificial intelligence, autonomous systems, and networked command structures to achieve information dominance. Today, China's drone ecosystem includes state-owned giants like the Aviation Industry Corporation of China (AVIC), Chengdu Aircraft Industry Group (CAIG), and the China Aerospace Science and Technology Corporation (CASC), which collectively form the backbone of both military and export UAV production.

> Several major state actors present proximate and enduring threats to the United States and its interests in the world, challenging U.S. military and economic strength, regionally and globally. **China stands out as the actor most capable of threatening U.S. interests globally**, though it is also more cautious than Russia, Iran, and North Korea about risking its economic and diplomatic image in the world by being too aggressive and disruptive. Growing cooperation among these actors expands the threat, increasing the risk that should hostilities with one occur, it may draw in others.
>
> - *2025 Annual Threat Assessment*

II. Technological and Doctrinal Trends

The PLA's UAV development follows a clear pattern of technological convergence with broader military modernization goals.

Autonomy and AI Integration

China emphasizes autonomy, machine learning, and real-time data fusion. Research institutions such as Northwestern Polytechnical University and CETC (China Electronics Technology Group Corporation) work on algorithms for cooperative control, object recognition, and adaptive mission planning. These advances aim to enable semi-autonomous or fully autonomous drone operations within “intelligentized” combat systems.

UAS Threat

Networked and Multi-Domain Operations

Chinese UAVs are increasingly linked with satellites, ground stations, and manned aircraft to create a comprehensive reconnaissance-strike complex. Drones serve as forward sensors for missile and artillery units, relay nodes for battlefield communications, and decoys or jammers against enemy air defenses. Integration across air, land, sea, and space domains reflects the PLA's shift toward system-of-systems warfare.

Maritime and Over-the-Horizon Surveillance

Maritime reconnaissance has become a key UAV mission area, especially in the South China Sea and the Taiwan Strait. The WZ-7 and other HALE drones are used for continuous monitoring of shipping lanes, foreign naval movements, and island outposts. UAV data is believed to feed directly into the PLA Rocket Force's targeting networks, enhancing China's ability to strike mobile maritime targets.

III. Primary UAV Systems

China fields a diverse portfolio of UAVs ranging from small tactical platforms to large, high-altitude, long-endurance (HALE) systems.

Small Unmanned Aircraft Systems (sUAS)

China's military small UAV (Unmanned Aerial Vehicle) program has rapidly expanded over the past decade, producing a wide range of reconnaissance, loitering, and strike-capable drones that form the backbone of the People's Liberation Army's (PLA) unmanned operations. These small UAVs—typically tactical or short-range systems—are designed for frontline use, providing surveillance, targeting support, and precision strike capabilities. Among the most widely recognized models are the ASN series (such as ASN-206 and ASN-209), developed by Xi'an Aisheng Technology Group, which provide real-time battlefield intelligence and artillery spotting.

(ASN-209) 2023-06-10) An unmanned aerial vehicle (UAV) attached to a brigade under the PLA 82nd Group Army is launched for a recent flight training exercise. (eng.chinamil.com.cn/Photo by Liu Kui)

In recent years, the PLA has increasingly emphasized swarm technology and AI coordination among small UAVs, aiming to integrate large numbers of expendable drones into its battlefield network for saturation attacks and real-time situational awareness. Systems like the CH-901 “suicide drone”, WS-43, and Feihong (FH)-901 can be launched from vehicles, aircraft, or naval platforms, allowing flexible, distributed firepower. These UAVs have also seen extensive export and adaptation by

IV. The Crossover of China's Civilian Drone Industry and Its Military Implications

China's booming civilian drone industry has become a cornerstone of its technological rise—and an increasingly important contributor to national defense. Over the past decade, companies like DJI, Autel Robotics, and EHang have turned China into the world's largest producer of commercial drones, commanding more than 70 percent of the global market. While these drones were initially designed for photography, logistics, and agriculture, the technologies underpinning them—autonomous flight control, miniaturized sensors, data transmission, and swarming algorithms—have found natural applications in the military domain.

This intersection between civilian innovation and defense needs exemplifies Beijing's policy of "military-civil fusion" (MCF), a national strategy that seeks to blur the lines between commercial and military technology development. Under MCF, research conducted by private drone firms and academic institutions is routinely leveraged by the People's Liberation Army (PLA).

The Ukraine conflict and other recent wars have demonstrated the battlefield value of small drones for reconnaissance, targeting, and improvised attacks. China has taken note: while major defense contractors continue to produce large UAVs like the Wing Loong II and CH-5, smaller civilian-derived platforms are being adapted for tactical missions. These include modified quadcopters capable of dropping grenades, drones designed for swarm reconnaissance, and systems fitted with jamming-resistant links. State media and military research institutes have increasingly showcased joint projects where civilian manufacturers collaborate with PLA units to create "dual-use" designs.

This crossover raises global concerns. Many Chinese civilian drones sold abroad could theoretically be repurposed for military or intelligence functions, raising export-control and cybersecurity issues. Western governments have responded with restrictions on the use of Chinese drones in sensitive sectors, citing data-security risks and potential ties to China's defense apparatus. For Beijing, however, the convergence of commercial and military drone ecosystems represents a strategic advantage—one that multiplies its research base, reduces costs, and accelerates deployment cycles.

> Several countries, particularly the United States and its allies, are considering or have enacted restrictions on DJI and other Chinese drone manufacturers due to national security and data privacy concerns. Critics argue that Chinese-made drones could transmit sensitive imagery or flight data back to servers in China, potentially allowing surveillance or mapping of critical infrastructure. Supporters of a ban emphasize that DJI's dominance in the global commercial drone market—estimated at over 70%—poses a strategic risk if supply chains or software updates can be controlled by Beijing. However, opponents of a full ban caution that such measures could disrupt industries reliant on affordable, high-performance drones for agriculture, construction, and emergency response, and note that evidence of systemic data misuse remains limited. The debate thus centers on balancing technological dependence, economic practicality, and national security assurance.

In essence, China's civilian drone success has evolved into a form of soft power and strategic leverage. The same innovation driving consumer markets and logistics efficiency is also reshaping modern warfare. As the line between hobbyist technology and military hardware continues to fade, the global community faces a new reality: the future of aerial warfare may be built as much in Shenzhen's factories as in the PLA's test ranges.

partner states and private defense firms, helping China refine both technology and tactics through external feedback. Collectively, China's small UAV ecosystem demonstrates a deliberate strategy of scalability and redundancy—producing low-cost, easily replaceable drones in massive quantities to overwhelm defenses, extend battlefield awareness, and enhance the PLA's combined-arms operations in future conflicts.

The CH-901, also known as the FH-901, is a type of loitering munition developed by the China Aerospace Science and Technology Corporation in 2016. (Credit: Mike1979 Russia, CC BY-SA 4.0 via Wikimedia Commons.)

The CH-4 and CH-5 (Cai Hong Series)

Developed by CASC, the CH-4 has become China's most recognizable UAV, similar in appearance and capability to the U.S. MQ-9 Reaper. It can carry precision-guided munitions such as the AR-1 missile and has an endurance of around 30 hours. The upgraded CH-5 offers extended range, heavier payload, and satellite communications for beyond-line-of-sight operations. These drones are actively deployed by the PLA and widely exported to countries across the Middle East, Africa, and Asia, providing China with both financial and strategic benefits.

CH-5 at Airshow China Zhuhai 2022. (Credit: Infinty 0, CC BY-SA 4.0 via Wikimedia Commons.)

Wing Loong II and Derivatives

Produced by CAIG, the Wing Loong family represents China's most commercially successful MALE (Medium Altitude, Long Endurance) UCAV line. The Wing Loong II can carry up to 12 missiles and operate for more than 20 hours. Its export versions are in service with Egypt, the UAE, Saudi Arabia, and Pakistan, among others. The PLA Air Force (PLAAF) uses the Wing Loong series for reconnaissance-strike missions, border patrols, and joint exercises simulating contested airspace.

Wing Loong II side view, Dubai Air Show 2017 (Credit: Mztourist CC BY-SA 4.0 via Wikimedia Commons).

High-Altitude and Strategic Drones: WZ-7 Soaring Dragon and WZ-8

The WZ-7, developed by Guizhou Aircraft Industry Corporation, is a HALE UAV featuring a distinctive "joined-wing" design optimized for long-range intelligence, surveillance, and reconnaissance (ISR) missions. It provides the PLA with persistent coverage over maritime zones and potential targets such as U.S. carrier groups. The WZ-8, by contrast, is a rocket-powered, high-speed reconnaissance drone launched from H-6 bombers, capable of penetrating deep into enemy airspace. These platforms are believed to support the targeting of long-range anti-ship ballistic missiles such as the DF-21D and DF-26.

WZ-7 at Airshow China Zhuhai 2022. (Credit: Infinty 0, CC BY-SA 4.0 via Wikimedia Commons.)

V. Challenges and Limitations

Despite rapid progress, China's UAV program faces several constraints. First, sensor and data-link technology—particularly satellite communication bandwidth and secure transmission—still lags behind that of leading Western systems. Second, reliability and maintainability issues have been reported by some export customers, reflecting differences in manufacturing quality and operational experience. Third, the PLA must still refine joint command and control structures to fully integrate drones with manned aircraft and ground forces in complex combat environments.

Additionally, the proliferation of counter-UAV technologies (jamming, laser defenses, kinetic interceptors) presents new challenges. As adversaries improve their defenses, Chinese drone survivability in high-threat airspace may depend on continued advances in stealth, autonomy, and electronic warfare resilience.

VI. Chinese Military Drone and UAV Support for an Attack on Taiwan

China's expanding fleet of military drones and unmanned aerial vehicles (UAVs) would play a critical role in any potential campaign against Taiwan, transforming the operational dynamics of such a conflict. Over the past decade, the People's Liberation Army (PLA) has invested heavily in a layered UAV architecture that integrates surveillance, strike, electronic warfare, and swarm operations. This unmanned network gives Beijing a significant advantage in intelligence, targeting, and battlefield coordination, which could enable more precise and sustained pressure on Taiwanese defenses while minimizing the exposure of manned aircraft.

At the Strategic Level

At the strategic level, high-altitude, long-endurance (HALE) platforms like the WZ-7 Soaring Dragon and WZ-8 would provide persistent reconnaissance over the Taiwan Strait, tracking Taiwanese air and naval movements and relaying data to missile forces. These UAVs could also support China's anti-access/area denial (A2/AD) strategy by helping guide long-range missiles against key targets such as radar stations, command centers, and air bases. Operating from mainland airfields well beyond Taiwan's defensive range, they would give the PLA continuous situational awareness and near-real-time battle updates—a critical advantage in the opening hours of any conflict.

Medium-altitude, long-endurance (MALE) drones like the Wing Loong II and CH-5 would likely serve as strike and reconnaissance assets during initial assaults, targeting air defenses, artillery, and troop concentrations. Their ability to loiter for extended periods and conduct precision strikes makes them ideal for suppressing Taiwan's coastal defenses or supporting amphibious landings. In conjunction with electronic warfare drones, they could also jam radar systems or communications, creating confusion and degrading command-and-control networks across the island.

At the Tactical Level

At the tactical level, small and swarming drones would be deployed for reconnaissance, decoy operations, and saturation attacks. Swarm-capable platforms—potentially launched from larger "drone carrier" UAVs such as the experimental Jiutian SS-UAV—could overwhelm Taiwanese air defenses through sheer numbers, saturating radar systems and creating false targets. These tactics could buy time and space for manned aircraft and missile salvos to penetrate deeper into Taiwanese airspace.

The civil-military integration of China's drone industry further amplifies this threat. Commercial drones from companies like DJI could be rapidly repurposed for battlefield reconnaissance or logistics support, providing low-cost, expendable surveillance coverage in urban and coastal areas. Combined with the PLA's extensive satellite and cyber capabilities, these drones would form part of a comprehensive intelligence network designed to identify and exploit weaknesses in Taiwan's defense posture.

For Taiwan and its partners, the implications are profound. The island's air defenses, already stretched by frequent Chinese incursions, would face a massive influx of unmanned threats that are cheaper, more expendable, and more numerous than traditional aircraft. Counter-drone strategies—ranging from electronic warfare to kinetic interceptors—would need to be integrated at every level of defense. Moreover, the PLA's reliance on drones underscores a broader shift in warfare: toward information dominance, persistent ISR, and unmanned precision strike, where the side that sees and reacts first holds decisive advantage. In any future Taiwan contingency, China's UAV force would not merely support its operations—it could define them.

Emerging Systems: Jiutian SS-UAV and Swarm Concepts

A newer concept known as the Jiutian SS-UAV—sometimes described as a "flying aircraft carrier for drones"—represents China's ambition to integrate swarm tactics into operational doctrine. This large UAV can reportedly carry and deploy dozens of smaller drones for reconnaissance, jamming, or strike missions. In parallel, Chinese defense industries are developing AI-driven swarm coordination technologies, integrating multi-UAV operations with electronic warfare and cyber capabilities.

The Jiutian SS-UAV represents a potential leap in unmanned aerial systems: a high-altitude, long-range platform not just for individual missions, but as a mothership that carries and deploys large numbers of smaller UAVs, loitering munitions, or sensors. If it enters full operational service under the People's Liberation Army or associated forces, it could significantly alter the calculus of unmanned warfare — especially in contested regions like the Taiwan Strait or South China Sea. That said, many of its claimed capabilities are still under development, and real-world effectiveness in contested scenarios remains unproven.

The "mothership" concept: rather than just being a conventional UAV, the Jiutian is designed to act as a platform that carries, launches, coordinates or supports many smaller unmanned systems (drones, loitering munitions) as part of a distributed system. For example, it might carry dozens of mini-drones, drop them into a theatre of operations, and serve as a command, relay, or coordination node.

Key Features & Specifications

- Wingspan reported around 25 meters (≈ 82 ft) and a maximum take-off weight in the 16-ton class.
- Operational ceiling of up to 15,000 meters (≈ 49,000 ft) and a range of up to ~7,000 km (≈ 4,350 miles) in some claims.
- Payload capacity reportedly around 6 tons (≈ 6,000 kg) for weapons, or the equivalent weight in smaller drones/loitering munitions.
- It features an internal bay or "hive" module for deploying large numbers of smaller UAVs/loitering munitions — some sources say up to 100+ such drones.
- It also reportedly carries under-wing hardpoints (e.g., 8 hardpoints in some sources) for additional weapons, sensors, or external payloads.

VII. Ukraine's Dependence on Chinese Drones

Ukraine loses an estimated 10,000 drones per month. Ukraine is heavily dependent on Chinese commercial drones and Chinese-made components for reconnaissance and many indigenous armed-drone programs. This dependence spans everything from small quadcopters (DJI-style MAVIC clones) up to parts used in larger tactical and loitering munitions — making China a critical node in Kyiv's unmanned systems supply chain and a potential strategic lever.

The most common battlefield workhorses are light commercial multirotor drones (sub-15 lb class) — almost all manufactured in China (estimates as high as ≈90–98% by unit count in that weight class).

Key components (flight controllers, motors, navigation cameras, batteries, antennas and certain RF/navigation modules) are also produced in China and integrate into "Ukrainian" systems — meaning many Ukrainian-assembled drones remain reliant on Chinese hardware.

Recent Shifts & Export controls

China introduced export controls and restrictions (late-2024 onward) on certain drone components and dual-use items. Those controls, plus international pressure and sanctions concerns, have intermittently disrupted shipments and forced Ukrainian firms and units to stockpile, re-engineer, or search for alternative suppliers. At the same time, investigative reporting shows Chinese components continue to flow into the conflict zone (including in ways that evade scrutiny), illustrating the difficulty of fully severing those supply chains.

Operational and Strategic Consequences

- **Vulnerability to supply shock / leverage**: Reliance on a small number of component suppliers creates a chokepoint. If China tightens controls, Ukraine's attrition-heavy drone operations (many drones are expendable) face immediate shortages.
- **Adaptation & innovation**: Kyiv's industry has accelerated domestic modifications, indigenization of flight controllers and software, and development of upgraded variants to resist EW (electronic warfare). Ukraine is also diversifying procurement, using parts from Turkey, Europe, and improvised local solutions.
- **Dual-use diffusion**: The same Chinese components power Russian and Iranian systems; this diffusion complicates sanctions and verification and provides adversaries with similar technological building blocks.

Example - Fiber-Optic Guidance Cables

Recent trade-data analyses reveal that the Chinese export of Fiber-Optic (guidance) cables has surged dramatically to Russia, while shipments to Ukraine remain minimal. According to a detailed review by the Washington Post of Chinese customs statistics, China exported a record ~328,000 miles of Fiber-Optic cable to Russia in August 2025, compared with only 72 miles to Ukraine in the same month. This stark imbalance provides concrete indication that China's role in the drone-supply chain is tilting in favour of Russia's unmanned-systems capability (particularly those using Fiber-Optic guidance wires) rather than supporting Ukraine's needs.

The implications of this disparity are significant for Ukraine's drone-campaign resilience. While Ukraine has relied heavily on Chinese-origin commercial UAVs and components, the minimal supply of critical guidance-cable materials means Kyiv faces an effective bottleneck in building or sustaining advanced loitering drones with Fiber-Optic control wires, whereas Russia is able to upscale production.

UAS Threat Overview (Systems & Capabilities)

II(b). Russia

I. Current Landscape

Over the course of the war in Ukraine, the Russian Federation has dramatically expanded its unmanned aerial systems (UAS) capabilities — shifting from mostly reconnaissance drones to massed strike and saturation-attack platforms. Russia is now manufacturing at a scale that challenges traditional air defences. For example, Russia reportedly produced more than 34,000 strike drones and decoys by mid-September 2025, nearly nine times the output of the same period in 2024.

A major factory in Yelabuga (Tatarstan) is described as possibly the world's largest drone factory, producing variants of the Geran-2 (and Shahed-derived) loitering munitions around the clock.

Industrial buildup is being supported by a domestic drone supply-chain drive: in 2025 Russia launched a lithium-ion battery facility in Tolyatti expected to produce over 100,000 small UAV units per year, and a rotor/blade factory in Novosibirsk for both military and civilian drones.

At the same time, Russia is moving to field more advanced navigation systems designed for GPS-denied environments, reflecting the electronic warfare (EW) environment on the Ukraine battlefield.

Tactically, Russia is employing swarms of low-cost attack drones to saturate Ukrainian air-defences and impose logistic/infra costs. A clear example: in one night Russia launched ~860 attack drones and missiles, and subsequent incursions saw 19 drones cross into Polish airspace, prompting NATO alert responses.

Russia's rapid scale-up of UAS production and deployment is shifting the paradigm of aerial warfare. Cheap, mass-produced strike drones allow saturation attacks that force defenders to expend high-cost interceptors for low-cost threats. The combination of loitering munitions, tactical FPV drones, and large numbers of quad/winged strikes imposes material and psychological costs. Meanwhile, increased industrial investment and the creation of dedicated drone units (e.g., the Russian drone-warfare specialised centers) signal that unmanned systems are now embedded institutionally.

II. Use of UAVs/Drones in the Russo-Ukrainian war (2022–present)

During the war between Russia and Ukraine, both sides have employed drones for reconnaissance, long-range strikes and support of ground operations. The Ukrainian military made extensive use of Turkish-made TB2 drones in their operations against Russian forces. On the Russian side, waves of Iranian-designed SHAHEED-136 drones were used in the October 2022 missile strikes on Ukraine. Drones' principal roles, however, have been in reconnaissance and artillery spotting, rather than purely independent bombing missions.

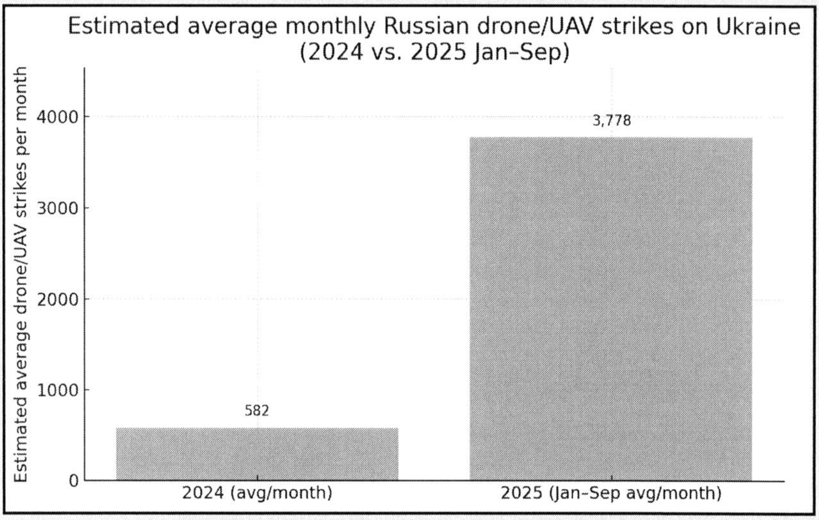

Estimated drone strike rates rose sharply from 2024 to 2025. Ukrainian officials reported roughly 6,987 attack UAVs launched at Ukraine in 2024 (≈582/month on average). By contrast, open-source reporting and U.S./Ukrainian officials recorded a massive surge in 2025 — journalists and analysts cite roughly 34,000 kamikaze drones and decoys sent at Ukraine through September 2025 (≈3,778/month for Jan–Sep), driven by rapid Russian domestic production, use of decoys, and mass-launch saturation tactics; analysts caution these counts mix attack drones and decoys and sources vary, so figures are best read as order-of-magnitude estimates rather than precise counts.

Russian sources claim to have employed a "Stupor anti-drone rifle" to jam the radio controls of Ukrainian drones, and there have been multiple encounters between drones of both sides. For example, above the Donetsk region a Ukrainian DJI Mavic quadcopter reportedly rammed a Russian drone of the same type, causing it to crash. On 24 November 2022 another collision was recorded, this time with a Russian Mavic plunging to earth after a Ukrainian drone rammed it. On 9 May 2023 a Russian conscript is said to have surrendered via a Ukrainian drone, rather than through conventional ground contact. The cost dynamic is striking: some sources estimate the SHAHEED-136 drone at about US$20,000 each, whereas the Ukrainian air defence IRIS-T missile costs roughly US$430,000—highlighting the economic asymmetry of the drone vs missile/higher-tier defence equation.

In the maritime arena since at least September 2022, Ukraine has reportedly deployed black naval drones, equipped with satellite-internet connectivity (via Starlink) to carry out attacks on the Russian Black Sea fleet at the Sevastopol naval base. What began as reconnaissance drones appear also to be weaponized, and observers note that the sensors on the front of these naval drones could double as a laser-range finder to assist in targeting. In late October 2022 seven of these drones were used in a successful attack on the Sevastopol base.

On 13 October 2022 a Ukrainian-flown MiG-29 reportedly became the first manned aircraft to be brought down in combat as a consequence of a drone's involvement: the pilot is claimed to have destroyed a SHAHEED-136 drone with its cannon, and the explosion apparently caused the plane to crash and the pilot to be hospitalized. (Specific independent verification is limited.)

By September 2023, Ukrainian forces were reportedly using improvised "cardboard" drones fitted with GoPro cameras for aerial reconnaissance, illustrating the extent of adaptation and resourcefulness in unmanned systems.

Since the 2022 invasion, Ukraine has seen the emergence of roughly 30 companies to mass-produce drones for the war effort. The Ukrainian Ministry of Digital Transformation initiated the "Army of Drones" project, with plans to purchase up to 200,000 drones in 2023 in an effort to offset Russia's advantages in conventional military equipment. The government also sponsored competitions inviting dozens of drone-developers across Ukraine to simulate strikes, pursue fixed-wing drones, or participate in "drone dogfight" contests.

By April 2024 Ukraine had begun deploying manned aircraft such as the Aeroprakt A-22 Foxbat and Yakovlev Yak-52, outfitted with snipers or machine-guns to shoot down Russian drones. Ukrainian drones were also being modified for new roles: for instance, in July 2024 an FPV (First-Person View) drone armed with a simple stick reportedly rammed and destroyed a ZALA 421-16E reconnaissance drone. Meanwhile, tether-controlled drones using optical fiber (to avoid radio jamming) became more common.

On 31 July 2024 a Russian Mi-8 helicopter was shot down over occupied Donetsk by a Ukrainian FPV drone—the first time a helicopter in combat was destroyed by a drone. The Mi-8 is believed to have been hit while on the ground during landing or take-off.

Russia launched what was reported as the war's largest drone strike on 25 November 2024, when 188 drones attacked Ukraine. Ukrainian forces shot down 76 and lost track of 96—likely due to electronic jamming. Other drones struck Ukraine's energy grid and civilian homes.

On 8 February 2025, Russian authorities reportedly uncovered a plot in which FPV drone headsets loaded with explosives (10-15 g each) were sent to Russian soldiers, programmed to detonate on activation. Officials compared it to electronic device attacks in Lebanon in 2024. Subsequent reports claimed eight Russian FPV pilots lost their eyesight between 4 and 7 February following explosions in Belgorod oblast, Kursk, Luhansk and Donetsk regions.

The largest combined drone-and-missile attack by Russia since the full-scale invasion occurred between 18–25 May 2025. Russian forces launched approximately 1,000 aerial weapons—comprising about 355 SHAHEED drones, ~23 Iskander ballistic missiles and over 60 air-launched cruise missiles (including the X-101 and X-555). The assault struck Kyiv and multiple other regions, killed more than 15 civilians and injured dozens, while sharply increasing pressure on Ukraine's air defenses and supply chains.

On 1 June 2025 Ukrainian FPV drones, smuggled into Russia and launched from trucks, reportedly struck Russian airbases (including Belaya and Olenya), destroying multiple Russian strategic bombers. Beginning in August 2025, Ukrainian drone attacks on Russian oil refineries are said to have triggered a fuel crisis in Russia.

III. Primary UAV Systems

These HALE/MALE systems are relatively expensive and produced at lower rates than small tactical drones, but they provide persistent ISR, target cueing for long-range fires, maritime surveillance, and a growing strike role when armed.

Sukhoi S-70 "Okhotnik-B"

Role & capability: stealthy, flying-wing heavy UCAV intended for long-range strike, ISR and "loyal wingman" operations with manned fighters. Large payload, turbofan propulsion and advanced avionics; prototype flight tests and some limited operational testing reported. Its scale and ambition place it in the HALE/UCAV class. An out of control S-70 was deliberately shot down by a Russian Su-57 over Ukraine in October 2024.

The Sukhoi S-70 Okhotnik-B ("Hunter-V"), a Russian heavy stealth unmanned combat aerial vehicle (UCAV). (Photo credit: Russian Ministry of Defence.)

Altius / Altius-M / "Altair" family

The Sokol Altius (also Altair) is a medium-altitude long-endurance Russian unmanned combat aerial vehicle developed by OKB Sokol (formerly known as Simonov Design Bureau) and Tranzas to perform reconnaissance, strike and electronic attack missions on behalf of the Russian Aerospace Forces and Navy. The Altius is comparable to the striking and reconnaissance capabilities of the US-made MQ-9 Reaper and RQ-4 Global Hawk UAVs.

Altius-U drone (Photo credit: Russian Ministry of Defence.)

Small Unmanned Aircraft Systems (sUAS) and Suicide /Loitering Munitions

Russia fields a growing array of low-cost tactical drones, drop-in loitering munitions, and simple winged decoys; combined employment (mixing Lancet, Geran-2, FPV and decoys) is a hallmark of saturation raids to overwhelm layered air defences.

Orlan-10 (Tactical ISR / Multi-Role Small UAS)

Role & capability: ubiquitous fixed-wing tactical ISR drone used in multi-aircraft groups for reconnaissance, EW support and data relay. Cheap, fielded in large numbers and a backbone of Russian tactical sensing. Over 1,000 produced and heavily used in Ukraine and other theaters.

Preparation for the launch of Russian Orlan reconnaissance drones. (Photo credit: Russian Ministry of Defence.)

ZALA Lancet & KUB series

Role & capability: loitering munitions (Lancet, KUB/BLA family) designed to seek and strike point targets (vehicles, radars, artillery). Lancet is a prominent precision loitering munition used operationally; ZALA (Kalashnikov group) produces several variants. These are tactical-to-operational strike tools.

Geran-2 / Shahed-derivatives and Gerbera

Role & capability: Geran-2 (Russian designation for Shahed-136 derivatives and Russian-built variants) is a delta-wing loitering munition used for deep strikes and saturation attacks against infrastructure. Russia has also developed lower-cost decoy/relay variants (e.g., “Gerbera”) to overwhelm air defences. These systems are central to Russia’s mass-launch saturation tactics.

The newly unveiled Russian mobile Geran-2 loitering munition launcher, mounted on a KamAZ-6350 8x8 truck, spotted during the Victory Day 2025 parade rehearsal in Alabino. (Picture source: Russian Social Network VK.)

Small quadcopters / FPV attack drone

Role & capability: modified commercial quadcopters and purpose-built FPV racing frames carrying shaped charges or explosives for short-range precision attacks.

Forpost-R

The Forpost-RE is a Russian reconnaissance and combat drone, based on the Israeli Searcher Mk II and developed into an "absolutely new and already Russian vehicle" with key components replaced. This armed drone can perform long-endurance surveillance and strike missions, carrying up to 100 kg of guided missiles or bombs, such as the KAB-20.

A Forpost in 2011, this is most likely an Israeli Searcher Mk II as it predates full domestication of the production line. (Credit: Letagur, CC BY-SA 3.0 via Wikimedia Commons.)

IV. Limitations

Russia's drone development faces growing limitations. Heavy combat losses in Ukraine are depleting both drone inventories and trained operators faster than replacements can be produced. Western sanctions have cut access to advanced microelectronics, optics, and avionics, forcing reliance on lower-quality domestic or Iranian components. The intense electronic-warfare environment demands costlier, more autonomous systems, increasing expenses. Russia can still build large numbers of simple loitering munitions, but sustaining advanced, long-endurance, or autonomous drone programs will remain a serious challenge.

Russia's Challenges

Even as Russia has proven resilient, it faces a myriad of challenges to remaining an indispensable global player, maintaining a sphere of influence, and upholding stability at home—its highest strategic aims—suggesting limits on its confidence dealing with the United States and the international community. Russia has paid a heavy price in blood, treasure, and loss of international reputation and foreign policy options because of its large-scale invasion of Ukraine. President Putin upended two decades of Russia's geopolitical resurgence, created new threats to its external and internal security, and strained its economic and military potential, making it more reliant on China and other like-minded partners like North Korea.

- Russia's military has suffered more casualties in Ukraine than in all of its other wars since World War II (750,000-plus dead and wounded), and its economy faces significant long-term macroeconomic headwinds and is increasingly dependent on China.
- Russia's aggression has strengthened European unity and prompted Finland and Sweden to join NATO. Efforts by Armenia, Moldova, and some Central Asian states to seek alternative partners highlight how the war has hurt Moscow's influence, even in the post-Soviet space, and derailed Putin's vision of a greater Eurasian union.

- *2025 Annual Threat Assessment*

V. Spillover into NATO (Drone Incursions)

The recent drone incursions into Poland and Denmark highlight a new phase of aerial challenge for NATO: high-volume unmanned systems, cross-border ambiguity, and critical-infrastructure targeting. The alliance must accelerate improvements in detection, engagement, and coordination across member states — while clarifying rules of engagement, deterrence signalling, and collective response options. NATO's ability to adapt to these drone-borne threats will shape its deterrence posture in Europe for years ahead.

Poland

On the night of 9-10 September 2025, about 19 to 23 drones entered Polish airspace during a major Russian strike on Ukraine. At least three drones were confirmed shot down. The incursion was the first time Russian drones (or objects attributed to Russia) flew from the direction of Belarus into NATO territory. In response, Poland invoked Article 4 of the NATO Treaty, requesting consultations with its allies.

Denmark

In late September 2025, Denmark reported a series of drone sightings over airports and military bases. The busiest airport in Copenhagen was shut for several hours after drones were observed, and a second airport in Aalborg was closed around midnight. Authorities described the incidents as a "systematic" or "professional" hybrid attack targeting critical infrastructure. NATO announced it would bolster its Baltic Sea region surveillance and defence posture in response.

Implications for NATO

Escalation and Deterrence Test

These incursions mark a shift in the scale and boldness of unmanned aerial threats penetrating NATO territory — from reconnaissance to actual engagement and air-defence responses. They test how the alliance reacts under ambiguous aggression and whether the collective defence guarantee remains credible.

Defense Gaps Exposed

Both incidents reveal vulnerabilities in ally airspace — particularly low-altitude detection, rapid engagement, and counter-drone capabilities. Analysts argue NATO's air-defence mesh needs denser radars, better integration across borders, and delegated decision authority for shoot-down.

Alliance Coordination & Burden-Sharing

Poland and Denmark have signalled the need for more allied resources and permanent posture enhancements. NATO assets are being repositioned (fighters, frigates, ISR platforms) to reinforce eastern and Baltic flanks.

Hybrid Warfare Dimension

These drone activities, especially when linked to critical infrastructure and civilian airports, blur the line between militarized raids and hybrid/grey-zone operations. NATO must adjust strategy to address a spectrum of threats below full conventional war.

Political and Legal Consequences

Invocation of Article 4 (consultation mechanism) by Poland signals serious concern. If such incursions increase, discussions may shift to Article 5 (collective defence) thresholds, how to attribute responsibility, and how NATO responds without triggering wider war.

VI. Russia's Drone Dependence and Mass-Production Surge

Russia's recent drone campaign blends imported designs and components with an accelerating domestic industrial effort. Early in the war Moscow relied heavily on Iranian loitering-munitions (Shahed/Geran-series) for long-range strike capacity; that transfer jump-started Russian operational use of delta-wing kamikaze UAVs and remains an important technological and doctrinal influence on Russia's mass-launch tactics.

Parallel to imports, Russia has prioritized scale. Large new facilities — most prominently the Yelabuga plant in Tatarstan — were established or repurposed to produce Geran-2/Shahed-type drones and simpler reconnaissance airframes around the clock. Open-source reporting and satellite imagery indicate high production rates at Yelabuga and other sites, enabling Moscow to field sustained, high-volume strike sorties that aim to saturate Ukrainian air defences.

China has emerged as a key supplier of dual-use components that materially ease Russian production constraints. Investigations and reporting document Chinese shipments of engines, fiber-optic cable, and lithium-ion batteries — sometimes concealed under benign descriptions — that are critical to sustaining Russian sortie rates and to improving range and reliability of attack drones. Beijing officially denies exporting lethal items, but the flow of industrial inputs has been linked to the Kremlin's ability to increase output and diversify platforms.

Third-country facilitation — through intermediaries, intermediated parts markets, and permissive commercial channels — has also helped Moscow mitigate Western sanctions on avionics and sensors. Russia has supplemented imports with reverse engineering, local substitution (including reworking Chinese modules), and sourcing via sympathetic or opaque trading networks. These stop-gaps permit mass production of lower-cost loitering munitions and enable higher sortie volumes, but they often come with quality and reliability trade-offs.

That trade-off between quantity and quality is central to Moscow's strategy and its limits. Mass production (and reported output in the tens of thousands of strike drones in 2025) gives Russia the ability to conduct saturation raids that strain defenders' economics and logistics, yet reliance on foreign components and sanctioned supply chains creates strategic fragility. Targeted sanctions, export controls, and interdictions can raise costs and slow improvements in autonomy, sensors, and propulsion that are needed for high-end MALE/HALE capabilities. Meanwhile, large factories and lengthy supply chains create attractive targets: strikes on manufacturing hubs (e.g., Yelabuga) and efforts to choke key inputs could progressively degrade both quantity and quality.

Russia's current drone posture is a hybrid of foreign-sourced know-how and components (notably Iranian designs and Chinese parts) plus an aggressive domestic scaling effort. That mix enables high sortie rates and saturation tactics today, but it also embeds vulnerabilities — dependence on external suppliers, quality control issues from rapid scaling, and exposure of manufacturing hubs — that could be exploited to blunt Moscow's longer-term shift to more capable, resilient unmanned systems.

UAS Threat Overview (Systems & Capabilities)

II(c). Iran & Middle East

Ref: Roster of Iran's Drones (dated April 12, 2024), United States Institute of Peace, The Iran Primer, https://iranprimer.usip.org/blog/2024/feb/02/roster-iran%E2%80%99s-drones (accessed 10 Oct 25).

I. History

Iran first developed unmanned aerial vehicles (UAVs) during the 1980-1988 with Iraq, when it was unable to replace U.S.-made warplanes shot down by Iraq. Getting access to spare parts to maintain its warplanes also grew harder because of U.S. sanctions. As the war dragged on, Iran turned to UAVs as alternatives to manned aircraft.

Iran's early models had vintage technology. Its first drone, the Mohajer-1 had only one camera—with film that had to be developed after recovery—for surveillance. Its radios to communicate with operators on the ground were comparable to a ham radio; they could be easily jammed, and thus disabled, by Iraqi forces. In the 1990s, after the war, Iran installed more sophisticated navigation systems that used GPS and higher resolution cameras that had longer ranges.

Iran developed UAVs with longer ranges that could carry bombs and missiles in the 2010s. In 2012, the new Shahed-129 could fly up to 24 hours and had a range of 1,700 kilometers (more than 1,000 miles). It was about the size of a U.S. Predator drone and had similar capabilities. Subsequent versions could carry up to eight precision-guided missiles or bombs. The Shahed-129 became the backbone of Iran's drone fleet. In 2014, Iran reportedly deployed the Shahed-129 with its forces in Syria to use against rebels and ISIS during the civil war.

In the early 2020s, Iran developed a new generation of suicide drones with advantages over ballistic missiles that had comparable ranges. They were cheaper. One Shahed-136 reportedly cost $20,000, while a Russian Kalibre cruise missile cost $1 million. They were more portable. The Arash, about 10 feet long, could reportedly be launched from a regular car. Their parts were easier to smuggle. And suicide drones were harder to detect--either visually or by radar—than ballistic missiles. They were not as noisy. They do not produce high thermal signatures, which makes them harder than ballistic missiles to track by satellites.

Spanning more than 4.6 million square miles, the Middle East has for millennia been a geographic and geopolitical crossroads and site of cooperation, competition, and conflict. The greatest single day-to-day threat to regional security and stability remains Iran, which challenges the U.S. and its allies by pursuing regional hegemony, breaching its JCPOA commitments, and posing a conventional threat to partner nations while facilitating and conducting coercive and malign activities. Iran will continue to use Syrian (and likely Iraqi) territory as a critical hub and resupply route for maintaining its campaign against Israel. Wielding the most advanced UASs and ballistic and cruise missiles Iran can design, build, and smuggle into Yemen, the Houthis have targeted Saudi Arabia's largest cities and its critical oil infrastructure.

II. Recent Developments

By 2024, Iran's military-industrial complex had produced thousands of advanced drones used for surveillance, reconnaissance, and combat used against U.S. forces as well as American allies in the Middle East. The drones (unmanned aerial vehicles) have become one of the greatest assets of the Iranian-backed "Axis of Resistance"—a network of militias in Iraq, Syria, Lebanon, and Yemen—and one of the greatest dangers to Iran's regional rivals. Among notable attacks that marked turning points in diverse Middle East conflicts:

- On Jan. 28, 2024, an Iranian-made drone attacked Tower 22, a U.S. outpost in Jordan along the Iraq and Syrian borders, that killed three American soldiers and injured more than 40.
- In January 2024, 18 Iranian-designed drones were fired by the Houthis in Yemen on ships in the Red Sea.
- In September 2019, drones and cruise missiles fired from Iran attacked Saudi Arabia's largest oil processing facility, temporarily crippling almost half of the kingdom's petroleum output.

Iranian-backed militias began using drones against U.S. military and diplomatic targets in Iraq and Syria in 2021. In 2021 and 2022, Iranian-designed drones were linked to at least 20 percent of attacks launched by militias linked to Tehran, according to War on the Rocks. Drone attacks on targets in Syria, Iraq and Jordan increased again after war erupted between Hamas and Israel on Oct. 7, 2023. Iran has also supplied hundreds of drones to Russia, beginning in 2022, that have been used against civilian and military targets in Ukraine. Russia established its own industry to replicate Iranian drone models.

Drones "pose the most immediate threat to Middle East security because of their low cost, widespread availability, and potential deniability—since their point of origin can be disguised by employing a convoluted flight path," Gen. Kenneth F. McKenzie Jr., former head of U.S. Central Command, wrote in a paper for the Washington Institute. The technology has produced "an inflection point in aerial warfare—comparable to the introduction of manned flight more than a century ago—that has ended the guarantee of U.S. air superiority over its forces and bases," he warned in February 2023.

See facing page for continued discussion.

III. Primary UAV Systems

Iran has more than a dozen models of combat drones that can attack ground, sea or air targets and then return to base. Larger models, such as the Shahed-149, have ranges of 2,000 kilometers (1,240 miles) and can carry up to 500 kilograms (1,100 pounds) of munitions or electronic equipment.

Most of Iran's combat drones, including long-range models, also have surveillance capabilities. Their functions range from taking photographs and video footage to marking targets for bombers, fighters or other drones. Iran also has smaller surveillance-only drones, such as the Hodhod-1, which have ranges as short as 30 kilometers (18 miles). They can only remain in the air for an hour or two at a time.

Some of Iran's more advanced UAVs/UCAVs include:

Mohajer-6

Multi-role UAV capable of reconnaissance and strike; can carry precision-guided munitions. The Ppehpad Mohajer-6 (literally: "Immigrant Drone") is an Iranian single-engine multirole ISTAR UAV capable of carrying a multispectral surveillance payload and/or up to four precision-guided munitions. It complements the larger Shahed 129 operated by the IRGC Aerospace Force. The drone has also been delivered to the Iranian Army.

In September 2022, a Russian Telegram post claimed, that Russia used a Mohajer-6 to guide Geran-2 suicide drones to their targets in Ukraine during their Invasion of

Examples of Iranian Drone/UAS Use

Since 2015, Iran has reportedly launched drones on Saudi oilfields, dissident groups in Pakistan and Iraqi Kurdistan, and jihadi groups in Syria. In 2018, the Islamic Revolutionary Guard Corps (IRGC) claimed to have conducted 700 drone strikes against ISIS targets in Syria alone. Iran has deployed armed drones beyond its borders. Tehran began exporting drone technology to Hezbollah, a Shiite militia in Lebanon, in the 2000s. In 2004, Hezbollah became the world's first non-state actor to use military drones. And in 2006, Hezbollah – with Iran's help – used armed drones during its 34-day war with Israel. Iran has exported drones or the technology to six other proxy militias and five governments, including Russia, Venezuela, Sudan, Ethiopia and Tajikistan.

Iran–Israel War (13 June – 25 June 2025)

The Iran–Israel war (13 June – 25 June 2025), also known as the Twelve-Day War, was an armed conflict in the Middle East. The war began when Israel bombed military and nuclear facilities in Iran in a surprise attack, assassinating prominent military leaders, nuclear scientists, and politicians, and damaging or destroying air defenses. Iran retaliated with over 550 ballistic missiles and over 1,000 suicide drones, hitting civilian population centers, one hospital and at least twelve military, energy, and government sites. The war followed a decades-long proxy war. Iran has called for Israel's destruction, and armed Hamas in Gaza and Hezbollah in Lebanon.

2025 Drone Attacks across Iraq

The 2025 Iraq drone attacks are a series of ongoing drone attacks throughout Iraq, likely carried out by the Iranian aligned Iraqi militias. The targets of the drone attacks include mostly the Iraqi and foreign military facilities and oil refineries.

The attacks on the oilfields marked significant escalation in regional tensions and imposed substantial threat to Iraqi energy infrastructure. The assaults hit multiple facilities including those operated by international companies and lead to the suspension of production and considerable production losses. No definite group took responsibility of the attacks, however the Kurdistan regional Government accuses the Iranian backed militias in Iraq for the repeated assaults.

Hamas Drone Attacks

On 7 October 2023, Hamas launched incursions of southern Israel, using commercial drones to bomb Israeli guard towers before breaching the border wall. Videos of Israeli troops and a Merkava IV tank being taken out by drones surfaced on the internet.

Houthi Drone Attacks

On 25 March 2022, the Houthis launched coordinated drone and missile strikes on multiple Saudi energy facilities, including sites in Jeddah and Ras Tanura. The attacks temporarily disrupted oil production and caused fires at Aramco facilities. Between late 2023 and July 2024, during the Gaza War, the Houthis carried out over 100 attacks on commercial ships in the Red Sea and Bab al-Mandab Strait. These attacks used drones, missiles, and fast boats, targeting vessels linked to Israel, the U.S., and its allies.

Use via Proxies

Iranian drones and drone-technology have been supplied (or used) by allied non-state actors (e.g., in Yemen, Iraq, Syria) for reconnaissance and strike missions.

Testing/Exports

The proliferation of Iranian drones means that they have been used in distant conflicts or supplied to other actors (which amplifies Iran's influence beyond its borders).

Ukraine. In the same month, Iran had also used the drone to attack Kurdish political party headquarters in Iraqi Kurdistan. Iran has supplied these drones to the Sudanese Army during the Sudanese civil war.

An Iranian Qods Mohajer-6 drone and two Qaem glide bombs are seen during a delivery of new equipment to the Islamic Revolutionary Guard Corps Ground Force. (Fars Media Corporation, CC BY-SA 4.0 via Wikimedia Commons.)

Shahed-149 Gaza

The Shahed 149 Gaza is an Iranian unmanned combat aerial vehicle (UCAV) operated by the Iranian Revolutionary Guard. The Gaza drone is a high-altitude long-endurance UAV similar in size, shape and role to the American MQ-9 Reaper.[7][8] It is a larger, heavier, and more capable aircraft than the earlier Shahed 129 which was similar to the MQ-1 Predator. It has a flight duration of 24 hours, a maximum operating radius of 2,500km (1,562 miles), 21m wingspan, 340 km/h maximum speed and is capable of carrying 13 bombs and 500 kg of electronic equipment. It is the first Iranian UAV powered by a turboprop engine.

Shahed-149 Gaza: A high-end UCAV, large wingspan (~21 m), range of ~2,000 km, endurance ~24 hours, capacity for significant payloads. (Student News Agency)

Shahed Saegheh (also "Shahed-191")

The Shahed Saegheh (Witness Thunderbolt), also called the Shahed 191, is an Iranian turbofan/piston-powered flying wing unmanned combat aerial vehicle (UCAV) produced by Shahed Aviation Industries. It is based on, but smaller than and substantially different from, a Lockheed Martin RQ-170 Sentinel UAV that was captured by Iran in 2011 and then reverse-engineered. It is one of two Iranian flying wing UAVs based on the RQ-170, along with the Shahed 171 Simorgh, a larger version. The drones can carry two Sadid-1 missiles, externally for the Saegheh-1, and internally for the Saegheh-2.

Small Unmanned Aircraft Systems (sUAS) and Suicide /Loitering Munitions

Iran's small and micro UAV capabilities focus on short-range reconnaissance, tactical support, and low-cost attack missions, providing flexibility for both state and proxy operations. Systems like the Ababil-3, Mohajer-6, and various quadrotor-style drones are used for battlefield surveillance, artillery spotting, and localized strike roles, often equipped with electro-optical sensors or lightweight munitions. At the micro scale, Iran has developed compact drones suitable for urban and asymmetric warfare, featuring commercial components adapted for military use. While these platforms lack endurance and sophistication compared to Western equivalents, their ease of production, portability, and expendability make them effective for saturation attacks, intelligence gathering, and covert operations across the Middle East.

Loitering Munitions ("Suicide Drones")

Iran has at least 10 different models of suicide drones, which explode on impact. They can be as accurate as a ballistic missile, but they can fly lower to evade radar. Smaller models, such as the Shahed-136, carry less than 45 kilograms (100 pounds) of explosives. One of the smallest models, the Meraj-521, carries only 3 kgs (6.6 pounds) of explosives. Iran's largest suicide drones, such as the Arash series, can carry nearly 260 kgs (575 pounds) of explosives. They have ranges of 2,000 kms (1,240 miles).

Arash-2 (Kian 2)

According to the Iranian media, the Arash-2 (Kian 2) has a wingspan of up to 4 meters, a length of up to 4.5 meters, and a range of up to 1,000 km. (Illustrative photo from open sources.)

HESA Shahed-136

Russia has made much use of the Shahed 136/ Geran-2 in its invasion of Ukraine, especially in strikes against Ukrainian infrastructure, and mass-produces its own version. (Photo credit: Iran Ministry of Defense.)

UAS Threat

As of 2017, 10 Saegheh drones were in production, and Iran planned to procure at least 50 by 2025. (Fars Media Corporation, CC BY-SA 4.0 via Wikimedia Commons.)

HESA Karrar

Originally a target drone, but modified with weapons/hardpoints; reportedly equipped with air-to-air or surface-to-air missiles in later iterations. The HESA Karrar is an Iranian jet-powered target drone manufactured by Iran Aircraft Manufacturing Industrial Company (HESA) since 2010. The Karrar is a derivative of the American 1970s-era Beechcraft MQM-107 Streaker target drone, probably incorporating elements from the South African Skua, with hardpoints added for munitions. It was framed as a "long-range bomber drone,"[19] and is the first long-range UAV manufactured in Iran. Multiple sources report that the Karrar has been exported to Hezbollah. The Karrar has supposedly been used in the Syrian Civil War.

A HESA Karrar UAV during Iran's "Sacred Defense Week 2010" with a Mk. 82 bomb mounted on the centerline. (By Fars Media Corporation, CC BY-SA 4.0 via Wikimedia Commons.)

IV. Regional (Middle East and North Africa) and Global Implications

Iran's growing fleet of drones and unmanned aircraft systems (UAS) has reshaped the security environment of the Middle East and extended its influence far beyond the region. In the Middle East and North Africa (MENA), the proliferation of Iranian drones has become a defining feature of Tehran's asymmetric strategy—enabling it and its proxies to project power, challenge regional adversaries, and strike high-value targets without direct conventional confrontation.

Groups such as Hezbollah in Lebanon, the Houthis in Yemen, and Shi'a militias in Iraq and Syria have all fielded Iranian-supplied or locally assembled UAVs for surveillance and precision strikes against Saudi, Emirati, and Israeli assets.

Iran's drone and UAV capabilities pose a persistent and growing **threat to maritime shipping in the Persian Gulf, Strait of Hormuz, and Gulf of Oman**—key chokepoints for global energy and trade. The Islamic Revolutionary Guard Corps (IRGC) routinely employs surveillance drones to monitor commercial and naval traffic, gather intelligence, and identify potential targets. Armed or loitering drones can be launched from land, ships, or converted "drone carrier" vessels to threaten tankers, oil platforms, and foreign warships.

Regional Impact

Regionally, Iranian UAVs have altered the balance of risk and response. Their affordability and operational versatility allow for sustained, low-cost harassment of adversaries, from maritime surveillance in the Persian Gulf to attacks on energy infrastructure in Saudi Arabia and the United Arab Emirates. By flooding the airspace with cheap, expendable systems, Iran can impose strategic and financial burdens on technologically superior militaries. This asymmetric pressure undermines traditional air dominance and challenges the long-standing assumption that expensive missile defenses and advanced fighter aircraft alone can ensure security. As a result, regional powers are now investing heavily in counter-UAV radars, directed-energy weapons, and electronic warfare systems—creating a regional arms race centered on drone defense rather than deterrence through firepower.

Global Impact

Globally, the implications are equally profound. The export of Iranian drones, most notably the Shahed-131 and Shahed-136 loitering munitions used by Russia in Ukraine, has demonstrated how low-cost, mass-produced UAVs can alter modern warfare and circumvent international sanctions. This proliferation blurs the lines between state and non-state actors, democratizing access to aerial strike capabilities once monopolized by advanced militaries. Moreover, Iran's use of commercial components and modular designs enables easy replication and complicates export-control enforcement. The result is a diffusion of drone warfare that destabilizes traditional power balances and challenges international norms on arms transfers and accountability.

Plausible Deniability and Asymmetric Leverage

Ultimately, the Iranian drone program represents a strategic blend of plausible deniability and asymmetric leverage. By supplying and operating UAVs through proxy networks, Tehran can strike adversaries while avoiding direct escalation. At the same time, its success in exporting drones to both state and non-state actors has made Iran a global model for low-cost, high-impact warfare. The MENA region—and increasingly the broader international community—now faces a future in which Iranian-style drone tactics, saturation attacks, and proxy-enabled airpower will continue to redefine the nature of conflict and deterrence.

V. Recent Iranian Drone Attacks on Israel (Outcomes and Costs)

On multiple occasions in 2024–2025 Tehran launched mass drone and missile barrages aimed at Israeli territory, employing a mix of long-range loitering munitions, cruise missiles and shorter-range UAVs intended to overwhelm air-defences and inflict damage on military and civilian infrastructure. These operations—framed by Iranian leaders as retaliation for strikes inside Iran or against Iranian proxies—ranged from the April 2024 direct strike that included explosive drones and missiles to the much larger June 2025 campaign in which hundreds of air vehicles were launched.

Despite the scale and intent of these attacks, the vast majority of incoming drones and missiles were neutralized before they reached critical targets. Israeli forces, supported operationally and materially by allied systems and platforms (notably U.S. air-defence assets in the region), achieved a very high interception rate—public reporting put successful interceptions at roughly the 90-plus percent level in the largest events, with Israeli official claims of about 99% of incoming projectiles stopped during the principal June barrage. This defensive success reflects layered systems (short-, medium- and long-range interceptors), integrated sensors, pre-emptive warning networks, and multinational tactical cooperation.

However, "shot down" is not the same as "cost-free." Intercepting hundreds of UAVs and missiles placed enormous operational strain on air-defence inventories and required a rapid mobilization of expensive interceptors, radars, electronic warfare measures, and allied assets repositioned to the theatre. Media and defence analysts estimated that the defensive effort to thwart the larger barrages ran into the hundreds of millions of dollars (one analysis estimated costs on the order of several hundred million pounds/dollars for a single major interception campaign), and it consumed limited interceptor stocks and generated readiness/tactical risks for follow-on threats. The expense and attrition incurred by defensive systems therefore represent a significant indirect success for the attacker even when physical damage is minimal.

Strategically, the episodes illustrate a few clear lessons. First, inexpensive massed UAS tactics can impose outsized economic and operational costs on higher-tech defenders: the attacker pays little per drone, the defender pays much more per interceptor. Second, multinational coordination (sharing radar, munitions, and rules of engagement) materially improves interception outcomes—U.S. contributions in the region were explicitly credited with helping stop inbound missiles and drones during peak exchanges. Third, while high interception rates blunt immediate physical damage and civilian casualties, they do not eliminate escalation risks, logistical strain, or political pressure to retaliate, all of which reshape regional deterrence calculations.

In sum: recent Iranian drone attacks against Israel were largely ineffective in causing mass destruction because layered Israeli and allied defences shot down almost all incoming threats, but that defensive success came at substantial financial cost, consumption of scarce interceptors and a heightened regional tension—precisely the indirect effects Iran's cheap, proliferated UAV arsenal is intended to produce.

UAS Threat Overview (Systems & Capabilities)

II(d). North Korea

I. Background

The Democratic People's Republic of Korea (DPRK) has, in recent years, placed increasing emphasis on unmanned aerial systems (UAS) and drones as part of its military modernization efforts. These systems span reconnaissance, strike (including "suicide" or loitering-munitions), and early-warning platforms. Several public milestones and expert assessments shed light on the state of this program, its ambitions, its limitations, and the implications for regional security.

North Korea's push into unmanned aerial systems is a clear indicator that the regime views drones as a significant component of modern military power — particularly given its asymmetrical posture and limited conventional resources. While the DPRK's UAS capabilities are still assessed to be well behind those of leading states, the combination of ambition, visible prototypes (e.g., Saebyeol-4/9), mass-production intent, AI/automation focus, and regional strategic context means the program cannot be ignored.

For South Korea, the U.S., and other regional stakeholders, the DPRK's developments pose both an intelligence challenge (to understand the real capabilities) and a military-defence challenge (to guard against new forms of unmanned threat). From surveillance and reconnaissance to loitering munitions and autonomous strike systems, drones are becoming a key domain of competition on the Korean Peninsula.

II. Strategic Drivers & Military Policy

Several factors are motivating North Korea's investment in drone and unmanned-aerial-vehicle (UAV) capabilities:

Modernization Push

The DPRK leadership under Kim Jong Un has declared unmanned systems and artificial intelligence (AI) as top priorities in military development. For example, as according to Korea Joongang Daily, Kim personally oversaw testing of "various kinds of reconnaissance and suicide drones … produced by [the] Unmanned Aeronautical Technology Complex.

Cost-Effective Asymmetric Capability

Drones offer a relatively low-cost way for Pyongyang to enhance surveillance, target acquisition, and strike options versus more expensive platforms (such as manned aircraft). The DPRK has ordered "mass production" of suicide drones.

Experience and Inspiration from Other Conflicts

Analysts suggest North Korea is influenced by drone employment in conflicts such as in Ukraine, and may be learning or procuring technologies and tactics via cooperation with other states (e.g., Russia) or through piecemeal acquisition.

Surveillance & Targeting Enhancement

The development of high altitude reconnaissance UAVs and even early-warning aircraft suggests an intention to improve the DPRK's ability to monitor the peninsula and potentially beyond, in contrast to its historically radar- and ground-based surveillance focus.

III. Primary UAV Systems

Here are several of the more prominent systems or lines of development that have been identified:

Saebyeol-4 (Morning Star-4)

A large reconnaissance UAV first revealed around July 2023. It resembles the U.S. RQ-4B Global Hawk in form, though analysts believe its capability is significantly inferior. It is estimated to have a wingspan of ~30-35 m, uses parts derived from older MiG-21/J-7 airframes, and is likely intended for strategic reconnaissance rather than tactical strike.

A close-up view of a variant of the Saetbyol-4 class UAV seen on the runway of Panghyon Airbase. March 27, 2025, (Photo source: Rodong Sinmun).

The Saebyeol-4 (Morning Star-4) first appeared publicly in July 2023, during North Korea's "Weapons and Equipment Exhibition 2023" held in Pyongyang to commemorate the 70th anniversary of the Korean War armistice (the "Victory Day" celebration). It was announced and unveiled through North Korean state media, notably the Korean Central News Agency (KCNA) and Rodong Sinmun, which released photographs showing Kim Jong Un touring the exhibition alongside visiting Russian and Chinese delegations.

Analysts immediately noted that one of the large UAVs on display strongly resembled the U.S. RQ-4 Global Hawk—a high-altitude, long-endurance reconnaissance drone—prompting the designation "Saebyeol-4." State media described it as a "strategic reconnaissance unmanned aerial vehicle" designed for long-range surveillance missions, although no technical details were provided. Follow-up imagery and analysis from 38 North, CSIS, and Defense Security Asia confirmed that the airframe's

Small Unmanned Aircraft Systems (sUAS) and Suicide /Loitering Munitions

At the 2024 "Weaponry Exhibition" in Pyongyang North Korea unveiled a family of at least six small tactical UAVs — a mix of light reconnaissance drones and one-way "suicide"/loitering-munition types — that analysts say are intended for short-range surveillance, targeting and massed attack/saturation tactics rather than long-endurance strategic ISR. These are harder to detect, cheaper to field, and more proliferable — making them potentially a more immediate challenge for neighboring forces.

Drones on show at the Weaponry Expo 2024 in Pyongyang, seen in a KCNA image from November 22, 2024. (Source: Korean Central News Agency via 38 North.)

North Korea's Kumsong Drones

North Korea's Kumsong-series drones ("Golden Star") are a family of loitering munitions and "suicide" attack UAVs first publicly tested in 2024 and showcased during the 2025 military parades, marking a major step in Pyongyang's push toward operational unmanned strike capabilities. These drones are designed for one-way attack missions — crashing into targets with explosive payloads — and closely resemble Israeli IAI Harop-style systems in both form and function. Mounted on truck-based six-cell launchers, Kumsong drones appear to provide North Korea with a low-cost, mobile, and scalable precision-strike option intended to overwhelm enemy air defenses through saturation tactics. While their range, autonomy, and guidance sophistication remain uncertain, the Kumsong line demonstrates Pyongyang's focus on asymmetric, mass-producible unmanned weapons to supplement its missile and artillery forces.

At the October 10, 2025 parade, DPRK exhibited a truck-based drone launching system carrying six loitering munitions each. (Source: Korean Central Television via 38 North.)

layout (wide wingspan, V-tail, top-mounted engine intake) mirrored Global Hawk characteristics but was likely far less capable, suggesting North Korea intended the debut to signal entry into the realm of large strategic drones rather than to reveal an operational fleet.

Saebyeol-9 (Morning Star-9)

First publicly revealed in July 2023, this UAV appears to mimic the U.S. MQ-9 Reaper — albeit at reduced scale and with less advanced avionics. It is described as a "multi-purpose attack drone" capable of both reconnaissance and strike missions.

Saetbyol-9 attack UAVs on transporters during the military parade in Pyongyang, July 27, 2023 (Photo source: KCNA).

There are no independently verified production totals for the Saebyeol-9. Analysts say reliable figures aren't available; open-source observers have only documented a handful of individual Saebyeol-9 airframes (exhibited at a weapons fair and seen in parade footage). Some Korean press reporting puts the observable count at 4-6..

MM-1 Attack UAV

The MM-1 is an early North Korean tactical attack/unmanned aerial vehicle — a relatively small, low-signature attack/reconnaissance platform that analysts say represents one of Pyongyang's first attempts at fielding purpose-built attack UAVs and experimenting with loitering/one-way-attack concepts; it appears to be more of a proof-of-concept and training platform than a mature, long-endurance system (limited sensors, modest payload and range, and heavy reliance on simpler navigation and off-the-shelf components), and it helped inform later, larger DPRK designs and the regime's push toward mass-produced suicide drones.

The North Korean MM-1 attack UAV displayed during the April 2012 Pyongyang Military Parade (Photo source: KCNA).

IV. Implications for the Korean Peninsula and Wider Region

The DPRK's drone developments carry a number of potential implications:

Increased Surveillance and ISR Capability

Should systems like Saebyeol-4 become operational, North Korea may gain improved ability to monitor South Korea's military posture, island outposts, and maritime zones. This could complicate South Korean and U.S. efforts at stealth or surprise. Small UAVs offer the DPRK a way into improved ISR (intelligence-surveillance-reconnaissance) at tactical levels (forward positions, the DMZ zone, maritime/coastal areas) without requiring the heavy infrastructure of large strategic drones.

Threat of Precision or Semi-Precision Strike By Drones

Although DPRK's strike drones are less capable than Western equivalents, massed employment of suicide drones might introduce new vulnerabilities for fixed or vulnerable assets (e.g., airfields, command posts, coastal defenses).

The loitering munitions provide the DPRK a tactical strike tool that can be used to harass or hit critical assets — radars, artillery batteries, fixed infrastructure, coastal vessels — with relatively inexpensive systems that may evade traditional air defenses.

Deterrence Escalation Dynamics

The DPRK may view unmanned systems as force multipliers and as another layer in its deterrence/war-fighting posture (alongside missiles, artillery, nuclear weapons).

Operational and Doctrinal Shift

Emphasis on unmanned systems suggests the DPRK is adapting its doctrine to the evolving character of warfare—less reliance solely on massed manpower and heavy conventional firepower, more on unmanned, automated, attritional tools.

Saturation and Asymmetric Threat

Because smaller drones are cheaper and simpler, the DPRK can conceivably deploy them in numbers (mass production is reportedly being ordered) to complicate detection and interception for South Korea and its allies. For example, one article estimates the DPRK may have 500 to 1,000 UAVs of various types, of which a portion will be smaller models.

Lower Profile Infiltration

The DPRK has already demonstrated the ability to fly smaller UAVs into South Korean airspace, evading detection or interception. One report: "drones … less than 3 meters long" flew into South Korea and returned.

Counter-Measures Challenge for South Korea and the U.S.

As the DPRK improves drone capabilities, counter-drone, electronic-warfare, air-defence and surveillance systems in the South and allied forces must adapt. The DPRK may exploit blind-spots (maritime, low altitude, night).

Proliferation and Export Concerns

The DPRK's progress could also raise concerns about proliferation of unmanned technologies or expertise to other states or non-state actors.

V. AI / Autonomous Features and Serial Production Push

Kim Jong Un has explicitly directed that the development of unmanned equipment and artificial intelligence (AI) be made a "top-priority" area within North Korea's defense modernization strategy, signaling a decisive shift toward automation and next-generation warfare technologies. This directive, first publicized in 2024 and reiterated during multiple weapons exhibitions and field inspections, emphasizes mass production of drones, loitering munitions, and AI-assisted targeting systems to enhance reconnaissance, precision-strike, and battlefield autonomy. It reflects Kim's recognition that unmanned and intelligent systems offer a cost-effective means to offset North Korea's technological and economic limitations, allowing Pyongyang to pursue asymmetric parity with more advanced militaries. Analysts view this as a clear attempt to institutionalize AI-driven and unmanned warfare as a permanent feature of the Korean People's Army's future doctrine and industrial focus.

VI. Limitations

While the DPRK has made visible progress, analysts remain cautious about the actual maturity of its systems. Public displays show prototypes and mock-ups; the key question is how many are produced, deployed, and sustained in operations.

The push for suicide drones and mass-production suggests a doctrine that emphasizes attrition, saturation, surprise, and asymmetric strike rather than purely high-tech precision.

Analysts at the Center for Strategic and International Studies (CSIS) note that while the DPRK's drones "mimic" advanced U.S. models in appearance, they lack the advanced sensors, satellites, communication links, and navigation systems that underpin true Global Hawk or Reaper-class performance. Signature reduction, communications security (jamming resilience), sensor quality, and battle-proven integration remain uncertain.

Issues such as remote-satellite communications, beyond-line-of-sight control, high-end sensors (EO/IR), data links, reliable propulsion, and maintainability remain significant challenges for North Korea. Smaller UAVs (especially loitering munitions) still require a ground control link, maintenance infrastructure and wartime logistics — areas in which the DPRK may be restricted due to sanctions.

The mass-production call is relatively recent and operational fielding at scale remains uncertain. Given the scale of sanctions and technological isolation on the DPRK, external sourcing or reverse engineering may be patchwork and uneven.

II(e). Hybrid, Terrorist, & Irregular Threat

I. Hybrid Threats

Hybrid threats are innovative, adaptive, globally connected, networked, and embedded in the clutter of local populations. They can possess a wide range of old, adapted and advanced technologies—including the possibility of weapons of mass destruction (WMD). They can operate conventionally and unconventionally, employing adaptive and asymmetric combinations of traditional, irregular, and criminal tactics and using traditional military capabilities in old and new ways.

A **hybrid threat** is the diverse and dynamic combination of regular forces, irregular forces, and/or criminal elements all unified to achieve mutually benefitting effects.

Hybrid threats are characterized by the combination of regular and irregular forces. Regular forces are governed by international law, military tradition, and custom. **Irregular forces are unregulated and as a result act with no restrictions on violence or targets for violence.**

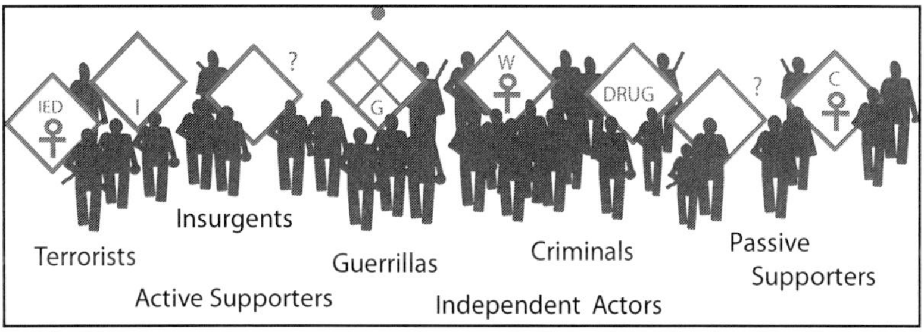

Ref: TC 7-100.3 (Jan '14), fig. I-1. Irregular force actors.

II. Irregular Forces

Irregular forces are armed individuals or groups who are not members of the regular armed forces, police, or other internal security forces (JP 3-24). The distinction of being armed as an individual or group can include a wide range of people who can be categorized correctly or incorrectly as irregular forces. Excluding members of regular

A hybrid threat is the diverse and dynamic combination of regular forces, irregular forces, and/or criminal elements all unified to achieve mutually benefitting effects. Irregular forces are armed individuals or groups who are not members of the regular armed forces, police, or other internal security forces. Irregular forces are unregulated and as a result act with no restrictions on violence or targets for violence. OPFOR5 topics and chapters include irregular and hybrid threat (components, organizations, strategy, operations, tactics), insurgents and guerillas forces, terrorists (motivations, behaviors, organizations, operations and tactics), criminals (characteristics, organizations, activities), noncombatants (armed & unarmed), foreign security forces (FSF) threats, and functional tactics.

III. Impact of UAS on Hybrid & Irregular Forces

UAS compress time, reduce required force size, and democratize capabilities that once distinguished states from irregular actors. Across military, paramilitary, insurgent, guerrilla, and criminal organizations, drones have become a force multiplier — altering tactics, eroding clear categorical distinctions, and forcing both kinetic and nonkinetic adaptations. Common effects of UAS on irregular and hybrid actors include:

Lowered cost and Lowered Entry Barrier

Commercial off-the-shelf (COTS) drones and inexpensive "kamikaze" loitering munitions put reconnaissance and strike capability within reach of small units, criminal groups, and non-state actors. This compresses capability gaps between regular forces and irregular actors.

Disaggregation and Distributed Lethality

Small units and even individuals can carry intelligence, surveillance, target acquisition, and precision strike (ISTAR) or terminal effects remotely, reducing the need for centralized platforms or large formations.

Faster Decision Loop and Increased Reach

UAS accelerate the observe-orient-decide-act (OODA) cycle: remote sensors feed near-real-time targeting and strike, enabling rapid local action that previously required larger systems.

Blurring of Legal/Operational Categories

When state security personnel or proxy forces operate UAS covertly, or when criminal groups adopt military-grade loitering munitions, distinctions among military, paramilitary, insurgent, guerrilla, and criminal activities become harder to sustain.

Information and Psychological Effects

Persistent overhead surveillance and publicized UAS strikes are potent tools for propaganda, intimidation, and shaping civilian behavior.

Countermeasure Cascade

Widespread UAS use drives demand for jamming, kinetic interceptors, nets, detection systems, and electronic warfare — but many irregular groups exploit the expense and complexity of counter-UAS to their advantage.

Effects and Examples by Component

Nation-State Paramilitary Forces (Internal Security, Police, Border Guards)

Paramilitary units rapidly integrate small UAS for border surveillance, crowd monitoring, and force protection. The affordability and deniability of certain drone operations allow states to pursue coercive or clandestine actions through proxies or non-transparent operations.

Examples: Paramilitary police units use quadcopters for crowd monitoring and tracking fugitives; border guards use UAS for long-distance observation and to interdict smuggling routes. In hybrid contexts, paramilitary units may operate armed or weaponized UAS in a gray zone below open warfare.

Insurgent Groups

Insurgents leverage UAS for reconnaissance, improvised targeting, and occasionally direct strikes. UAS enable insurgents to emulate conventional reconnaissance and strike cycles without needing airfields or heavy logistics. They also facilitate combined operations with guerrilla or criminal networks.

Examples: Insurgents have used off-the-shelf drones for roadside-bomb targeting, overwatch of movement routes, and to deliver small explosive payloads against checkpoints or convoys. Public release of drone reconnaissance footage is often used to amplify psychological effects.

Guerrilla Units (Irregular Indigenous Forces in Occupied Territory)

Guerrillas use UAS to bypass frontline observation, coordinate ambushes, and verify strikes using small loitering munitions. UAS allow small, dispersed guerrilla cells to conduct precision attacks and to maintain situational awareness over contested terrain.

Examples: Guerrilla units deploy quadcopters for reconnaissance of occupation forces' patrol routes and to aim small loitering munitions at exposed vehicles, fuel sites, or command posts — increasing effectiveness of hit-and-fade tactics.

Criminal Organizations (Gangs, Cartels, Hackers)

Criminals exploit UAS for surveillance (tracking targets, spotting police), smuggling (payload delivery across borders or into prisons), and as a platform for improvised weapons or intimidation. Drone-assisted criminal activity is often synergistic with cyber capabilities (remote control, encrypted comms) and corrupt supply chains.

Examples: Cartels have used drones to drop contraband across borders and to surveil rival movements and law enforcement. Gangs have used drones to observe police tactics or to drop small incendiaries or contraband into controlled areas. Cybercriminals may also attempt to hijack or spoof drones for criminal ends.

How UAS Blur Categories and Empower Hybrid Strategies

- **Proxy and deniable operations**. States can provide UAS or UAS-capabilities to proxies or paramilitary groups, creating plausible deniability while achieving political objectives.
- **Fusion of criminal and insurgent finance/logistics**. Criminal networks supply materiel, launch sites, and funding for insurgent UAS operations; in turn, insurgents provide protection and intelligence.
- **Convergence of tactics**. Terrorism, criminality, and military-style attacks may converge in a single operation (e.g., criminally funded drone strike executed by politically motivated guerrillas).

Operational Implications and Mitigation

- **Affordable counter-UAS is critical**. Low-cost detection, layered defenses (RF detection, EO/IR cueing, jammers, capture methods), and hardened procedures reduce UAS effects.
- **Policy and legal frameworks need updating.** Law enforcement, customs, and armed forces must clarify rules for attribution, evidence, and proportional response in gray-zone UAS incidents.
- **Intelligence fusion and supply-chain interdiction.** Targeting the procurement, training, and logistics networks that supply UAS (components, encrypted links, ground control stations) is often more effective than focusing solely on individual drones.

armed forces, police, or internal security forces from being considered irregular forces may appear to add some clarity. However, such exclusion is inappropriate when a soldier of a regular armed force, policeman, or internal security force member is concurrently operating in support of insurgent, guerrilla, or criminal activities.

Irregular forces can be insurgent, guerrilla, or criminal organizations or any combination thereof. Any of those forces can be affiliated with mercenaries, corrupt governing authority officials, compromised commercial and public entities, active or covert supporters, and willing or coerced members of a populace. Independent actors can also act on agendas separate from those of irregular forces.

Blurring of Categories

Although three basic types of forces can be part of the irregular OPFOR, the distinctions among insurgents, guerrillas, and criminals are sometimes blurred. That is because they may have more in common than they have that is different. From the viewpoint of the existing government authority, for instance, the activities of all three types are illegal, that is, criminal. Not just criminals but also insurgents and guerrillas can engage in criminal activities. Some insurgent organizations can include guerrilla units (developed from within or affiliated) and some guerrilla units may be part of an insurgency. In advanced phases of an insurgency, guerrilla units may begin to look and act more like regular military units.

There are three general tactics available to the irregular OPFOR—

- Military-like functional tactics.
- Criminal activity.
- Terrorism.

At any given time, the irregular OPFOR could use any of these means. The differences among these three can become blurred.

III. UAS Threats to the Homeland

The United States is confronting an increasingly complex national security threat environment. In addition to traditional military modernization, developments in artificial intelligence (AI), biotechnology, quantum sciences, microelectronics, space, cyber, and **unmanned systems** are rapidly transforming the nature of conflict and the global threat landscape.

I. Domestic Counter-UAS Authorities

DOD is one of four federal agencies—the others are the Department of Energy (DOE), Department of Homeland Security (DHS), and Department of Justice (DOJ)—granted limited authority by Congress to take actions against drones.

In December 2016, Department of Defense (DOD) was granted limited authority to protect assets related to nuclear deterrence against drone threats. This authority allows DOD to identify, monitor, and track drones; warn operators; disrupt or take control of drones; or use reasonable force to disable, damage, and seize or destroy drones deemed to pose a threat. The authority has since been broadened to include protections of other assets located in the United States, including those related to missile defense, national security space, protection of the President and Vice President, air defense missions, combat support, special operations activities, and major range and test facilities. Similar authority was also given to the Department of Energy to protect certain facilities that store or use special nuclear material.

In 2018, the **Department of Homeland Security (DHS) and the Department of Justice (DOJ)** were given similar authorities to protect certain covered facilities and assets from threats posed by drones. Covered facilities and assets include those related to Customs and Border Protection functions; Secret Service protection operations; FBI activities; and operations of the U.S. Marshals Service, including protection of jurists, court officers, witnesses, and other threatened persons, as well as facilities operated by the Federal Bureau of Prisons and federal courts and DOJ buildings.

While other law enforcement entities lack explicit authority to interdict drones, FAA developed a drone response playbook to guide in responding to suspected prohibited drone activity and teamed with other federal agencies to publish an advisory on the use of technology to detect and mitigate drone threats.

Refer to our series of related Homeland Defense, DSCA, Disaster and National Response SMARTbooks for further discussion. The US Armed Forces have a historic precedent and enduring role in supporting civil authorities during times of emergency, and this role is codified in national defense strategy as a primary mission of DOD.

UAS Threat

II. Implications of the UAS Threat: A Homeland Perspective

Based on the DHS Office of Intelligence and Analysis Homeland Threat Assessment, 2025.

Unmanned Aircraft Systems (UAS) are now integral to the evolving homeland threat landscape described by the Department of Homeland Security (DHS). As technology proliferates and becomes cheaper, the potential misuse of UAS intersects with every major threat category—terrorism, transnational crime, border security, infrastructure vulnerability, and economic espionage—multiplying both the scope and immediacy of risk.

Across all categories of the homeland threat environment, UAS represent a unifying technology that amplifies traditional risks while eroding the effectiveness of legacy defenses. Mitigating this challenge will require a coordinated approach that combines airspace awareness, data security, counter-UAS technologies, and legal frameworks capable of addressing rapid innovation, binding together the diverse and evolving dangers facing the modern Homeland.

Public Safety and Security

Small UAS can serve as force multipliers for violent extremists and lone offenders. Their ability to conduct reconnaissance, deliver small payloads, or disrupt large public events creates asymmetric advantages for attackers with minimal resources. Domestic violent extremists could exploit drones to surveil potential targets, intimidate election workers, or deliver improvised explosives. Foreign terrorist organizations, while geographically distant, may attempt to inspire or enable such tactics through online dissemination of drone modification techniques and operational guidance. The convergence of political polarization, election-related tensions, and easily accessible drone technology compounds this risk to public order.

Border and Immigration Security

At U.S. borders, criminal cartels and smuggling networks increasingly use UAS to transport contraband, conduct counter-surveillance of law enforcement, and coordinate migrant movement. The integration of drones with ground networks blurs traditional security perimeters, providing TCOs with real-time intelligence and low-cost delivery mechanisms for narcotics and other illicit goods. The potential for nation-state proxies to exploit these same routes for surveillance or covert delivery of materials also cannot be dismissed.

Critical Infrastructure Security

The threat to critical infrastructure from UAS is acute. Drones can be used to collect imagery of sensitive sites, interfere with air traffic operations, or carry payloads capable of disrupting electrical grids, refineries, or communication hubs. Adversarial nation-states such as the PRC, Russia, and Iran could integrate drones into broader hybrid operations combining cyber, physical, and psychological attacks. Even unsophisticated drone incursions can trigger costly shutdowns, erode public confidence, and reveal procedural weaknesses in site security.

Economic Security

In the economic sphere, the UAS threat extends beyond sabotage to include espionage and intellectual property theft. Commercial drones equipped with advanced imaging or data collection systems can be used to map industrial sites, monitor proprietary processes, or gather geospatial data valuable to foreign competitors. The People's Republic of China, identified as the leading economic threat in the DHS assessment, is also the world's primary producer of civilian drone technology—raising persistent concerns about embedded data links, supply chain vulnerabilities, and potential backdoor access to sensitive U.S. networks.

UAS Threat Implications: Insights from the DIA Worldwide Threat Assessment

Based on the 2025 Statement for the Record by the Director, Defense Intelligence Agency (May '25).

The Defense Intelligence Agency's *2025 Worldwide Threat Assessment* underscores how Unmanned Systems (UxS)—including air, maritime, and ground variants—have become defining features of modern conflict and intelligence competition. The report details a sharp expansion in the accessibility, lethality, and strategic integration of UAS across all major threat categories, from state adversaries to non-state actors, reshaping both the global and homeland security environments.

The 2025 DIA assessment portrays the UAS threat as a cross-domain phenomenon—strategic, operational, and tactical—driven by technological democratization and adversary adaptation. Drones now serve not only as tools of warfare but as instruments of coercion, espionage, and influence. Countering their proliferation and misuse will require synchronized intelligence, regulatory, and defense measures that evolve as fast as the technology itself.

Strategic Adversaries

Nation-states—principally China, Russia, Iran, and North Korea—are leveraging UAS for reconnaissance, strike, and denial-of-access operations. China continues to integrate drones into its joint maritime and air exercises, employing them in swarms and as extensions of its space and cyber-enabled targeting networks. Russia's heavy use of Iranian-supplied UAVs in Ukraine demonstrates how drones enable sustained attrition warfare and strategic terror campaigns at low cost. Iran's drone exports now serve as diplomatic currency, traded for advanced technology such as fighter aircraft and missile components. North Korea, meanwhile, employs UAS for espionage and proliferation, sharing designs with Russia in exchange for conventional arms.

Irregular and Proxy Forces

Iran's regional proxies—including Hizballah, HAMAS, and the Huthis—have normalized drone warfare as part of hybrid campaigns. The Huthis' repeated maritime UAV and missile attacks on commercial shipping illustrate how small actors can impose disproportionate global economic and strategic effects. These systems provide deniable strike options that blur traditional lines between state and non-state aggression, complicating deterrence and response.

Terrorism and Transnational Crime

The DIA warns that terrorist networks such as ISIS and al-Qa'ida are seeking to weaponize commercial drones and integrate artificial intelligence for autonomous targeting. The widespread availability of low-cost platforms enables actors with minimal resources to conduct surveillance, precision strikes, or symbolic attacks. Similarly, cartels along the U.S. southern border use UAS for smuggling, counter-surveillance, and narcotics delivery, highlighting the convergence of criminal and military-grade drone applications.

Technology Diffusion and Homeland Vulnerability

Rapid commercialization and the global supply chain—dominated by Chinese manufacturers—allow both adversaries and opportunistic actors to exploit civilian technologies for military purposes. The DIA emphasizes that attribution challenges and AI-enhanced autonomy make detection and response increasingly complex. Future UAS swarms, integrated with 5G and big data analytics, could overwhelm traditional defense systems and undermine situational awareness at home and abroad.

III. Case Study: 2024 U.S. Drone Sightings

The 2024 United States drone sightings, also referred to as the New Jersey drone sightings, were a series of reports involving large, unidentified drones observed at night across multiple regions of the United States between November and December 2024. The phenomenon originated in New Jersey before spreading to neighboring states like New York and Pennsylvania, and eventually across the Northeastern United States and other parts of the country. These sightings, often occurring over residential areas and critical infrastructure, prompted investigations by local, state, and federal authorities.

Airspace Restrictions

Small drones are generally restricted to altitudes below 400 feet and must remain within visual line of sight of the operator, although some commercial and governmental operators have obtained waivers to operate beyond visual line of sight. Drones may fly at night if equipped with approved anticollision lights visible for three statute miles.

Notable incidents included reports of drones over the military bases Picatinny Arsenal and Naval Weapons Station Earle in New Jersey. Some sightings, including one resulting in a delayed medical evacuation and another reported by Coast Guardsmen, turned out to be planes approaching airports. Still others were debunked as the constellation Orion or the planet Venus.

In response to the sightings, government officials called for increased transparency and resources for investigations. The Federal Aviation Administration (FAA) implemented temporary flight restrictions over sensitive areas in New York and New Jersey. Overflights of certain U.S. military bases led officials to request expanded authority to counter drones, but the Pentagon noted that drone flyovers are common and typically not malicious, and experts like Jamey Jacob suggested the incidents were likely the result of careless actors.

Remote Identification Requirements

FAA regulations generally require most nonmilitary drones to broadcast position information and a unique identifier, referred to as Remote ID. This can be accomplished through built-in capabilities generally required for drones manufactured since September 2022, via Remote ID modules affixed to drones without bult-in capabilities, or by operating within the confines of an FAA-recognized identification area (like a model aircraft park). FAA requires drone operators to register every three years. Recreational operators must complete online training and pass a safety test, and commercial drones must be operated by FAA-certificated remote pilots.

Investigations by civilian and military agencies and independent experts concluded that the reported sightings largely consisted of authorized drones and misidentified manned aircraft, celestial bodies, and other routine aerial objects. Commentators also attributed the sightings to widespread confirmation bias and mass hysteria, comparing them to a traditional UFO flap. Alternative explanations such as military operations received limited support.

Unidentified drones were reported over U.S. bases in the United Kingdom in November 2024, followed by sightings over Ramstein Air Base and arms factories in Germany in December 2024. Major General Patrick S. Ryder commented that while private drones periodically fly over military bases, most are not considered threats and do not affect operations. The FAA reported receiving over 100 drone sighting reports monthly in 2024. In a February 2025 Senate Armed Services Committee hearing, General Gregory Guillot, head of NORAD and NORTHCOM, stated that 350 drone incursions were reported over 100 U.S. military bases during 2024. He

IV. Protection of Domestic Military Installations

Ref: CRS Report R48477, Department of Defense Counter Unmanned Aircraft Systems, Daniel M. Gettinger (Mar '25), pp. 28-29.

In December 2016, Department of Defense (DOD) was granted limited authority to protect assets related to nuclear deterrence against drone threats. This authority allows DOD to identify, monitor, and track drones; warn operators; disrupt or take control of drones; or use reasonable force to disable, damage, and seize or destroy drones deemed to pose a threat. The authority has since been broadened to include protections of other assets located in the United States, including those related to missile defense, national security space, protection of the President and Vice President, air defense missions, combat support, special operations activities, and major range and test facilities. Similar authority was also given to the Department of Energy to protect certain facilities that store or use special nuclear material.

Reports of potentially illegal or unauthorized drone activity near or around DOD installations have prompted congressional inquiries and expressions of concern by some Members of Congress. In November 2024, DOD's Office of Inspector General announced it would evaluate DOD actions to address UAS incursions over military installations. As discussed above, DOD has designated U.S. Northern Command to coordinate domestic counter-UAS activities; the Department has also announced that technological solutions for the defense of military installations, including for those located in the United States, is a focus of DOD's Replicator 2 initiative. DOD has proposed expanding its authorities to conduct counter-UAS operations in the United States in what the Department has described as an effort to align its authorities with those granted by Congress to other federal agencies.

Falcon Peak 2025

Teledyne Cerberus XL Counter-Small UAS static display at Falcon Peak on Fort Carson, Colorado, Oct. 30, 2024. Falcon Peak is a USNORTHCOM led counter-small unmanned aircraft system experiment and the first Department of Defense C-sUAS initiative focused on detecting, tracking, and mitigating sUAS incursions at DoD installations in the United States. Falcon Peak occurred between Oct. 19-30, 2024. (U.S. Department of Defense photos by Josh Armstrong)

V. Case Study: C-UAS at the Superbowl (National Special Security Event)

Ref: Feature article, "Keeping Watch on the Sky Above the Big Sky," Department of Homeland Security, Science and Technology Directorate (May 6, 2025).

The Super Bowl is one of the largest public events to take place in the United States. This year's matchup, held at the Caesars Superdome in New Orleans, was attended by more than 65,000 fans, with thousands more enjoying related festivities in and around the stadium. The Science and Technology Directorate's (S&T) Counter-Unmanned Aircraft Systems (C-UAS) program was on-site providing technical support to the United States Coast Guard (USCG) to help keep everyone safe from careless, clueless or nefarious Unmanned Aircraft Systems (UAS).

The Superdome before the game. (Photo credit: USCG.)

Security for an event like the Super Bowl must be airtight and seamless. Keeping everyone—including spectators, the competing teams, halftime performers, venue staff, media and surrounding establishments—safe is a top priority. No one wants to miss a moment of the festivities, pageantry, and action.

Providing this kind of security requires the professionalism and cooperation of countless local, state and federal agencies, and S&T was a big part of that, supporting USCG as part of a C-UAS team that also included U.S. Immigration and Customs Enforcement, the Transportation Security Agency, the Federal Bureau of Investigation and the Federal Aviation Administration (FAA) among others.

The FAA estimates that as of this month, more than 1 million UAS (or drones) are registered in the U.S. alone. The overwhelming majority are used for legitimate commercial and recreational purposes. But there is no avoiding that they can also be used for illicit and nefarious means. During the Super Bowl, the FAA designated the Superdome and the area around it as a No-Fly Zone, and this included UAS. However, this did not prevent more than 70 drone operators from attempting to enter the restricted area and approach the stadium. All of these attempts had to be taken seriously and were dealt with appropriately.

The C-UAS team positioned sensors in advance at multiple locations around the Superdome. The sensors provided coverage of the airspace that needed to be monitored, detecting any drones in flight approaching the restricted area.

"S&T provided our team with absolutely essential expertise that allowed us to network a lot of different equipment from different agencies and centrally locate and monitor all of it," said a USCG representative. "The five different sites that the Coast Guard set up around the city were all networked electronically back to the NFL Security Room. This network arrangement allowed us to monitor in real time exactly what every one of those sensors was detecting and manipulate and control them from the central location."

When a drone was detected, the first step determined if it was an authorized drone being flown by law enforcement or media covering the game. If not, then the C-UAS team treated it as a potential threat.

"Sometimes we can infer their intent based on the flight pattern, or where they took off from, or where they were intending to go," said S&T C-UAS Program Manager Shawn McDonald. "Normally we try to send it home, or fend it off, and see if the operator is going let it land and pack it up. If they keep coming, you know they're trying to go where they shouldn't be."

The team had options for dealing with unauthorized drones. In some instances, they could make the drone fly back to where it launched; the drone pilot would then see the UAS returning to them, and most people get the message and pack up to leave.

The C-UAS command center. (Photo credit: USCG.)

Another potential option would be to take control of the UAS. The C-UAS team could fly the drone to its operator, sometimes accompanied by local law enforcement. Or, if there is some concern about whether the UAS has hazardous payload or the pilot is in a crowded area, the C-UAS team could fly it to a pre-selected "safe" area away from people, like an empty field.

Then the drone could be examined without putting any bystanders at risk.

Unique situations also arose during the Super Bowl. For instance, two different attempts were made to fly drones towards the Super Bowl from a large ocean cruise liner docked on the Mississippi River in the French Quarter. The systems detected the drones, and a ground team was able to get over there and interview one of the pilots.

As it turned out, the unauthorized drones during this year's Super Bowl all fell into the "careless and clueless" category (an unofficial, but common term in the C-UAS world). They were not carrying any dangerous payloads but still had to be kept clear and checked out.

At the end of the day, those attending the big game and the millions watching across the globe had no idea that a very successful collaboration was taking place outside, keeping watch and making sure that it all went off without a hitch to keep the public safe and enjoy the game.

UAS Threat

VI. Case Study: Detecting and Mitigating Drones on the Border

Ref: Feature article, "Detecting and Mitigating Drones on the Border," Department of Homeland Security, Science and Technology Directorate (April 22, 2025).

The ability to locate and deal with drones operating in unauthorized areas is vital to maintaining secure borders and protecting assets. The Science and Technology Directorate (S&T) and Canadian partners are testing and deploying the latest technology to address these challenges.

In early autumn near Buffalo, New York, the weather was comfortable and warm, with a touch of color in the trees marking the change of seasons. As locals went about their business and tourists enjoyed the changing leaves, several nondescript vehicles traveled alongside them on the scenic byways along the Niagara River. Inside these trucks and vans were S&T technicians watching screens, tracking the movements of small Unmanned Aerial Systems (sUAS) flying across the U.S.-Canada border.

A mobile detection system mounted in the bed of a truck, one of many vehicles being used to practice detecting UAS. (Photo credit: S&T.)

S&T carried out the exercise in coordination with U.S. Border Patrol (USBP) Buffalo Sector and Canadian government counterparts. The goal was to practice locating and intercepting UAS flying near or crossing the Niagara River, which marks the international border. UAS are often used to move drugs and other contraband back and forth across the border. As technology advances, some have become large and powerful enough to potentially transport a person.

"We're only able to see our side of the border, and because the UAS move so fast, we don't have enough time to do an intercept," said Tony Hammerquist, Deputy Program Manager in S&T's C-UAS Program. "If we're able to see from the Canadian side, then we're able to prepare an intercept, if required, and the same thing goes for the Canadians on their side."

In addition to the mobile sensors being driven on local roads, and the existing sensors installed on poles and rooftops, S&T also tested some portable units in fixed positions. One such location was in Fort Niagara State Park on the top of a round hill covered with green grass. Members of the team set up a detection system and flew UAS in the surrounding areas.

One of the target drones being tracked across the international border. (Photo credit: S&T.)

This collaboration between the U.S. and Canada is not unique. "We host events between the two countries where we showcase technologies available in the United States, and the Canadians showcase technologies from Canada," said Hammerquist. "We also share UAS and C-UAS technology trends to prepare for the next generation of threats."

The chosen location along the border has had a S&T test sensor network in place for several years, providing an opportunity to test drone detection and mitigation. The S&T sensor network is used by USBP in securing the airspace along the Niagara River. The downside is that there is no time to waste. Favorable weather conditions maximize testing and flight time, and this part of the world has five solid months of winter, which can limit access to the sensor network and prevent target drones from flying due to bad weather.

Technicians monitor a target UAS as it flies near a portable detector. (Photo credit: S&T.)

UAS are a part of our world, and the vast majority of them are used for recreational and business purposes. However, there are nefarious actors out there using them in ways that threaten our safety and security. S&T is working on solutions to detect, and if necessary, intercept drones that are potentially dangerous. The kind of exercises that took place in Buffalo will help S&T develop better sensors and better techniques for use on the front lines of our security.

emphasized surveillance risks and advocated for expanded counter-drone authority for military installations.

Sightings by Armed Forces

Multiple drone sightings were reported over Picatinny Arsenal and Naval Weapons Station Earle, with some accounts coming from what military officials described as "highly trained security personnel." The first confirmed sighting, witnessed by a police or security officer, was reported on November 13, 2024. That day, a contractor at Picatinny Arsenal reported seeing "a light rising straight up from the tree line and toward the arsenal."

A month later, multiple drone incursions were again reported, this time over Naval Weapons Station Earle. In response, the Joint Chiefs of Staff issued a statement confirming some sightings. They stated that the Department of Defense (DoD) was unaware of the drones' operators and that there was no current indication of involvement by adversary nations. The statement added, "We have not been able to locate or identify the operators or the points of origin," and noted that the DoD lacks the necessary authority to trace the drones outside the bases, requiring assistance from law enforcement.

United States Coast Guard

Legislators Chris Smith and Paul Kanitra reported that multiple drones followed a United States Coast Guard 47-foot Motor Lifeboat. Coast Guard officer Luke Pinneo stated that "multiple low-altitude aircraft were observed in the vicinity" of one of their vessels. White House spokesman John Kirby announced that forensic analysis concluded the reported aircraft were "air traffic going into JFK International Airport, and not drones at all." Smith challenged this explanation, emphasizing the experience of the officers involved. Later, Transportation Security Administration (TSA) documents revealed that their investigators concluded the "swarms" were commercial jets executing S-shaped maneuvers while aligning for landing at JFK Airport, creating an illusion of hovering when viewed from certain perspectives.

Other Military Sightings

On December 9, 2024, federal officials arrested a Chinese national at San Francisco International Airport for flying a drone over Vandenberg Air Force Base. The individual was attempting to board a flight to China. The following day, a press release from Hill Air Force Base in Utah acknowledged recent observations of "unidentified drone swarms" over several U.S. military bases, confirming nearby sightings that did not affect operations. The United States Marine Corps confirmed that unidentified drones were observed over Marine Corps Base Camp Pendleton beginning on December 9 and into the following week.

Wright-Patterson Air Force Base in Ohio experienced airspace closures in mid-December due incursions by "small unmanned aerial systems". Shortly after, police in White Settlement, Texas, confirmed an investigation involving the FBI, Naval Criminal Investigative Service and Air Force into drone incidents at Naval Air Station Joint Reserve Base Fort Worth and a nearby National Guard armory.

I. Air & Missile Defense (AMD) Overview

Ref: FM 3-01, U.S. Army Air and Missile Defense Operations (Aug '25), chap. 1.

Understanding the integration of **air defense artillery (ADA) assets** is critical when planning and conducting counter–unmanned aircraft system (C-UAS) operations because ADA provides the structural foundation and doctrinal framework for layered, defense-in-depth protection against aerial threats. Effective C-UAS defense requires the synchronization of area and local defense assets to ensure redundancy, mutual support, and overlapping coverage that closes dead spaces and enhances survivability as threats approach protected areas. ADA units, as the Army's dedicated air and missile defense (AMD) force, bring specialized capabilities, sensors, and weapons systems that extend the joint force's ability to detect, track, identify, and defeat hostile UAS across multiple echelons. Integration ensures that tactical air defense measures at the division and brigade levels align with operational and strategic AMD actions, enabling a cohesive, multidomain defense.

The Army supports joint operations by conducting multidomain operations. While operating as part of the joint force and working with interorganizational and multinational partners, the Army forces gain, sustain, and exploit control over land to deny its use to an enemy. They do this using combined arms formations and capabilities to defeat an enemy and establish control of areas, resources, and populations. AMD is one of the Army's critical contributions to these efforts. Confronted by decentralized, networked, and adaptive enemies in dynamic and uncertain environments, the Army must possess a versatile mix of capabilities, formations, and equipment to conduct AMD. The ADA force must deter, defeat, and destroy air and missile threats in support of joint campaigns and assist in achieving air superiority to assure victory in a complex and uncertain world.

Air & Missile Defense (AMD)

Air and missile defense is the direct (active and passive) defensive actions taken to destroy, nullify, or reduce the effectiveness of hostile air and ballistic missile threats against friendly forces and assets (JP 3-01). AMD is the comprehensive role of ADA. AMD is embedded in the defensive counterair portion of the joint counterair operational framework. AMD refers to defensive counterair actions and to the capabilities of other Service and multinational partners involved in the counterair fight (for example, Navy AMD).

Air Defense

Air defense is the defensive measures designed to destroy attacking enemy aircraft or aerodynamic missiles, or to nullify or reduce the effectiveness of such attacks (JP 3-01). Air defense refers to active and passive defensive measures against air threats (such as manned and unmanned aircraft, cruise missiles, and air-to-surface missiles) to enable Army echelons freedom of action. Air defense operations occur at the tactical level (division and below maneuver formations). Air defense typically appears in discussions of the defense of corps-, division-, and brigade-level operations.

Air Defense Artillery (ADA)

In the joint lexicon, air defense artillery is defined as weapons and equipment for actively combating air targets from the ground (JP 3-01); more precisely, air defense artillery is the dedicated Army systems, personnel, and forces that provide active, land-based defense against air and missile attacks. ADA units execute AMD operations. ADA is used throughout the document as the Army's dedicated AMD force. Army AMD appears in lieu of ADA when more broadly applied to ADA actions and activities in AMD operations.

I. Joint Counterair Operational Framework (Offensive/Defensive Counterair)

Ref: FM 3-01, U.S. Army Air and Missile Defense Operations (Aug '25), pp. 1 to 3.

The joint counterair operational framework is based on the integration of offensive and defensive counterair operations.

Offensive Counterair

Offensive counterair are offensive operations to destroy or neutralize enemy aircraft, missile launch platforms, and their supporting structures and systems both before and after launch, and as close to their sources as possible (JP 3-01). Attack operations are the predominant offensive counterair operations. Attack operations include offensive action by any part of the joint force in support of the offensive counterair mission against surface targets which contribute to the enemy's air and missile capabilities (JP 3-01). Offensive actions can be conducted before, during, and after launch and, though focused on surface targets in the above definition, can include attacks against airborne command and control (C2) platforms. Defensive counterair are all defensive measures designed to neutralize or destroy enemy forces attempting to penetrate or attack through friendly airspace (JP 3-01).

Note. The operational framework for AMD had previously been known as the pillars of AMD. The pillars were active defense, passive defense, attack operations, and C2. Active defense, passive defense, and attack operations are now elements of the joint counterair operational framework. While C2 is not presented as a separate element, it serves as a baseline and enabler of all the other framework elements.

Defensive Counterair

Defensive counterair consists of two operational elements: active AMD and passive AMD.

Passive AMD *(See pp. 5-15 to 5-27.)*

Passive AMD operations are all measures, other than active AMD, taken to minimize the effectiveness of hostile air and ballistic missile threats against friendly forces and critical assets. These measures include detection, warning, camouflage, concealment, deception, dispersion, hardening, and the use of protective construction (JP 3-01).

Active AMD *(See pp. 5-28 to 5-30 and chap. 6.)*

Active AMD operations are direct defensive actions taken to destroy, nullify, or reduce the effectiveness of air and missile threats against friendly forces and assets. Active AMD deters, defeats, or destroys aerial threats, thus enabling freedom of action for friendly forces. Active AMD includes air defense and missile defense (defensive measures designed to destroy attacking enemy missiles or to nullify or reduce the effectiveness of such attack [JP 3-01]). Though not included in the definition of active AMD operations, countering indirect fire munitions is a fundamental part of active AMD. Counter-indirect fire actions are defensive measures to destroy, nullify, or reduce the effectiveness of rocket, artillery, and mortar threats. For simplicity's sake, air, missile, and indirect fire threats are generally referred to as "air and missile threats" hereafter.

Note. Fixed-wing aircraft, rotary-wing aircraft, cruise missiles, and unmanned aircraft systems (UAS) have traditionally been referred to as "air-breathing threats," or more commonly as "ABTs", by ADA Soldiers. The term "air-breathing threat" has evolved over time to "air threat" in joint doctrine. Neither term has been formally defined in joint or Service doctrine; they are described simply in terms of the threat set that they both encompass. Air threat is used consistently throughout this document to refer to these collective threats. Missile threat refers to ballistic missiles; cruise missiles are considered to be air threats.

Active AMD is typically the more prevalent counterair operational element in the early phase of campaigns; however, active AMD is finite and arrayed against those assets essential for achieving the supported commander's end state. Attack operations reduce the capacity of the enemy to launch air attacks and willingness to employ those capabilities, due to threat of losses, over time. Attack operations may thus decrease the demand for active AMD munitions in the later stages of the campaign. C2 systems link the planning and execution activities of the other operational elements. All defensive and offensive counterair operations are enabled by joint and Army C2 elements and further facilitated by mission command. A command and control system is the facilities, equipment, communications, procedures, and personnel essential for a commander to plan, direct, and control operations of forces pursuant to the missions assigned (JP 6-0). Mission command is the Army's approach to command and control that empowers subordinate decision making and decentralized execution appropriate to the situation (ADP 6-0).

AMD is conducted from the strategic to the tactical levels. Strategic-level AMD encompasses defense of assets of national interest in the homeland and allied nations. Air defense aspects of AMD are typically conducted by defensive counter-air aircraft and Army air defense systems, while missile defense is conducted by wide-area missile defense systems, such as the Ground-based Midcourse Defense system and the U.S. Navy's Aegis weapon system. All ADA systems may augment missile defense operations. Strategic missile defense operations can affect more than one combatant commander and is often referred to as global missile defense. Strategic-level AMD normally deals with countering the longer-range ballistic missile threats, such as intermediate-and intercontinental-ballistic missiles. It is important to remember that multinational partners approach AMD from a homeland perspective, so what ADA units categorize as theater AMD is strategic AMD to them. Multinational partners will generally employ their AMD systems to defend critical assets in accordance with their national priorities.

While all operational elements and levels are mentioned in this document, the focus throughout is active AMD and C2 as executed by ADA organizations primarily at the operational and tactical levels. Some discussion of strategic AMD capabilities is presented, but most is deferred to FM 3-27 and JP 3-27. The combination of active AMD to defeat, attack operations to prevent, and passive AMD to mitigate air and missile attacks and surveillance, enhanced by C2 capabilities, is critical to sustainable AMD operations, enabling freedom of action and the protection of forces, equipment, and other military and geopolitical assets.

AMD Aligned w/Fires Warfighting Function

The planning, preparations, and execution of active and passive AMD operations against all air and missile threats by all ADA echelons are aligned with the Fires warfighting function and provide the surface-to-air component of the Fires triad (surface-to-surface, air-to-surface, and surface-to-air). ADA and field artillery elements in fires cells exchange information on current and planned operations, support the coordination of airspace, and work to assure and enable commanders to have freedom of maneuver and action. Supported commanders and their staffs determine their AMD priorities which form the basis of the AMD plan. Protection elements at all echelons are informed of AMD plans and activities to ensure the synchronization of protection of critical assets against ground and air threats. The protection element coordinates for AMD support of a commander's critical assets, but defining and approving AMD priorities is ultimately the supported commander's responsibility. Representatives from ADA echelons actively participate in fires and protection cells and keep all cell personnel advised of pertinent AMD directives, actions, information collection efforts, and the overall AMD picture. *(See chap. 9.)*

II. Air Defense Artillery Role and Functions

The role of air defense artillery is to defeat aerial threats, through surface-to-air fires and effects, and provide early warning to enable supported commanders to complete their missions in multidomain operations.

ADA units conduct AMD operations to support U.S., joint, and multinational forces across the range of military operations, from military deterrence through large-scale combat operations. The primary ADA functions in executing AMD operations are—

- Defeat the full range of enemy air and missile threats encountered in current and future strategic, operational, and tactical fights.
- Integrate with Army, joint, interorganizational, and multinational elements.
- Provide early warning.
- Enhance situational awareness.
- Contribute to airspace management.

ADA capabilities are currently being enhanced by the fielding of the Integrated Air and Missile Defense Battle Command System (IBCS) and Maneuver Short-Range Air Defense (M-SHORAD) system and further facilitated by the continued contributions of space elements. Though currently limited to fielding in some Patriot units, IBCS constitutes the beginning of a common C2 system—the backbone of the integrated ADA force. IBCS provides commanders with tailorable sensor and weapon capabilities to meet mission needs. It integrates ADA C2 nodes, sensors and weapons systems into a meshed network that eliminates the stove-piped processes of previous generations, enabling any-sensor, preferred-shooter engagements. M-SHORAD is a divisional ADA system that provides dedicated defense of maneuvering forces in BCTs against rotary-and fixed-wing aircraft and UASs.

III. The Army Air and Missile Defense System

ADA is composed of composite systems (systems consisting of sensor, weapon, and C2 components) and singular systems (a sensor, weapon, or C2 system) that provide sense, engage, and interoperable capabilities to counter aerial threats to maneuver forces and other high priority assets. ADA systems are arrayed throughout a theater. Composite ADA systems are generically referred to as high-to medium-altitude air defense (HIMAD), and singular systems typically in the context of short-range air defense (SHORAD). Singular systems require sensors or weapons networked by and with C2 capabilities.

Short-Range Air Defense (SHORAD) *(See pp. 9-2 to 9-10.)*

Short-range air defense are those capabilities that provide air defense against low-altitude air threats. Short-range air defense (SHORAD) consists of dedicated ADA and non-dedicated air defense (known as combined arms for air defense) capabilities which deny an enemy the use of airspace and provide the supported commander freedom of action. SHORAD capabilities do this by deterring, defeating, or destroying low-altitude air threats to defended critical fixed and semi-fixed assets and maneuvering forces. SHORAD systems are M-SHORAD, Avenger, Stinger man-portable air defense (MANPAD), Counter-Small Unmanned Aircraft System (C-sUAS) variants, Land-Based Phalanx Weapon System (LPWS), Sentinel, and the forward area air defense (FAAD) C2 system.

High-to-Medium Altitude Air Defense (HIMAD) *(p. 9-11 to 9-12.)*

High-to-medium altitude air defense are those capabilities that provide defense against longer-range air and ballistic missile threats. HIMAD addresses capabilities that are primarily employed to counter ballistic missile threats, ranging from close-range to intermediate-range variants, in theater. Most HIMAD systems have diverse, system-specific C2 sub-systems that require conventional manual processes and methodology to perform integrated AMD operations, though, as noted above, IBCS is being fielded to some Patriot units. Patriot and THAAD are HIMAD systems.

IV. Air & Missile Defense Principles

Ref: FM 3-01.44, Short-Range Air Defense Operations (Jul '22), pp. 6 to 7.

Armed with a thorough understanding of the operational environment that is further focused through the lens of current mission, enemy, terrain and weather, troops and support available, time available, and civil considerations (METT-TC) conditions, commanders apply AMD principles when planning active AMD operations. The AMD principles are mass, mix, mobility, integration, flexibility, and agility. Mass, mix, mobility, and integration are traditional principles that have stood the test of time. Flexibility and agility are inherent considerations for how ADA forces organize and operate on future battlefields (FM 3-01).

Mass. Mass is the concentration of combat power sufficient to achieve the commander's intent. Mass, when applied to AMD, is achieved by allocating enough AMD firepower to successfully defend the force or the asset against aerial attack or surveillance. To mass AMD combat power in one area, commanders may have to accept risks in other areas of the battlefield.

Mix. Mix is the employment of a combination of weapons and sensors to protect the force and assets from the threat. Mix offsets the limitations of one system with the capabilities of another and complicates the situation for the attacker. Proper mix causes the enemy to adjust tactics. Enemy tactics designed to defeat one system may make the enemy vulnerable to another system. For instance, an enemy aircraft flying low to avoid Patriot may be vulnerable to engagement by an Avenger system.

Mobility. Mobility is a quality or capability of military forces, which permits them to move from place to place while retaining the ability to fulfill their primary mission (JP 3-36). M-SHORAD is capable of matching the cross-country mobility of maneuvering forces. Avenger systems can move with and maintain defense of the maneuver force's semi-fixed assets. Mobility of SHORAD increases their survivability as well as that of their supported assets.

Integration. Integration is the arrangement of military forces and their actions to create a force that operates by engaging as a whole (JP 1). Integration constitutes the combination of ADA and joint counterair forces, systems, functions, processes, and information acquisition and distribution required to efficiently and effectively perform the mission. Integration combines separate systems, capabilities, or functions in such a way that they can operate singly or in concert without adversely affecting individual elements.

Flexibility. Flexibility is the employment of a versatile mix of capabilities, formations, and equipment for conducting operations (ADP 3-0). Flexibility enables adaptive forces, facilitates collaborative planning and decentralized execution, and fosters individual initiative. The AMD principle of mix discusses the combination of ADA systems as task force tailored formations. The principle of flexibility is applied in AMD terms primarily by building METT-TC informed task organizations.

Agility. "Agility is the ability of friendly forces to react faster than the enemy" (ADP 3-90). For AMD, at the platoon level, it is the ability to leverage digital capabilities (e.g., air picture) to maneuver to defeat an air and missile threat. At the battalion level, it is the use of continuous air IPB and digital systems to employ forces to defeat the threat by operating inside the enemy's decision space.

V. Air Defense Artillery & the Air Threat

Ref: FM 3-01, U.S. Army Air and Missile Defense Operations (Aug '25), pp. 3 to 5.

As noted in FM 3-0, the operational environment is in a constant state of flux for each commander in their respective operational area, when friendly or enemy operations may be initiated or curtailed at any time. What is more of a given, regardless of the turmoil inherent in these environments, is that the United States will face a peer threat and must plan for large-scale combat operations against that threat. In the past, U.S. forces have enjoyed relative advantages over their adversaries with superior capabilities in each of the domains, particularly in the air domain. However, adversaries of the past have become peer threats, capable of challenging U.S. operations in all domains. Peer threats generate tactical, operational, and strategic challenges that are an order of magnitude more challenging militarily than other adversaries.

Potential adversaries are investing in antiaccess strategies and area denial capabilities to counter the U.S. ability to project military force into an operational area with sufficient freedom of action to accomplish assigned missions. The concept of antiaccess and area denial is to control the access into and movement within a region. Preclusion, the combination of antiaccess and area denial methods, seeks to influence the ability of an enemy, outside the region, to introduce forces into the theater and sustain combat power. Many countries are pursuing weapon systems/capabilities (such as ballistic or cruise missiles, UASs, and electromagnetic warfare) which enable antiaccess and area denial. To deter current and potential adversaries from achieving these objectives, the United States must maintain its ability to project power in areas where access and freedom to operate are challenged.

From a pure ADA perspective, operational environment considerations are focused on the air domain, while maintaining an awareness and understanding of the potential impacts of actions on AMD operations within the other domains. The air domain challenges to Army operations include:

- **Increased threats from traditional ballistic missile capabilities.** The ballistic missile threat has increased both quantitatively and qualitatively and is likely to continue to do so over the foreseeable future. Our known and potential adversaries will continue to challenge both the proficiency and sufficiency of U.S. ballistic missile defenses with improvements in ballistic missile technical sophistication and accuracy, increases in inventories and launch systems, advancement of countermeasures, and continued proliferation.
- **Expanding spectrum of air and missile threats.** The air and missile arsenal of today's adversaries has become more robust, diverse, and complex. Not only do our adversaries have access to the increasingly sophisticated ballistic missiles, but they also have a growing array of cruise missiles.
- T**he availability and use of UASs** has grown exponentially and technological trends are dramatically transforming legitimate applications of small-unmanned aircraft systems (sUASs) while simultaneously making them increasingly capable weapons in the hands of state actors, non-state actors, and criminals. They are effective for surveillance, targeting, electronic attack, and direct kinetic attack using onboard munitions or acting as precision guided munitions themselves. Manned fixed-and rotary-wing aircraft capabilities continue to evolve, and these aircraft remain as lethal threats to maneuver formations. Indirect fire munitions are easily attainable, relatively inexpensive, and extremely lethal. All air and missile threats must be addressed by Army and joint AMD forces within the context of antiaccess and area denial operations and of defeating countermeasures, electronic attack, and cyberspace efforts.

The threats to be countered by ADA systems are ballistic missiles, large-caliber rockets, cruise missiles, air-to-surface missiles, hypersonic weapons, manned fixed-and rotary-wing aircraft, UASs, and indirect fire (rocket, artillery, and mortar) weapons. Most aerial threats will be employed in the areas of operations in which ADA units operate.

Non-Dedicated Air Defense

Ref: FM 3-01, U.S. Army Air and Missile Defense Operations (Aug '25), pp. 51 to 52.

Given the limited amount of dedicated ADA resources available within the Army, **it is incumbent on non-ADA units to contribute to their own defense against air threats** using organic weapons when ADA units are not available or to supplement any dedicated ADA units allocated to them. In addition to organic crew-served and individual weapons, divisional forces have Stinger MANPAD systems and are being fielded with C-sUAS capabilities.

Engagements can be lethal or non-lethal, and they can be executed against air and missile threats that are airborne. Engagements are conducted in accordance with the promulgated ROE and the local commander's priorities. Engagements are conducted by the best available system or systems positioned and capable of negating an enemy's capabilities. Key to the engagements is the detection of the threat in sufficient time and with sufficient accuracy to allow friendly forces to execute necessary responses.

Figure 2-1 presents a doctrinal framework for C-UAS applicable to non-dedicated air defense and dedicated AMD Soldiers and units. It notes the non-dedicated protection task for air defense and considerations for C-UAS AMD missions. It also depicts overlaps in capabilities in engagements.

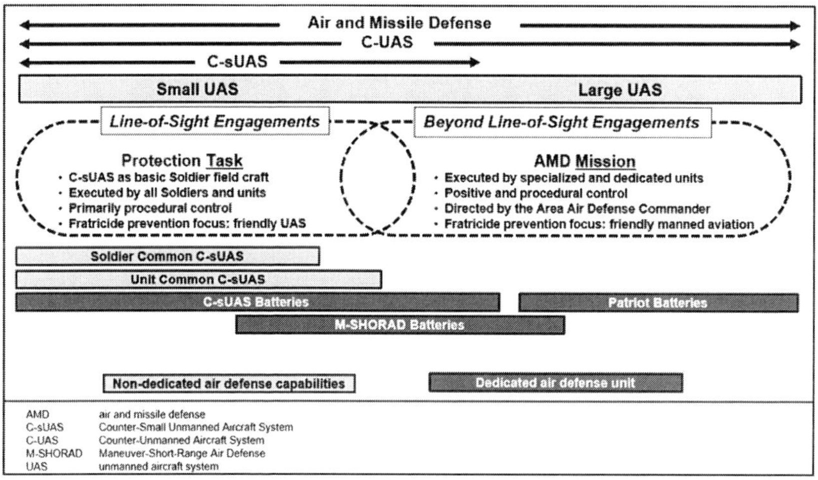

Ref: FM 3-01 (Aug '25), fig. 2-1. Doctrinal framework for counter-unmanned aircraft systems.

Maneuver forces in the close area use their organic weapons to engage surveilling or attacking rotary-or fixed-wing aircraft and UASs. Crew-served weapons are more capable against fixed-wing aircraft, rotary-wing aircraft, and sUAS; small arms are generally better against the group 1 UASs. All such efforts are considered self-defense actions. Threat systems that transit a unit's area of operation, but which are not attacking that unit, can also be engaged; however, they are more likely allowed to continue transiting given the relative short ranges and limited AMD effectiveness of maneuver force weapons. Transiting threats should be reported for early warning of and potential engagement by other units.

C-UAS "Concept of Operations"

See pp. 6-2 to 6-3 for an overview and further discussion of C-UAS concept of operations including air defense artillery and non-dedicated air defense assets.

Air & Missile Defense

VI. Air & Missile Defense Positioning (within the Operational Framework)

Ref: FM 3-01, U.S. Army Air and Missile Defense Operations (Aug '25), fig. 1-2. Air defense artillery forces arrayed across the operational framework, pp. 18 to 21.

The **operational framework** is a cognitive tool used to assist commanders and staffs in clearly visualizing and describing the application of combat power in time, space, purpose, and resources in the concept of operations (ADP 3-0). Commanders build their operational framework on their assessment of the operational environment, including all domains and dimensions. The joint force commander assigns land forces an operational area within a joint organizational construct. There are three types of assigned areas that a land component or ARFOR commander uses: area of operations, zone, and sector.

ADA commanders generally use **areas of operations** in addressing assigned areas. An area of operations is an operational area defined by a commander for the land or maritime force commander to accomplish their missions and protect their forces (JP 3-0). A land area of operations by definition does not include a volume of airspace to control. Control of the airspace above assigned land areas is not an inherent responsibility of the land component commander.

Airspace control authority resides with the joint force commander, who designates an airspace control authority to develop operational policies and procedures for airspace usage. Airspace control authorities delegate airspace control to Army commanders based on the situation. All commanders must thus be prepared to enable or coordinate airspace management. Airspace control must be delegated for operations below the coordinating altitude. The coordinating altitude is an airspace coordinating measure that uses altitude to separate users and as the transition between different airspace control elements (JP 3-52). The joint airspace control authority designates the coordinating altitude in coordination with the joint force land component commander and joint force air component commander (in the event the airspace control authority is not also the joint force air component commander). Within assigned areas, commanders organize their operations in terms of time, space, and purpose by synchronizing deep, close, and rear operations.

Figure 1-2 depicts the nominal locations of ADA units across the four areas of the strategic framework. The assigned operational area is further reflected as the rear, close, and deep areas.

ADA units are arrayed from the homeland to the forward areas of a theater of operations.

Homeland

Though not portrayed in the figure, ballistic missile defense systems defend the United States against ballistic missile attack, while AMD systems of all Services add defense against other air and missile threats.

Strategic Support / Joint Support Areas

The army air and missile defense command (AAMDC) orchestrates the allocation and positioning of Patriot and THAAD units, as well as the integration of joint and multinational AMD capabilities to defend high priority theater and corps assets in the strategic support and joint security areas against the cited threats. Patriot and THAAD can also support AMD efforts across theaters or regions as well as those within the theater.

Corps/Theater Deep Area

In support of the fight in the corps or theater deep area, the AAMDC or ADA brigade and battalion employ the IPOE process to identify prospective enemy air and missile operating and launch areas and nominate those areas for subsequent targeting by long-range artillery and aviation assets.

Division/Brigade Areas

DIVAD forces, such as M-SHORAD batteries and platoons, generally defend assets in the division and brigade areas. They are positioned with division or brigade maneuver formations in the close area, where most subordinate maneuver forces conduct close combat. They are oriented to the threat using techniques such as target lines, sectors, or aerial engagement areas. An aerial engagement area, for instance, identifies the engagement parameters of DIVAD units and is established/adjusted to ensure denial or defeat of enemy air platforms. The C-sUAS batteries in DIVAD battalions and teams assigned to divisions can mitigate surveillance and attack options by unmanned aircraft and can assist in thickening defenses. The air defense support element (ADSE) personnel in the BCT plan and coordinate the support of SHORAD or other ADA units and relay pertinent AMD information and early warning of enemy air activity to maneuver formations.

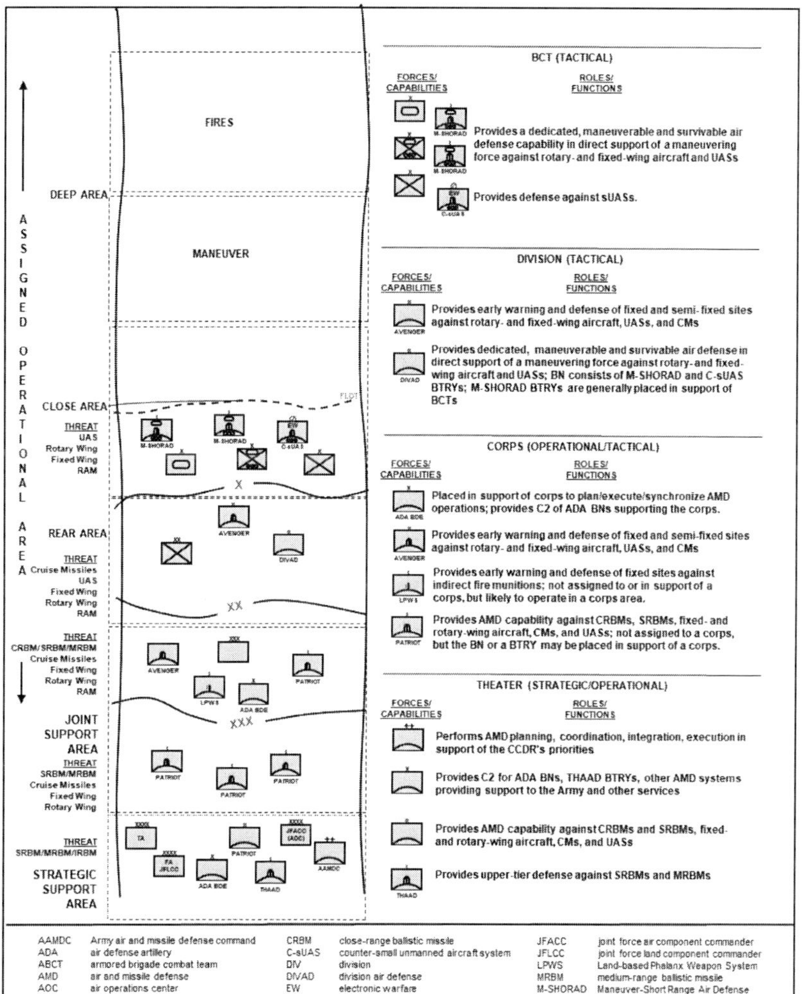

AAMDC	Army air and missile defense command	CRBM	close-range ballistic missile	JFACC	joint force air component commander
ADA	air defense artillery	C-sUAS	counter-small unmanned aircraft system	JFLCC	joint force land component commander
ABCT	armored brigade combat team	DIV	division	LPWS	Land-based Phalanx Weapon System
AMD	air and missile defense	DIVAD	division air defense	MRBM	medium-range ballistic missile
AOC	air operations center	EW	electronic warfare	M-SHORAD	Maneuver-Short Range Air Defense
BCT	brigade combat team	FA	field army	RAM	rocket, artillery, and mortar
BN	battalion	FY	fiscal year	SBCT	Stryker brigade combat team
BDE	brigade	FLOT	forward line of own troops	SRBM	short-range ballistic missile
BTRYS	batteries	HQ	headquarters	sUAS	small unmanned aircraft systems
C2	command and control	IBCT	infantry brigade combat team	TA	theater army
CCDR	combatant commander	IFPC	indirect fire protection capability	THAAD	Terminal High Altitude Area Defense
CM	cruise missile	IRBM	intermediate-range ballistic missile	UAS	unmanned aircraft system

VII. Air & Missile Defense Employment Tenets

Ref: FM 3-01.44, Short-Range Air Defense Operations (Jul '22), pp. 1-4 to 1-5.

While commanders should always start AMD employment planning by applying the principles described above, they should also strive to adhere to employment tenets when planning and positioning their ADA resources. The tenets are desirable attributes that should be built into all plans and operations and are directly related to how assets should be employed. There are seven tenets (figure 1-2).

Mutual Support

Weapons are positioned so that the fires of one weapon can engage targets within the dead zone of the adjacent weapon. For guns, this dead zone is usually small. For missiles, the dead zone may be large, and mutual support is a critical element. Mutual support can also cover nonoperational weapons or weapons at lower states of readiness. Mutual support, when applied to sensors has the same connotation; that is, sensors are deployed to cover the dead zone of adjacent sensors.

Overlapping Fires and Overlapping Coverage

Weapons are positioned so that their engagement envelopes overlap. Because of the many altitudes, directions, and ranges from which the enemy can attack or conduct surveillance operations, defense planners must apply mutual supporting and overlapping fires vertically and horizontally. Overlapping coverage is the positioning of sensors such that their coverage does not leave any seam in the defense that might be used by incoming threats. Overlapping fires and overlapping coverage should be planned during defense design.

Balanced Fire

Weapons are positioned to deliver an equal volume of fires in all directions. This is necessary for AMD in an area where the terrain does not canalize the threat or when the avenues of approach are unpredictable. Balanced fires is a desired characteristic of defense design against cruise missiles and other air threats.

Weighted Coverage

Weapons coverage is combined and concentrated toward the most likely threat air avenues of approach or directions of attack. Based on the tactical situation, a commander may risk leaving one direction of attack unprotected or lightly protected to weight coverage toward another direction. Weighted coverage and balanced fires are not mutually achievable, requiring the defense designer to give up most aspects of one to achieve the other.

Early Engagement

Early engagement generally requires extending the defense away from the defended asset. Sensors and weapons are positioned so they can engage the threat before ordnance release or detection of friendly forces. Early engagements enable destruction of enemy platforms over enemy forces and unoccupied areas, thereby reducing the possibility of friendly collateral damage and fratricide. As with weighted coverage, early engagement is achieved at the expense of balanced fires.

Defense in Depth

Sensors and weapons are positioned so that the threat is exposed to a continuously increasing volume of fire as it approaches the friendly protected asset or force. Defense in depth decreases the probability that attacking missiles, aircraft, or RAM will reach the defended asset or force.

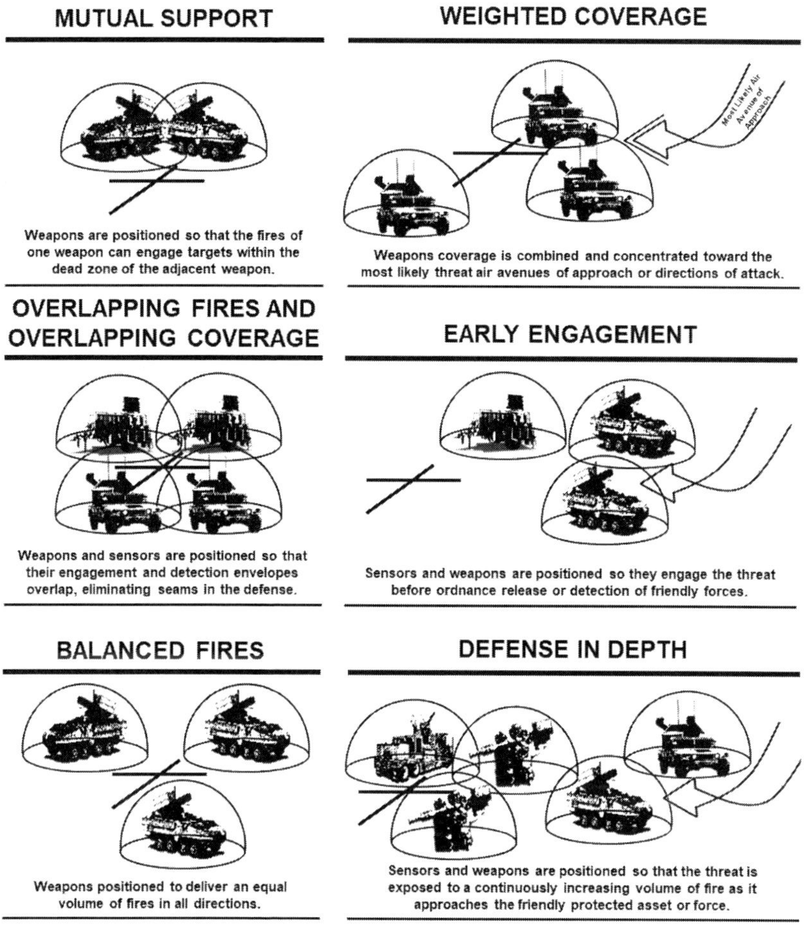

Ref: FM 3-01.44 (Jul '22) fig. 1-2. AMD employment tenets.

* Resilience

Resilience is the quality of the defense to maintain continuity of operations regardless of changes in or unanticipated tactics by enemy air or losses of critical air and missile defense components (FM 3-01). ADA planners must plan for defense design adjustment based on ADA system attrition. Resilience is a key determinant when considering which tenet (or tenets) to use in maintaining the defense.

VIII. Air Defense Artillery (ADA) in Support of Multidomain Operations

Ref: FM 3-01, U.S. Army Air and Missile Defense Operations (Aug '25), p. 17.

Army forces conduct operations in support of the joint force, with multinational allies and partners, and in coordination with other agencies and organizations. The Army's contribution to joint operations is multidomain operations.

Multidomain Operations *(See pp. 1-21 to 1-32.)*

Multidomain operations are the combined arms employment of joint and Army capabilities to create and exploit relative advantages that achieve objectives, defeat enemy forces, and consolidate gains on behalf of joint force commanders. Employing Army and joint capabilities makes use of all available combat power from each domain to accomplish missions at least cost. Multidomain operations are the Army's contribution to joint campaigns, spanning the competition continuum.

Multidomain operations span the competition continuum and seek to integrate and employ available capabilities to create and exploit relative advantages across the domains. All operations are multidomain operations. Army forces integrate land, maritime, air, space, and cyberspace capabilities in the operational environment to facilitate operations across the competition continuum. Within the context of an operational environment, a domain is a physically defined portion of an operational environment requiring a unique set of warfighting capabilities and skills (FM 3-0).

ADA units at the tactical level that support Army formations are positioned throughout the land domain to defend assets critical to the execution of operations within all domains, such as maneuver forces, naval and air bases and forces, cyberspace and space nodes, and C2 facilities. ADA units execute their missions primarily against aerial threats transiting the air domain; they also contribute to the joint targeting process and the application of offensive actions to defeat or negate air and missile threats through the use of air, land, maritime, and cyberspace capabilities. Using organic sensors, ADA units assist in collecting information on air threat capabilities which enable the acceptance or rejection of potential threat courses of action. ADA elements support air operations and contribute to airspace management. Engaging, deterring, or destroying threat air platforms support the air component's ability to achieve air superiority. ADA units collaborate with Navy AMD elements, typically in the littoral regions, to engage air and missile threats in the land domain.

The air battle will be prosecuted in all domains, with the objective of winning the battle. Where and how the threat aerial platform is defeated or negated is irrelevant. Land-based maneuver, aviation, and artillery forces provide security for ADA and other AMD forces against land-based threats and their support structures and can proactively target air and missile threats on the ground, greatly reducing their numbers and/or effectiveness before they can be employed.

Multidomain operations are aligned with the six phases of joint operations. However, they are executed under the rubric of competition below armed conflict, crisis, and armed conflict.

II. AMD in Large-Scale Combat Operations

Ref: FM 3-01, U.S. Army Air and Missile Defense Operations (Aug '25), chap. 6.

This section addresses operations during **armed conflict**, specifically large-scale combat operations. It provides an overview of large-scale combat operations, noting an enemy's generic approach to such operations and the AMD responses by the ADA force. This is followed by sections that describe deployment and entry, offensive and defensive operations and the actions of ADA forces in each. The focus of most of the discussions is the actions of the maneuver force and those of supporting ADA units.

Large-Scale Combat Operations (LSCO)

The focus of Army readiness is on large-scale combat operations. Large-scale combat operations are extensive joint combat operations in terms of scope and size of forces committed, conducted as a campaign aimed at achieving operational and strategic objectives (ADP 3-0). During ground combat, they typically involve operations by multiple corps and divisions, and they typically include substantial forces from the joint and multinational team. Large-scale combat operations often include both conventional and irregular forces on both sides.

Conflicts encompassing large-scale combat operations are more intense and destructive than limited contingencies, often rapidly amassing heavy casualties. Peer threats employ networks of sensors and long-range massed fires that exploit electromagnetic signatures and other detection methods to create high risk for ground forces, particularly when they are static. Army forces must account for constant enemy observation, including the threat from unmanned systems that saturate the operational environment. Army forces take measures to defeat the enemy's ability to effectively mass effects while creating exploitable advantages to mass effects against enemy capabilities and formations.

- FM 3-0, Operations (Mar '25).

Potential conflicts with state or non-state actors can ultimately lead to large-scale combat operations. Large-scale combat operations are extensive joint combat operations in terms of scope and size of forces committed, conducted as campaigns aimed at achieving operational and strategic objectives (ADP 3-0). The ability to prevail in ground combat is a decisive factor in breaking an enemy's will to continue a conflict.

The Army always fights as part of a joint force and usually as part of a multinational force during large-scale combat operations. The Army supports the joint force by employing capabilities from the land, maritime, air, space, and cyberspace domains in support of ground operations on land and employing ground-based capabilities to enable operations in the other domains.

Refer to AODS7-1: The Army Operations & Doctrine SMARTbook, w/SMARTupdate 1 (Multidomain Operations), pp. 1-87 to 1-26 for detailed discussion of ***armed conflict and large-scale combat operations*** *from FM 3-0, Operations (Mar '25). Armed conflict encompasses the conditions of a strategic relationship in which opponents use lethal force as the primary means for achieving objectives and imposing their will on the other.*

Antiaccess and Area Denial (AA/AD)

Antiaccess is an action, activity, or capability, usually long-range, designed to prevent an advancing enemy force from entering an operational area (JP 3-0). Area denial is an action, activity, or capability, usually short-range, designed to limit an enemy force's freedom of action within an operational area (JP 3-0). Enemies pursue these approaches with lethal means that significantly increase the risk to forward-stationed forces and the ability to deploy and stage additional forces into tactical assembly areas. Defeating antiaccess and area denial actions requires—

- A multidomain approach that includes Army forces retaining or seizing critical terrain to establish the depth necessary for defeating enemy forces.
- The employment of multiple attacks through multiple domains.
- The use of space and cyberspace effects to disrupt C2 networks, impacting human and automated communications between sensors, firing units, and command posts.
- The enabling of joint forcible entry operations and the movement of friendly forces from aerial ports of debarkation and seaports of debarkation to their tactical assembly areas.

Joint force commanders seek to achieve air, maritime, space, and cyberspace superiority early to allow the joint force to conduct land operations without prohibitive enemy interference. ADA has a critical role in countering antiaccess and area denial activities and assuring access into a given region. The joint warfighting force's ability to conduct force projection is reliant on ADA's ability to rapidly provide expeditionary AMD capabilities. When deployed, ADA units will be integrated with joint and multinational AMD forces to improve and extend surveillance and defend land, air, and sea bases that support operations in the antiaccess and area-denial environment.

I. Deployment and Entry Operations

ADA units are trained, structured, and postured for rapid deployment. Deployment planning tools allow commanders to adapt to rapidly changing circumstances. Lift assets are limited, but critical to the successful projection of the force. ADA commanders make every effort to integrate the capabilities of the host nation, joint and multinational forces, and forward presence forces with those of the deploying force. Commanders must balance the factors of METT-TC (I) against available lift assets to determine the composition of the initial response force. Each crisis has unique demands, causing commanders to balance requirements against time and lift availability and resolve shortfalls. In deployment, commanders must maintain flexibility and agility in force mix, their combat capability, sustainment, and lift, along with the need to forecast future events that call for decisions early in the deployment stage of the operation.

ADA commanders are responsible to provide AMD during deployment. They must tailor the force to accomplish the mission against the threat developed during IPOE. They may have to sacrifice mobility, redundant communications, and sustainability to bring in sufficient AMD fires to defend the force and designated assets during the initial phases of the deployment. Tailoring includes force refinement, which is the continuous process of adjustment conducted by commanders. Countering enemy surveillance activities is also a significant part of the responsibilities of the ADA commanders, especially with respect to the increasingly expanded roles of sUASs. AMD of joint priorities may outweigh defense of service priorities.

In preparations for and the conduct of deployments, ADA units must consider—

- Potential enemy air and missile capabilities, to include quantities and qualities.
- The presence and availability of joint and multinational AMD systems.

Air Defense Artillery Considerations (During Large-Scale Combat Operations)

Ref: FM 3-01, U.S. Army Air and Missile Defense Operations (Aug '25), p. 136.

ADA is a significant enabler in the battle to achieve air superiority, or at a minimum, maintain air parity throughout large-scale ground combat operations. Large-scale ground combat operations require the employment of a considerable amount of ADA forces to defend the most critical theater assets, as designated by the joint force commander. An army air and missile defense command (AAMDC) is positioned in the theater of operations to provide overall command of deployed ADA forces. An ADA brigade can be placed in supporting roles to an Army corps in accordance with METT-TC (I) conditions and the joint force commander's directives. A divisional air defense (DIVAD) battalion (or supporting ADA unit when DIVAD is not assigned) employs SHORAD units in support of the division commander's scheme of maneuver. The DIVAD's M-SHORAD batteries are normally deployed with BCTs.

ADA commanders task organizes units to execute an AMD operation plan, optimizing available ADA forces for defenses of the supported commander's priority assets. Battalion-level task forces are generated to defend theater, corps, and division assets, and task-organized batteries generally support brigade and below formations. Each task organization is constructed based upon mission-available ADA systems and commanders' intents and allows the matching of unit capabilities to tasks.

The phasing of operations determines where ADA forces are positioned in the theater of operations.

In the early phases of operations, Patriot and Terminal High Altitude Area Defense (THAAD) units are generally arrayed in the strategic support area to combat enemy ballistic missiles, cruise missiles, manned aircraft, and UASs, as the joint force commander sets the theater and builds combat power.

SHORAD units may remain with their assigned forces and defend their assets or can be task organized to other units, to provide an inner layer of defense against the air threats.

Once offensive ground operations begin, DIVAD units deploy with maneuver formations in divisions and BCTs, and Patriot units may be directed to provide AMD to a corps, as it maneuvers forward.

Defeating enemy forces in close combat is normally required to achieve campaign objectives and national strategic goals after the commencement of hostilities. Divisions and corps are the formations central to the conduct of large-scale combat operations, as they are organized, trained, and equipped for the deep, rear, and support operations that enable success during close combat. The ability to prevail in ground combat is a decisive factor in breaking an enemy's will to continue a conflict (FM 3-0).

- The presence of and coordination of efforts with forward-stationed ADA units.
- Identification of the supported commander's critical assets.
- ADA capabilities and organizations to be deployed.
- Priority of ADA forces and capabilities to be deployed.
- Coordination and integration with supported forces.
- Command and support relationships with supporting and supported Army and joint units.
- Enroute communications and internal/external C2 networks upon arrival at a lodgment.
- Initial and build-up sustainment needs.

Forces are most vulnerable, and the success of the operation is at greatest risk during initial entry. This vulnerability is intensified when the enemy possesses weapons of mass destruction. Defensive and offensive operations to counter these weapons will affect ADA, Army, joint, and multinational planning efforts. Defending the entry force is critical.

The requirements of entry operations vary. Each operation is different. Entry operations include rapid buildup and expansion that can require—

- Positioning the forces to achieve initial mission success and maximize future employment options.
- Establishing forward operating and logistics bases.
- Closing the remainder of the force.
- Expanding the lodgment.
- Linking up with other forces.
- Preparing for future operations.

Entry can be either opposed or unopposed. While it is conceivable that an entry may be unopposed, it is highly unlikely. Enemies of the United States have studied previous U.S. operations and understand that allowing U.S. forces to enter and build up allows them to mass technologically superior weapons and systems whose effects can rapidly change the operational environment. Thus, opposed, or forced entry, can be assumed to be the norm for all future entry operations. Forcible entry is the seizing and holding of a military lodgment in the face of armed opposition or forcing access into a denied area to allow movement and maneuver to accomplish the mission (JP 3-18). Forcible entry operations can be used to conduct operational movement and maneuver to attain positional advantage or as part of a deception.

The joint force is particularly vulnerable to attacks by enemy air platforms and missiles during the early stages of a forcible entry. AMD forces assist the joint force in gaining air superiority and protecting the assault force.

An ADA commander must ensure that the supported commander understands the need to deploy ADA forces with the supported force during entry operations. The ADA commander and staff develop an AMD plan to integrate AMD capabilities into the supported force commander's plan, facilitating a coordinated effort, providing defense of the supported commander's priority assets, and achieving the supported commander's goals and intent. The AMD plan must identify and allocate sufficient capabilities to deny, defeat, and destroy projected enemy air and missile systems. The ADA commander may create a task force with mixed long-and short-range capabilities to reduce risks to entry forces from enemy UASs, fixed-and rotary wing aircraft, and missiles.

The ADA commander must have the requisite authorities to execute AMD tasks in entry and follow-on operations. If operating as part of a joint entry force, the commander may require decentralized control of supporting ADA forces, as authorized by a joint engagement authority, to enable timely engagements of enemy air and missile systems. If operating as part of an Army-only force, the ADA commander requires identification and engagement authority within the entry area. Such authority must be identified, approved, and promulgated in the supported commander's operation order.

Forcible Entry Operations

Forcible entry operations are conducted in five phases: preparation and deployment, assault, stabilization of the lodgment, introduction of follow-on forces, and termination or transition of operations.

Commanders design their forcible entry operations to seize and hold a lodgment against armed opposition. A lodgment is a designated area in a hostile or potentially hostile operational area that, when seized and held, makes the continuous landing of troops and materiel possible and provides maneuver space for subsequent operations (JP 3-18). Friendly forces must be combat loaded and prepared for immediate combat operations upon arrival on the ground in the lodgment area. A force defends the perimeter of a lodgment until it has sufficient forces to break out and conduct offensive operations (FM 3-0).

Continuous intelligence support is critical. ADA commanders must have in-flight intelligence during deployment and entry operations. Once on the ground, ADA commanders and staffs must ensure that intelligence is timely, relevant, accurate, and predictive. The ADA S-2/G-2 gathers information on the enemy air and missile capabilities from joint and multinational elements and U.S. national intelligence agencies. Coordination is also established with supported unit, local units, and civilian entities to enhance knowledge of the immediate operational environment.

Early deployment of ADA units is crucial to the success of entry operations. Initial entry forces should be accompanied by organic, SHORAD systems, such as M-SHORAD or Avenger. More robust, HIMAD systems, such as Patriot, should be deployed with follow-on forces. Even in an apparently benign entry operation, defense of the force remains a critical command consideration. Theater missile defense units defend the lodgment, geopolitical assets, and debarking forces. As the joint force expands the lodgment, ADA units continue to defend the force and geopolitical assets and deny threat reconnaissance, surveillance, and target acquisition throughout the area of operations. This is particularly critical to deceiving the enemy on the U.S.'s intended course of action.

The AAMDC deploys its headquarters and subordinate units into theater to defend Army, joint, and multinational maneuver forces, critical assets, and designated support areas against enemy surveillance and air and missile attacks. The AAMDC is the area air defense commander's (AADC's) and ARFOR commander's primary resource and means to plan and prepare AMD assets during large-scale combat operations. The AAMDC directs ADA brigades, and other assigned or attached units, in the conduct of all operational-and tactical-level AMD missions within the joint area of operations. The AAMDC staff coordinates with the AADC and Service components to again define or confirm the critical assets requiring defense and array the ADA units. The staff coordinates AMD operations with the fires and protection cells in an Army corps to ensure that AMD actions are synchronized with those of the supported commander.

ADA brigades are deployed early to defend aerial and seaports of debarkation, early arriving forces, and critical sustainment elements. The ADA brigade, as a force provider for the AAMDC, ensures that required forces are available to meet mission objectives. The brigade coordinates the operations of subordinate ADA battalions, typically organized as task forces and assigned or attached units within the joint area of operations. The brigade plans, prepares, and executes operations based on the commander's intent as derived through guidance from the AAMDC. As a lodgment expands, the brigade may reposition its ADA units to better defend maneuver forces and other critical assets, such as communications nodes and transportation hubs.

II. Defensive Operations

A defensive operation is an operation to defeat an enemy attack, gain time, economize forces, and develop conditions favorable for offensive or stability operations (ADP 3-0). Reasons for conducting a defense include to—

- Create conditions for the offense that allow Army forces to regain the initiative.
- Retain decisive terrain or deny a vital area to an enemy force.
- Attrit or fix an enemy force as a prelude to the offense.
- Counter enemy action.
- Accept risk in one area to create offensive opportunities elsewhere.

While the offense is more decisive, the defense is typically stronger. However, a defense normally cannot determine the outcome of battles (FM 3-90).

Defenses for virtually all Army elements are focused on the potential or projected ground threat. Units and their weapons are positioned to deny or mitigate this threat. However, with the increasing proliferation of sUASs and potential for complex integrated attacks of any friendly concentration of forces, commanders must be cognizant of, plan, and allocate resources to defend against all air threats. Thus, defenses of installations, such as operating bases, or areas where brigade elements may concentrate, should consider the most likely air avenues of approach, surveillance orbits, and suspected areas from which UASs and rotary-and fixed-wing aircraft can be launched.

The Army conducts three types of defensive operations: area defense, mobile defense, and retrograde. For more information on these, see FM 3-90. Although the names of these types convey the overall aim of a selected operation, each type usually contains elements of the other and combines static and mobile elements (FM 3-90).

These types of defenses are predominately supported by DIVAD forces. They provide direct support of division forces. While Patriot and THAAD can also support these defensive operations, they provide defense from established positions in the rear or joint security areas, from where they focus on the defense of higher echelon (theater and corps) assets. Coverage is extended over maneuver force operations as a complementary capability (complementary to the defense of the designated assets) or per the direction of a joint/ARFOR commander.

Note. ADA units typically provide point and area defense of designated critical assets. A point defense is the employment of defense measures to protect limited areas (JP 3-01). A Maneuver Short-Range Air Defense (M-SHORAD) platoon may establish a point defense of a forward arming and refueling point or maneuver battalion assembly area. A Patriot battery may establish a point defense of an airfield. Included within this defense may be the airfield and control tower, operations center, and fuel area, given the range and altitude capabilities of the Patriot system. An area defense, in AMD terminology, is one that employs defensive measures to protect assets over an extended area. THAAD, with its extended range and altitude capabilities, is the only ADA system that can provide area defense. Subsequent discussions of area defense, however, are keyed to the Army definition in ADP 3-90.

A. Area Defense

An area defense is a type of defensive operation that concentrates on denying enemy forces access to designated terrain for a specific time rather than destroying the enemy outright (ADP 3-90). The purpose of an area defense is to retain the terrain where the bulk of the defending force positions itself in mutually supporting prepared positions.

DIVAD forces in an area defense are generally focused on defending designated fixed or semi-fixed assets, rather than denying access to terrain. ADA commanders and leaders implement defenses based on the IPOE, METT-TC (I), AMD employment tenets, and the maneuver commander's scheme of maneuver. Positions must be planned and prepared in coordination with the supported commander to ensure integration of efforts and optimal defenses against enemy air surveillance and attacks. During preparations of the maneuver forces' defensive positions, the unit preparing positions and obstacles is initially the air defense priority. The air defense priority can then shift, based on the supported commander's priorities. When maneuver is required, for instance, priority may be given to the maneuvering elements.

Position Areas for Air Defense (PAAD)

DIVAD units establish and coordinate position areas for air defense and aerial engagement areas to ensure that the supported force and adjacent forces are cognizant of DIVAD unit locations while preparing for and engaging enemy air platforms. Position areas for air defense are located near the defended asset based on METT-TC (I). M-SHORAD, Avenger, and Stinger units can move from primary to secondary and supplementary locations within the position area, as needed, for survivability and enhanced engagement opportunities. Deliberate planning and coordination are required for the emplacement and operation of position areas for air defense near the forward line of own troops to minimize the risk for fratricide between maneuver forces operating along there and the ADA units forward positioned to defend those forces. This can be extremely important with operating with Coalition forces. Position areas for air defense—

- Are planned for in coordination with the supported commander's schemes of action and maneuver.
- Are designated for a specific period of time.
- Clearly delineate areas where DIVAD units can operate while optimizing survivability.
- Ensure that assigned areas for DIVAD units are included in the common operating picture.
- Assist in ensuring DIVAD unit engagement areas support the maneuver force's offensive and defensive operations.

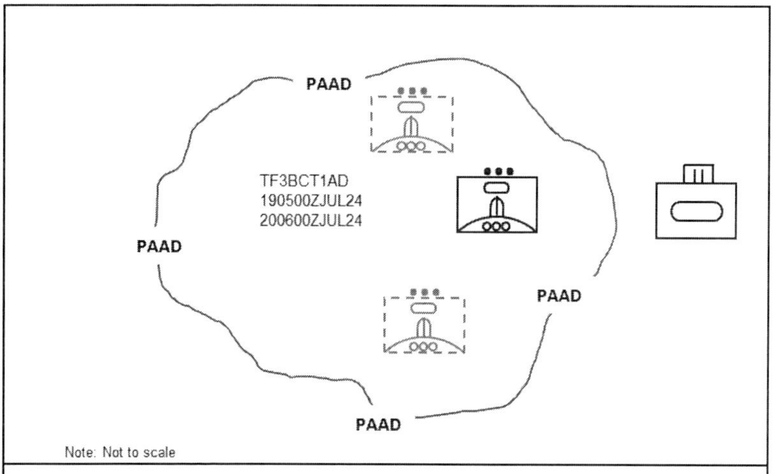

AD	armored division	PAAD	position area for air defense
BCT	brigade combat team	TF	task force

Ref: FM 3-01 (Aug '25),fig. 6-1. Nominal position area for air defense.

Figure 6-1 depicts a nominal position area for air defense for a M-SHORAD platoon defending an armor battalion-sized task force. The PAAD is irregularly shaped

Characteristics of the Defense

Ref: FM 3-01, U.S. Army Air and Missile Defense Operations (Aug '25), pp. 141 to 142.

Defensive operations are conducted throughout the area of operations, from the deep area to the strategic support area. Depending on METT-TC (I), maneuver units typically execute operations in portions of the close area—a main battle area, security area, or along the forward edge of the battle area or forward line of own troops—or in the rear area, The main battle area is the area where the commander intends to deploy the bulk of their unit to defeat an attacking enemy (FM 3-90). The security area is that area occupied by a unit's security elements and includes the areas of influence of those security elements (ADP 3-90). The forward edge of the battle area is the foremost limits of a series of areas in which ground combat units are deployed to coordinate fire support, the positioning of forces, or the maneuver of units, excluding areas in which covering or screening forces are operating (JP 3-09.3). The forward line of own troops is a line that indicates the most forward positions of friendly forces in any kind of military operation at a specific time (FM 3-90). The rear area constitutes that area behind the security and main battle areas. DIVAD units may defend their supported forces in each of these. HIMAD units defend key fixed assets in the rear area of committed corps or division forces, joint security area, and the strategic support area. Patriot may also be required to defend assets in a corps or division's sector where an operational activity is occurring.

See pp. 3-8 to 3-9 for the arrayal of ADA units (within the operational framework).

ADA units begin defensive preparations immediately upon moving into position. Priority for air defense in the defense may be to the battle positions in the main battle area, command and control elements, or logistics assets. ADA commanders and leaders must perform mission analyses, ensuring they understand the supported commander's intent and concept of the operation. Based on these considerations, they develop coverage plans to support the defensive concept of operations. Actual positioning of ADA systems should be based on the air IPOE, the commander's priorities for AMD, and the location of the supported unit's assets. ADA commanders and leaders must ensure that their plans for support of the defense are properly coordinated, within their unit and with adjacent and supported forces. Constant coordination and exchange of information is key. Leaders consider these planning aspects in the coordination process:

- The higher commander's intent and concept of operations.
- Updated locations and composition of the supported force.
- The tactics applied by the supported force.
- Control measures that enhance coordination between units.
- Planning for mutual support—air defense of the supported force by ADA units, and ground defense of ADA units by the supported force.
- Airspace coordination measures.
- Information collection and sensing of the airspace.
- ADA unit locations and coverage areas.
- Engagement operations and clearance of fires.
- Communications.

ADA forces can be allocated to reserve or counterattack forces, C2 and logistics facilities, and fire support units. If the supported unit is defending against an enemy main attack, the ADA commander may move ADA forces forward, to achieve the mass and early engagement needed to defeat potential enemy UAS and manned air platforms and take risks in the rear area. If defending against a supporting attack, more Avenger and Stinger teams can be allocated to static assets in the unit's rear. The AMD plan must be flexible enough, however, to quickly mass against enemy air surveillance and attacks should the enemy's main effort shift and occur in the supported unit's sector.

Successful defenses share some important characteristics. These characteristics and associated AMD actions are presented in table 6-2.

Characteristic	***Description***	***ADA Actions***
Disruption	Deceiving or destroying enemy reconnaissance forces, breaking up combat formations, separating echelons, and impeding an enemy force's ability to synchronize its combined arms.	ADA forces and non-dedicated air defense capabilities provide defense against UASs, manned attack aircraft, rockets, and missiles.
Flexibility	Developing plans that anticipate a range of enemy actions and allocate resources accordingly.	ADA commanders, staffs, and leaders conduct thorough IPOE analysis of potential enemy aerial capabilities and develop plans, with branches and sequels, that allocate the right-size forces that provide suitable defenses against them.
Maneuver	Achieving and exploiting a position of physical advantage over an enemy force.	DIVAD forces maneuver with their supported forces, synching their movements and activities to best defend the supported maneuvering units. Patriot units conduct movements, as ordered, to extend AMD coverage over maneuvering forces.
Mass and concentration	Creating overwhelming combat power at specific locations to support the main effort.	ADA forces mass capabilities to mitigate the effects of enemy air threats. Massing effects can be created through the use of multiple ADA units or systems or by multiple launches/shots fired. Massing effects does not imply massing units or systems at one location.
Depth	Engaging multiple enemy echelons, enemy long-range fires, sustainment, and C2.	ADA units seek to create defense in depth wherever feasible. The intent is to bring enemy aerial platforms and missiles under increasingly more AMD fires as they ingress to the defense.
Preparation	Preparing the defense before attacking enemy forces arrive.	AMD plans identify the allocation of ADA forces to designated critical assets. Commanders and leaders plan defenses of these using the ADA employment tenets and IPOE analyses. Once in position, ADA commanders and leaders adjust system component positions to best achieve the planned defense. All preparations are typically initiated prior to enemy air actions.
Security	Conducting security, protection, information activities, operations security, and cyberspace and electromagnetic warfare tasks.	ADA units conduct security operations with their supported units or assets. They effect coordination to prepare and execute security of positions and force protection. They synchronize AMD operations with offensive and defensive tasks conducted by cyberspace forces.

ADA	air defense artillery	DIVAD	divisional air defense
AMD	air and missile defense	IPOE	intelligence preparation of the operational environment
C2	command and control	UAS	unmanned aircraft system

Ref: FM 3-01 (Aug '25), table 6-2. Characteristics of the defense.

Refer to AODS7-1: The Army Operations & Doctrine SMARTbook, w/SMARTupdate 1 (Multidomain Operations), pp. 1-110 to 1-117 for detailed discussion of ***defensive operations*** *from FM 3-0, Operations (Mar '25). Defensive operations defeat an enemy attack, buy time, economize forces, hold key terrain, or develop conditions favorable for offensive operations.*

as M-SHORAD systems in the platoon can be positioned at various ranges and different locations from the task force, though all must focus on projected enemy air avenues of approach. The PAAD contains primary, alternate, and supplementary positions. Irrespective of locations, the M-SHORAD systems must be capable of maintaining the primary target line. The date-time group identifies the effective time that the platoon occupies the PAAD and the time the supported force is expected to stay in that location.

Aerial Engagement Areas

Aerial engagement areas are created over the defended maneuver formation or other asset in coordination and synchronization with the supported commander and other ADA/joint units providing AMD in or near the supported commander's area of operations. The use of aerial engagement areas can alleviate potential overkill and excessive expenditure of munitions in engaging air targets. Aerial engagement areas are adjusted as assets move or priority of assets change. ADA commanders implement the following steps of engagement area development adapted from FM 3-90:

- Identify likely enemy air avenues of approach.
- Identify the most likely enemy course of action for aerial attacks and surveillance.
- Determine where to engage and destroy the enemy before its air platforms and missiles can reach their target areas (keep-out range).
- Position subordinate forces and weapons systems to optimize defenses against enemy aerial surveillance and attack.
- Rehearse the execution of operations within the engagement area.

DIVAD assets in each defense focus on the templated air avenues of approach. M-SHORAD systems can be positioned to conduct early engagements of UASs. They can be collocated with counter-reconnaissance forces or independently along air avenues of approach that allow the enemy surveillance of defensive positions and preparations. While DIVAD systems are emplaced where they can best defend against air threats, maximizing their engagement capabilities, ADA commanders should mitigate exposure to direct fires as much as possible.

DIVAD units provide warning of threat actions, most notably UAS surveillance and attacks, to their supported forces. Detections by Sentinel radars or other sensors are broadcast to affected units or installations. Detections of sUASs, particularly groups 1 and 2, can trigger self-defense actions by non-dedicated air defense elements in Army units. Passive defense measures should be executed to mitigate indirect fire attacks.

Patriot and THAAD units typically operate at the strategic or operational level to provide defense of critical assets designated by joint force commander against the most likely threats to the joint force—longer-range ballistic and cruise missiles. These assets include geopolitical assets, aerial and seaports of debarkation, critical command posts, logistical facilities, assembly areas, forward operating bases, and maneuvering forces.

The HIMAD-based task force fights from stationary locations; therefore, tactics are a matter of the proper positioning, orientation and system configuration of subordinate components with respect to the defended assets and threat. Tactics for threat engagements will vary based on the specific threat platforms, such as ballistic missiles, cruise missiles, or fixed-wing aircraft.

Patriot units in task forces are generally employed as task-organized batteries. The task-organized battery affords continuous defense of assets against threats within its METT-TC (I)-tailored capabilities. A Patriot task force can also be missioned to defend a maneuver force. It conducts this defense using bounding overwatch techniques, as Patriot lacks the mobility to maintain pace with maneuvering forces.

Bounding overwatch is a movement technique used when contact with enemy forces is expected. The unit moves by bounds. One element is always halted in position to overwatch another element while it moves. The overwatching element is positioned to support the moving unit by fire or fire and movement (FM 3-90).

Figure 6-2 presents a nominal area defense. THAAD conducts the area defense, with surveillance and engagement coverage overlaying the point defenses provided by Patriot and M-SHORAD. Patriot could be positioned to extend its coverage over the tactical assembly area if directed by the controlling commander— in this case, the corps commander.

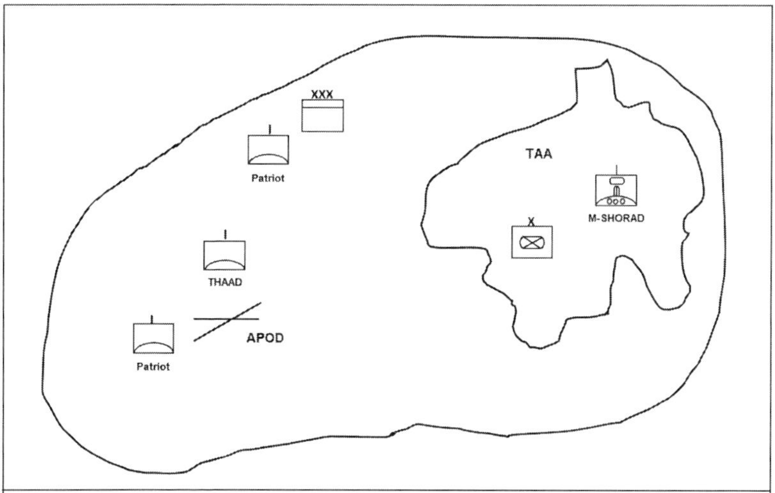

APOD	aerial port of debarkation	TAA	tactical area assembly
M-SHORAD	Maneuver-Short-Range Air Defense	THAAD	terminal high altitude area defense

Ref: FM 3-01 (Aug '25), fig. 6-2. Air defense artillery units in a nominal area defense.

B. Mobile Defense

Mobile defense is a type of defensive operation that concentrates on the destruction or defeat of the enemy through a decisive attack by a striking force (ADP 3-90). The mobile defense focuses on defeating or destroying enemy forces by allowing them to advance to a point where they are exposed to a decisive counterattack by a striking force. A mobile defense includes a fixing force and a striking force. A fixing force is a force designated to supplement the striking force by preventing the enemy from moving from a specific area for a specific time (ADP 3-90). The striking force is a dedicated counterattack force in a mobile defense constituted with the bulk of available combat power (ADP 3-90). A fixing force supplements the striking force by holding attacking enemy forces in position, by canalizing attacking enemy forces into ambush areas, and by retaining areas from which to launch the striking force.

M-SHORAD units may accompany the striking force. M-SHORAD systems are positioned with the maneuver force formations to defend against manned aircraft attacks and UAS surveillance. Defense against manned aircraft is focused on the likely air avenues of approach. Defense against UASs requires 360-degree coverage. M-SHORAD systems may also be placed in defense of a site, such as a wet gap crossing or terrain obstacle. Avenger platoons and C-sUAS/Stinger teams may support the fixing force by engaging aerial surveillance and attack platforms from select, suitable locations. Positioning is determined in accordance with the projected enemy air and ground capabilities and activities and the supported commander's objectives.

Figure 6-3 presents a nominal mobile defense after the commitment of the strike force. A M-SHORAD platoon accompanies each of the maneuver battalions in the strike force. Another M-SHORAD platoon defends a fixing force battalion. The M-SHORAD platoons are placed in direct support of the maneuver units. C-sUAS teams are positioned to deny or defeat sUAS surveillance and attack options. C-sUAS teams are in general support of the fixing force units.

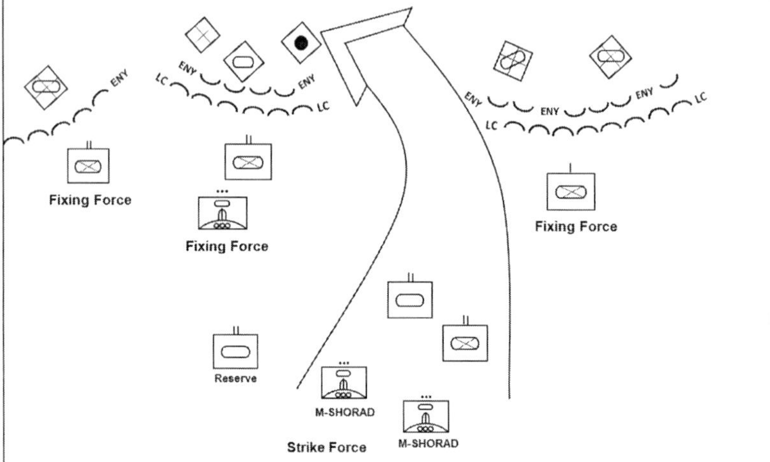

Note: Though the M-SHORAD platoons are depicted following the attacking battalions, they would typically provide defense against air threats from overwatch positions

C-sUAS	counter-small unmanned aircraft system	LC	line of contact
ENY	enemy	M-SHORAD	Maneuver-Short-Range Air Defense
EW	electromagnetic warfare		

Ref: FM 3-01 (Aug '25),fig. 6-3. Divisional air defense units in a nominal mobile defense.

The key aspects of supporting mobile defense are mobility and survivability. The moving elements in a mobile defense are the most vulnerable to attack and surveillance. Counterintelligence, surveillance, and reconnaissance are critical because the plan relies on surprise.

C. Retrograde

A retrograde is a type of defensive operation that involves organized movement away from the enemy (ADP 3-90). A retrograde is a transitional operation. It is always part of a larger scheme of maneuver designed to regain the initiative and defeat the enemy.

DIVAD units can be missioned to defend the covering force, assuming that the covering force's mission is not to destroy the lead elements of the enemy force. A covering force provides the main body of the maneuver force with early warning, protects its movement, and develops the situation before the main body is committed. M-SHORAD units can maneuver with the covering force and provide the required defense against air threats. Avenger and Stinger can provide overwatching fires of a covering force, that is used as a screening force, providing information on enemy air activity and engaging enemy air targets as required.

Since the covering force units are typically not the air defense priority, the covering force units normally rely on passive air defense.

See facing page for an overview and further discussion of passive air defense.

Passive Air and Missile Defense

Ref: FM 3-01, U.S. Army Air and Missile Defense Operations (Aug '25), pp. 147 to 148.

Passive actions are a unit's first line of defense and improve survivability by reducing the likelihood of being detected and targeted from the air and by mitigating the potential effects of air surveillance and attack. Passive air and missile defense measures include detecting air and missile launches, predicting impact points, providing threat identification, and disseminating early warning. Passive measures also include those initiated to reduce vulnerability and minimize the effect of damage caused by missile attack. Five fundamental methods can be used to camouflage and conceal individual weapons, units, installations, and activities:

Hiding. Hiding is the complete concealment of an object by some form of physical screen. For example, the overhead canopy of trees hides the objects beneath from aerial observation; tunnels hide objects located within them; a building's roof and walls, camouflage net, or tarpaulin hides objects beneath them; and a defilade position hides objects from ground observation. In some cases, the screen may be invisible. In other instances, the screen may be visible, but it hides the activity behind it.

Blending. Blending is arranging or applying camouflage materials on, over, and around an object so that it appears to be part of the background. Examples include applying face paint to the exposed areas of skin, and adding burlap, paint, and live vegetation to helmets and clothing to closely resemble or blend into the background. Units can apply the same technique for equipment or structures.

Disrupting. Disrupting is the technique that alters or eliminates regular patterns and target characteristics. Disrupting techniques include pattern painting, deploying camouflage nets over selected portions of a vehicle, and using shape disrupters (such live vegetation) to eliminate regular target patterns.

Disguising. Disguising is the intentional misleading of the enemy about the friendly force's identity, strength, and intention, and may draw enemy fire from real assets. Therefore, the simulation of objects, pieces of equipment, or activities may have military significance. Decoys, such as inflatable tanks or ADA radars, can look like the real thing to an aerial observer.

Decoying. Decoying is the technique that deploys a false or simulated target within a target scene or in a position where the enemy might conclude that it has found the correct target. Decoys generally draw fire away from real targets. Depending on their fidelity and deployment, decoys greatly enhance survivability.

Other Considerations

In addition to camouflage and movement control, units can plan for and institute electromagnetic hardening, electronic masking, and emission control by such measures as shielding sensitive equipment components, controlling signatures to protect against threat signals intelligence operations, and selecting and controlling the use of sensors and C2 emitters, respectively (FM 3-12). Units on installations should continually harden their locations and incorporate all means to lessen the potential for threat observation and mitigate the effects of air attacks.

Non-Dedicated Air Defense Capabilities *(See p. 6-2 to 6-3.)*

Non-dedicated air defense capabilities should also be integrated into defenses, as complements to supporting ADA units and systems or as standalone resident capabilities if ADA units are not available. A ground force's organic weapons are its primary defense against short-range air threats.

For more information on passive AMD measures, refer to ATP 3-01.8.

III. Offensive Operations

An offensive operation is an operation to defeat or destroy enemy forces and gain control of terrain, resources, and population centers (ADP 3-0). The main purpose of the offense is to defeat or destroy enemy forces and to gain control of terrain, resources, or population centers (FM 3-90). Other purposes are to—

- Secure decisive terrain.
- Deprive the enemy of resources.
- Gain information.
- Fix an enemy force in position.
- Disrupt an enemy force's attack.
- Set the conditions for successful future operations.

Commanders seize, retain, and exploit the initiative when conducting the offense. Taking the initiative from an enemy force requires conducting offensive operations, even in the defense. Defeating enemy air and missile threats is necessary to create opportunities for offensive maneuver.

ADA units generally do not conduct offensive operations in the traditional sense; they support offensive operations. However, in certain circumstances, they can set air defense ambushes, with positions forward along the likely avenues of approach to engage enemy air—this may be considered an offensive action.

Support of offensive operations is primarily executed by M-SHORAD and Avenger batteries/platoons, C-sUAS teams, and Stinger man-portable air defense (MANPAD) teams. DIVAD battery commanders and platoon leaders organize and position their units to support the maneuver commander's mission and scheme of maneuver. They should organize their units for maximum flexibility and plan for decentralized execution. Patriot and THAAD units, if so directed, can also provide defense of maneuvering forces from their primary defensive positions or relocate to extend such coverage. Any repositioning of Patriot or THAAD must be coordinated with the AADC and senior ADA commander to ensure that the relocation does not result in a subsequent gap in a defense. If a Patriot unit is placed in direct support of a corps, the corps commander sets the priorities for the Patriot unit. Patriot units placed in direct support of a maneuver force notify the controlling joint C2 element, through the ADAFCO, and their parent ADA brigade when their batteries are executing movements in accordance with the supported commander's scheme of maneuver. In this case, the supported commander dictates movements as the supported unit is the defended asset.

AMD tasks to enable offensive operation include—

- Planning air defense coverage, to include denying surveillance by threat UASs, to provide freedom of maneuver for supported forces and their critical assets.
- Developing air aspects of the IPOE.
- Developing and coordinating IPOE information to support attack operations of enemy air bases and other aerial launching sites.
- Developing air defense priorities.
- Defending forward-based infrastructure, such as lines of communications and command nodes, from aerial attack.
- Planning early warning and surveillance.
- Maneuvering to maintain air defense coverage.
- Developing engagement guidance for ground targets.

There are four types of offensive operations: movement to contact, attack, exploitation, and pursuit.

Characteristics of the Offense

Ref: FM 3-01, U.S. Army Air and Missile Defense Operations (Aug '25), p. 150.

Successful offenses share four characteristics: audacity, concentration (mass), surprise, and tempo. These characteristics and associated AMD considerations are presented in table 6-3.

Characteristic	*Description*	*ADA Actions*
Audacity	The ability to assume risk to create opportunity with bold action.	DIVAD units support maneuver commanders and forces in aggressively seizing opportunities to maintain the initiative and force enemy forces to react.
Concentration (mass)	Orchestrating forces or effects to create and exploit opportunity.	As in defensive operations, ADA forces mass capabilities to mitigate the effects of enemy air threats. Massing effects may be created through the use of multiple ADA units or systems, at different locations, to execute multiple launches/shots fired from a single system or multiple systems.
Surprise	Taking action that catches enemy forces off guard.	SHORAD systems may assume positions away from defended forces or other assets, and along likely air avenues of approach, to conduct early engagements of enemy air transiting the area.
Tempo	Maintaining a pace of operations that is faster than that of the enemy, but not so fast that it cannot be sustained for as long as necessary to achieve all assigned objectives.	M-SHORAD units maintain pace with their supported maneuvering forces to mitigate or deny enemy air surveillance and attacks.

ADA air defense artillery
DIVAD divisional air defense
M-SHORAD maneuver short-range air defense
SHORAD short-range air defense

Ref: FM 3-01 (Aug '25), table. 6-3. Characteristics of the offense.

*Refer to AODS7-1: The Army Operations & Doctrine SMARTbook, w/SMARTupdate 1 (Multidomain Operations), pp. 1-118 to 1-126 for detailed discussion of **offensive operations** from FM 3-0, Operations (Mar '25). The key to successful offensive operations is to achieve all desired objectives prior to culmination. This requires the force in the offense to have some combination of relative advantage in the physical, information, or human dimensions.*

A. Movement To Contact

The goal of a movement to contact is to make initial contact with a small element while retaining enough combat power to develop the situation and mitigate the associated risk. Commanders conduct a movement to contact when an enemy situation is vague or not specific enough to conduct an attack. Once contact is made with an enemy force, the friendly force quickly transitions either to an attack, a hasty defense, or bypasses enroute to their objective.

In a movement to contact, forces deploy from assembly areas in attack formations that facilitate movement and mass sufficient combat power to ensure decisive victory. Commanders adjust their tempo appropriately as they anticipate closing with enemy forces. They decide where their forces can deploy into formations that facilitate the initial contact and still provide freedom of action for the bulk of their forces.

A movement to contact can result in a meeting engagement. A meeting engagement is a combat action that occurs when a moving force, incompletely deployed for battle, engages an enemy at an unexpected time and place (ADP 3-90). Once contact is made and a meeting engagement ensues, M-SHORAD units move to planned positions, observe and focus on enemy air avenues of approach, and activate planned target areas of interest.

M-SHORAD platoons are integrated with the maneuvering forces. M-SHORAD platoons must be able to posture themselves quickly to counter any enemy surveillance or attack efforts. They are positioned based on the maneuver force's type of travel (for example, deliberate movement when contact with enemy forces is likely) and formation and other METT-TC (I) factors. The M-SHORAD battery commander or platoon leader makes positioning decisions based on preventing enemy aircraft from successfully observing or attacking the vulnerable moving elements. M-SHORAD systems are generally placed behind the lead elements of the supported force during movement to provide overwatch of enemy air avenues of approach. Air defense coverage is extended forward of the lead elements. C-sUAS teams can also be positioned with the M-SHORAD systems or as standalone capabilities to strengthen defenses against small surveilling and attacking unmanned aircraft.

Avengers can be deployed with the covering force, along the flanks, or in overwatch positions of the maneuver force. Avenger fire units are light-skinned vehicles and should not be integrated into the maneuver force when contact is expected. The Avenger can be positioned on the maneuver force flank and rear where it is best suited to defend enemy air avenues of approach. Avengers may also be employed to protect the supported unit's critical assets, such as C2 elements and artillery radars. Avengers and Stinger MANPAD teams can provide coverage for choke points along the march route or to defend the maneuver force as it moves prior to contact. Stinger teams are primarily used to protect maneuver support assets, such as fire support systems and C2 nodes. Once contact has been initiated, Stinger teams need to assume overwatch positions.

B. Attack

Attacks incorporate coordinated movement supported by fires. An attack differs from a movement to contact because in an attack commanders know at least part of an enemy's dispositions. A commander may describe an attack as hasty or deliberate, depending on the time available for assessing the situation, planning, and preparing. A commander may decide to conduct an attack using only fires, based on an analysis of METT-TC (I).

See following pages (pp. 3-30 to 3-31) for an overview and further discussion.

C. Exploitation

Exploitations seek to disintegrate enemy forces to the point where they have no alternative but to surrender or retreat. Exploitation follows any successful attack and is conducted to take advantage of the success. The exploiting force must have mobility and balanced firepower.

DIVAD units supporting an exploiting force must be as mobile as the maneuver forces. They must be able to move with the exploiting force to provide defense against aerial surveillance and attacks. A M-SHORAD platoon, for instance, may move with the main body (exploitation force), or it may be task organized to provide support for both the exploitation force and the logistics elements. Avenger units can be used to keep maneuver force lines of communications open and defend the exploiting force's less mobile assets. Stinger MANPAD teams can provide coverage of key terrain and choke points and protect C2 posts. C2 is critical to support constantly moving units.

D. Pursuit

There are two variations of the pursuit: frontal and combination. A pursuit normally follows a successful exploitation. However, if enemy resistance breaks down and enemy forces begin fleeing the battlefield, any type of offensive operation can transition into a pursuit. Pursuits entail rapid movement and decentralized control. The pursuit is oriented on the enemy rather than on terrain objectives.

As in an exploitation, DIVAD units deployed with the encircling forces must be as mobile as the defended asset. Flexibility, decentralized execution, and the direct fire threat to ADA fire units are key considerations in the employment of DIVAD units. M-SHORAD units maneuver with the pursuing forces, normally traveling at their flanks. M-SHORAD battery commanders and platoon leaders employ their M-SHORAD systems to prevent the enemy from using attack helicopters to disrupt, delay, or deny the pursuing force. Avengers are generally positioned at the rear of a maneuvering force to defend the maneuver support, C2, fire support, and maneuver reserve forces. Stinger MANPAD teams can be best used to provide area defense along main supply routes or logistics support areas. The ability of the DIVAD units to maintain ADA coverage over the maneuver force ensures that the maneuver force commander can retain the initiative during pursuit operations.

IV. Consolidate Gains

Consolidate gains are activities to make enduring any temporary operational success and to set the conditions for a sustainable security environment, allowing for a transition of control to other legitimate authorities (ADP 3-0). Though presented as part of the offensive operations section, consolidate gains are also part of defensive and stability operations.

Consolidation of gains generally signifies a greater focus on security and stability tasks than on combat operations. ADA forces support operations to consolidate gains by maintaining defense of friendly forces and critical assets as areas are secured. ADA formations, in accordance with the priorities for defense and available assets, can be positioned in corps or division consolidation areas to defend against enemy residual air and missile capabilities. While an enemy may possess few of these capabilities—the majority having been destroyed or disabled during the dominate phase (large-scale ground combat) of an operation, an attack by a single missile can have a catastrophic impact on a maneuver formation, C2 facility, or geopolitical asset.

Transitions occur throughout the theater or various areas of operation. Offensive, defensive, and stability operations can be conducted by different echelons simultaneously in various areas. This transition can occur in one part of a theater while combat operations are still underway in other parts. Post-conflict operations focus on restoring order, minimizing confusion following the operation, reestablishing host

Attack

Ref: FM 3-01, U.S. Army Air and Missile Defense Operations (Aug '25), pp. 151 to 153.

Attacks incorporate coordinated movement supported by fires. An attack differs from a movement to contact because in an attack commanders know at least part of an enemy's dispositions. A commander may describe an attack as hasty or deliberate, depending on the time available for assessing the situation, planning, and preparing. A commander may decide to conduct an attack using only fires, based on an analysis of METT-TC (I).

Figure 6-4 presents a nominal attack with supporting DIVAD units. Two M-SHORAD batteries accompany the main attack and one battery the supporting attack. The M-SHORAD batteries are portrayed as following the attacking maneuver units for illustrative purposes. They are generally integrated with the maneuver forces, positioned to defend against air threats from overwatch positions. M-SHORAD systems are positioned in accordance with the projected air threat and mission demands. C-sUAS teams, though not shown, are likely employed with the M-SHORAD systems.

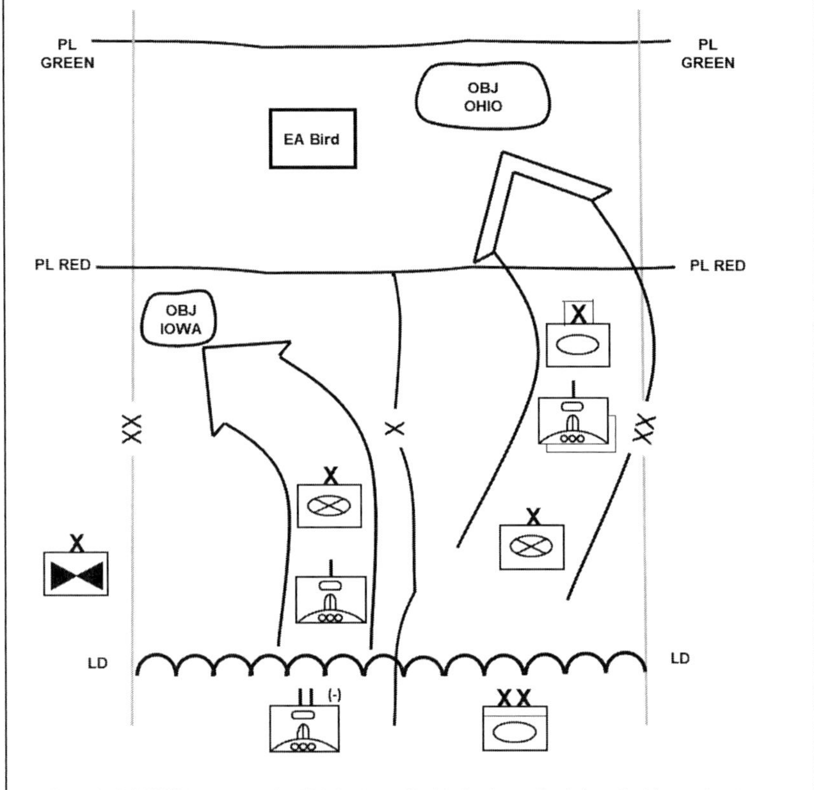

Note: Though the M-SHORAD batteries are depicted following the attacking brigades, they would typically provide defense against air threats from overwatch positions

EA	engagement area	OBJ	objective
LD	line of departure	PL	phase line
M-SHORAD	Maneuver-Short-Range Air Defense		

Ref: FM 3-01 (Aug '25), fig. 6-4. Divisional air defense units in a nominal attack.

Hasty Attack

In a hasty attack, the commander must move quickly to gain the advantage. Speed and violence of action can overcome lack of preparation. A deliberate attack is much more thoroughly planned and prepared. Hasty attacks are not based on detailed planning, but the commander must anticipate and plan for them as much as possible. They are conducted to catch the enemy off-guard. The maneuver commander can use assets available and synchronize all available support and combat multipliers provided they can be committed without significant delay. Hasty attack plans are disseminated through fragmentary orders. The use of standard operating procedures and battle drills are critical for effective execution of hasty attacks. The success of DIVAD units in a hasty attack is dependent on thorough prior planning. Prior to the mission, M-SHORAD, Avenger, and Stinger MANPAD systems may be positioned along suspected enemy air avenues of approach. ADA platoon leaders plan operations based on the supported commander's scheme of action and designated priorities.

Deliberate Attack

The deliberate attack requires more time for a detailed analysis of the operational environment and preplanned positions throughout the expectant zones of advance. For a deliberate attack, air defense coverage must be extensive and well-coordinated. This requires a thorough IPOE analysis that includes—

- Enemy aerial types, quantities, and capabilities.
- Potential launch sites.
- Ground and air avenues of approach.
- Aerial surveillance positions and routes.
- Aerial attack profiles.

ADA platoon leaders use system-specific graphic control measures and those directed by the supported unit to control M-SHORAD, Avenger, and any Stinger team movements. All movements are conducted in synchronization with the supported unit's scheme of maneuver. During an attack, M-SHORAD systems are generally positioned to support the main effort. Avengers are positioned to protect maneuver support, C2, and fire support assets. However, M-SHORAD or Avenger units may be redirected to shift fires or positions and change priorities in accordance with the supported commander's directives,

An attack can require a forward passage of lines through a stationary force. In a passage of lines, the stationary force's SHORAD units (C-sUAS teams from a DIVAD battalion or Avenger systems/Stinger MANPAD teams) protect the passage lanes. The attacking force's M-SHORAD units accompany the passing forces. Airspace coordinating measures and weapons control measures must be planned and disseminated by either the division AMD section or brigade air defense support element (ADSE). Similar procedures apply for a rearward passage of lines.

nation infrastructure, preparing forces for redeployment, and continuing a presence to allow other elements of national power to achieve the overall strategic aims. Post-conflict operations place demands on every level of command.

The post-conflict stage may be interrupted by the resumption of hostilities. The enemy may continue to challenge the establishment of normal peacetime operations or stability operations, with the objective of creating a favorable political settlement, set conditions for a protracted resistance, and alter the nature of the conflict to gain relative advantages. Thus, units must rapidly consolidate, reconstitute, train, and prepare to remain in theater should the fighting resume. During this time, force protection is vital to prevent isolated attacks. ADA units continue to execute AMD operations to prevent surprise, deter aerial attacks, permit unimpeded reconstitution, and support the joint force commander's end states for stability operations. ADA units may remain in theater after the end of hostilities to perform stability and support operations.

Stability Operations

A stability operation is an operation conducted outside the United States in coordination with other instruments of national power, to establish or maintain a secure environment and provide essential governmental services, emergency infrastructure reconstruction, and humanitarian relief (ADP 3-0). While each stability task is different, AMD coordination, planning, and execution to meet the mission requirements of the maneuver commander remain the same.

The ability of ADA commanders to stabilize a crisis is directly related to their ability to defend the supported force and civilian critical assets, as necessary. ADA units can be called upon to conduct humanitarian assistance and population control. ADA personnel may be required to control prisoners, handle refugees, and perform other related humanitarian assistance and control activities. ADA units, in conjunction with joint AMD forces, will likely be required to continue to maintain presence in contested areas as armed conflict and crisis end and the environment returns to one of competition.

Refer to TAA2: Military Engagement, Security Cooperation & Stability SMARTbook (Foreign Train, Advise, & Assist) for further discussion. Topics include the Range of Military Operations (JP 3-0), Security Cooperation & Security Assistance (Train, Advise, & Assist), Stability Operations (ADRP 3-07), Peace Operations (JP 3-07.3), Counterinsurgency Operations (JP & FM 3-24), Civil-Military Operations (JP 3-57), Multinational Operations (JP 3-16), Interorganizational Cooperation (JP 3-08), and more.

Redeployment

Redeployments occur as forces that are no longer needed in theater. Post-conflict requirements have a direct effect on the redeployment flow. Commanders contend with the same challenge as in deployment, which is balancing the factors of METT-TC (I) against available lift assets. Forces not required for subsequent operations return to home station and prepare for future missions. Protection of the force during redeployment is as critical as during deployment or any other stage of the operation. While the most significant aerial attack capabilities may have been eliminated, ADA forces must continue to be prepared to counter desperation or retaliatory air and missile attacks.

*Refer to SMFLS6: The Sustainment & Multifunctional Logistics SMARTbook, 6th Ed. (Operational & Tactical Level Sustainment Operations), pp. 7-39 to 7-44 for detailed discussion of **redeployment operations** from ATP 3-35, Army Deployment and Redeployment (Mar '23). Redeployment involves the return of personnel, equipment, and materiel to home and/or demobilization stations and is considered as an operational movement critical in reestablishing force readiness.*

Planning & Preparing (Overview)

Ref: ATP 3-01.81, Counter-Unmanned Aircraft System (C-UAS) Operations (Mar '25), chap. 4.

This chapter describes the planning considerations and integration efforts needed to seamlessly incorporate and react during C-UAS operations into the overarching layered defense respective to the air & missile defense (AMD) mission set. Planning to defeat or mitigate impacts of threat, UAS must be included as part of an overall protection plan. *(See also chap. 3, Air & Missile Defense.)*

Protection planning occurs as a holistic analysis of all threats, hazards, vulnerabilities, and criticalities. Commanders at the company and battery level must consider most likely threats, hazards, and the commander's protection priorities. *(See pp. 4-21 to 4-30.)*

C-UAS Planning Considerations

Countering the UAS threat can be accomplished through a combination of individual Soldier actions and survivability activities, blended with employment of equipment and systems ranging from organic weapons, to specialized handheld EW and lethal devices, to complex systems designed specifically to detect, identify, and defeat UAS. Many units may not be equipped with equipment designed specifically to counter the threat. Countering the threat may not focus solely on defeating the threat UAS; much of the activities to counter the threat may center on the mitigation efforts to lesson or eliminate the impact of a UAS mission against U.S. forces. These efforts should be conducted as part of a unit's overall protection and not as a standalone C-UAS plans and activities.

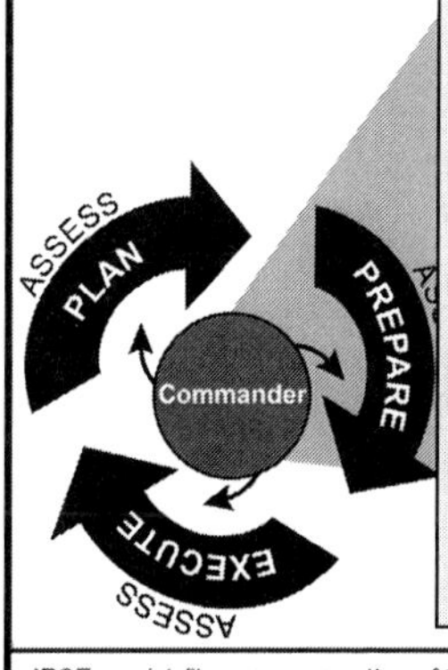

- Refine defended asset list based on IPOE, risk, and commander's assessment
- Define the UAS rules of engagement (hostile criteria/hostile acts
- Disseminate weapons control status (can be different for UAS, fixed-wing, or rotary-wing
- Establish the level of control
- Determine the identification authority
- Determine the engagement authority (lowest level)
- Inform Soldiers of C-UAS criteria
- Establish air defense warnings (based on running estimates of the air threat)
- Coordinate coverage that may extend beyond organic sensor coverage
- Coordinate with friendly mission command nodes and airspace users to reduce fratricide
- Establish notification procedures

IPOE	intelligence preparation of the operational environment
C-UAS	counter-unmanned aircraft system
UAS	unmanned aircraft system

ATP 3-01.81 (Mar '25), fig. 4-1. C-UAS planning considerations.

Planning is the first step toward effectively countering UAS threats. Combating the UAS threat requires a combined arms approach that leverages capabilities from all the warfighting functions. Combined arms operations use the capabilities of each function, along with leadership and information, in complementary and reinforcing capabilities (ADP 3-0). Planning effective actions to counter the UAS threat must synchronize activities across all the warfighting functions to ensure complementary and redundant capabilities executed as part of a holistic, layered plan.

The challenge for commanders at all levels is integrating available resources during the planning process to ensure their effective utilization during mission execution. Some systems are available that are specifically designed and/or optimized to address the UAS threat. These systems are employed based on their own unique capabilities to counter UAS threats. While specialized systems exist, commanders and planners must understand the UAS threat is not a singular problem requiring its own processes and solutions. The UAS threat is one of many threats that commanders face. Many of the actions used to mitigate the UAS threat are the same actions used to mitigate other threats and may already be employed. Commanders and leaders should employ basic planning methods and concepts, including the military decision-making process and troop leading procedures, integrating the UAS threat into the overall analysis and threat mitigation efforts.

Commanders have the responsibility to understand the ROE in order to take the appropriate actions to protect their forces. Planning and understanding the six AMD ROE categories will aid in the protection of forces and equipment. The ROE do not inhibit commanders at all echelons from exercising the inherent right of self-defense to protect their forces from UAS.

Planning for C-UAS activities should be part of a holistic protection planning effort. Successful efforts involve the tailored selection and application of multilayered active and passive measures commensurate with the level of risk. Countering UAS threats includes:

Passive Defense *(See pp. 5-15 to 5-27.)*

A measure taken to reduce the probability of and to minimize the effects of damage caused by hostile action without the intention of taking the initiative (JP 3-60). It includes measures, other than active defense, to minimize, mitigate, or recover from the consequences of threats against friendly forces and assets. These measures include camouflage, concealment, deception, dispersion, reconstitution, redundancy, detection and warning systems, and the use of protective construction.

Active Defense *(See pp. 5-28 to 5-30 and chap. 6.)*

The employment of limited offensive action and counterattacks to deny a contested area or position to the enemy (JP 3-60). It includes all direct defensive actions taken to destroy, nullify, or reduce the effectiveness of hostile threats against friendly forces and assets.

I. Assessment

Planning occurs as a holistic analysis of all threats, hazards, vulnerabilities, and criticalities. Commanders and staff must consider-

- Most likely threats and hazards.
- Protection priorities.
- Which personnel, physical assets, and information to protect.

See facing page for an overview and further discussion. (See pp. 4-21 to 4-30 for an overview and discussion of protection planning.)

Assessment Categories

Ref: ATP 3-01.81, Counter-Unmanned Aircraft System (C-UAS) Operations (Mar '25), pp. 23 to 24.

Commanders and staff must address C-UAS relevant areas within the protection planning process. Initial planning requires various assessments to establish protection priorities including-

- Threat Assessments.
- Hazard Assessments.
- Criticality Assessments.
- Vulnerability Assessments.

Initial protection planning requires various assessments to establish protection priorities. These assessments include criticality assessments, vulnerability assessments, and threat and hazard assessments. During these assessments commanders and staff must ensure C-UAS relevant areas are addressed.

Threat and Hazard Analysis *(See pp. 4-23, 4-24 to 4-25.)*

Threat and hazard analysis comprises a thorough, in-depth compilation and examination of information and intelligence that addresses potential threats and hazards. During threat and hazard analysis commanders and staffs must ensure-

- UAS threats are included: actors, entities, or forces that have the capability and intent to harm U.S. forces, U.S. national interests, or the homeland are considered threats.
- Potential hazards from threat UAS are identified; a threat is a condition with the potential to cause injury, illness, or death of personnel; damage to or loss of equipment or property; or mission degradation.

Analysis of the results of criticality assessments, vulnerability assessments, and threat and hazard assessments feed into the protection priority list. All assets nominated for the protection priority list are usually prioritized based on a methodology of assessing the three major factors of criticality, vulnerability, and threat. The criticality, vulnerability, and threat process are subjective and considers intelligence, air operations, ground combat operations, support operations, and host nation (HN) considerations. Each asset is evaluated against defined criteria, and these criteria are weighed based commander's priorities and intent.

Criticality Assessments *(See p. 4-23.)*

A criticality assessment identifies key assets that are required to accomplish a mission:

- It addresses the impact of a temporary or permanent loss of key assets or the unit's ability to conduct a mission.
- It should also include high-population facilities (recreational centers, theaters, sports venues) that may not be mission essential.

Vulnerability Assessments *(See p. 4-26.)*

A vulnerability assessment is an evaluation to determine the magnitude of a threat's or hazard's effect on an installation, personnel, a unit, an exercise, a port, a ship, a residence, a facility, or other site.

The vulnerability assessment determines an asset's susceptibility or the degree to which an asset is susceptible to surveillance, attack or damage.

The vulnerability assessment also addresses recoverability of an asset, if the asset is attacked and damaged; this assists in determining the degree of potential damage and ability to recover/reconstitute from inflicted damage in terms of time, equipment, and manpower and to continue the mission.

II. Planning and Preparing Components

Commanders utilize the operations process to drive the conceptual and detailed planning necessary in order to better understand the OE. This allows them the ability to visualize and describe the operations end state and operational approach; make and articulate decisions; and direct, lead, and assess military operations. The major command and control activities performed during operations are planning, preparing, executing, and continuously assessing the operation (ADP 5-0).

Framework	*Operational Phase*	*End State*
Deterrence	Plan/Prepare	Enemy dissuaded from employing UAS
Preparation		Understand the OE, threat, TTPs, and countermeasures
Detection		Enemy UAS identified at the earliest opportunity
Response	Execute	Neutralize or evade the enemy threat
Site exploitation	Assess	Enemy UAS rendered safe and information extracted
Reports		Timely, detailed incident report submitted
Synthesis		Incident drives adaptive response to future incidents
OE operational environment UAS unmanned aircraft system TTPs tactics, techniques, and procedures		

Ref: ATP 3-01.81 (Mar '25), table 4-2. Framework components

Risk management is the process to identify, assess, and mitigate risks and make decisions that balance risk and cost with mission benefits (JP 3-0). The C-UAS planning process assists commanders in making informed decisions to balance risk with mission benefits. The process itself evaluates the threat problem, collateral concerns, environment or medical hazards, impacts to capabilities, risk to force, and risk to mission, before or during execution of actions.

Refer to ATP 5-19 for additional information on risk management.

Table 4-2 depicts the relationship between framework components, the operational phase, and the desired end state related to—

Deterrence (Pre-Incident)

- Shape OE through policy, influence activities, and understanding of civilian UAS manufacture geofence software.
- Prevent or limit neutral or non-combatant UAS in the OE.
- Limit enemy UAS procurement, distribution and employment networks.

Preparation (Pre-Incident)

- Identify enemy UAS TTP, payload capabilities, and methods of UAS control, expected number of UAS in the OE, flight profiles, and evasion capability.
- Develop TTPs for detection, response, site exploitation, and reporting based on the IPOE and potential enemy UAS activity.
- Implement ROE and rules of use of force for UAS, focused with a proportionate response to avoid premature forced evolution of enemy TTPs.
- Conduct UAS battle drills based on enemy UAS capabilities, ROE/rules for the use of force, and C-UAS capabilities for fixed and dismounted scenarios.
- Conduct spectrum coordination and deconfliction with friendly actors in the OE.
- Analyze and understand the impact friendly C-UAS capabilities will have on other friendly assets operating in the spectrum.
- Format standardized, disseminated, and rehearsed reports for C-UAS.
- Train and rehearse for C-UAS responses.
- Identify potential launch sites and controller locations

Detection

- Identify enemy UAS as early as possible.
- Send hasty initial report of UAS sighting.
- Conduct C-UAS response informed by IPOE and preparation.

Response (trans-incident)

- Secure the site of a UAS incident in coordination with tempo of land operations.
- Conduct UAS retrieval in accordance with unit TTP developed in the preparation component.
- Maintain security of enemy UAS and submit to appropriate lab for analysis.

Report (post-incident)

- Submit full report in accordance with unit standard operating procedure.

Synthesis (post-incident)

- Leverage intelligence obtained from a UAS incident to inform potential methods of deterrence and engagement, thus enhancing C-UAS actions against future attacks.
- Use intelligence obtained from UAS incident to strengthen preparation for future C-UAS operations.

III. Planning Responsibilities

Brigade and Higher

Brigade and higher headquarters are responsible for the integration of C-UAS into military decision-making process and working groups to include targeting and protection processes. In addition, the ADAM/BAE and EW officer will manage the airspace and C-UAS sites and mobile systems' employment. The C-UAS fight requires a multi-echelon approach, synchronizing resources of higher echelons to support mitigation of the tactical UAS threat. Brigade and lower echelons execute protection and survivability activities to mitigate UAS threats on their positions and conduct direct engagement of immediate UAS threats (in flight).

Higher Echelons

Higher echelons conduct analysis and planning to mitigate the UAS threat and direct resources to enhance survivability of subordinate friendly forces. They direct resources to ensure U.S. forces and our allies maintain a real-time common operating picture relevant to the UAS threat and identify UAS production and associated sustainment resources used by threat elements. These activities may lead commands above brigade level to direct and coordinate direct action against those threats when required. Actions against large UAS platforms are conducted at echelons leveraging joint capabilities.

Division and Brigade-Level

Division and brigade-level ADA/EW staffs should advise lower echelons with respect to integration of planning C-UAS activities into their military decision-making, targeting, and protection processes.

Company-Level

Company-level mission command requires an understanding of the tasks relative to C-UAS operations using a quick reference guide or pre-combat checklist. Company will be responsible for the employment of the handheld equipment to detect and/or defeat Group 1 UAS. Company-level security operations will be enhanced by integrating air guards into their security planning. Platoon leaders conduct rehearsals to plan air guard tasks, locations, UAS spot report procedures, and communication plans. Soldiers will incorporate air guard principles into existing search and scan techniques throughout all phases of the operation.

IV. Electromagnetic Warfare (EW) Considerations

Ref: ATP 3-01.81, Counter-Unmanned Aircraft System (C-UAS) Operations (Mar '25), pp. 27 to 28.

Electromagnetic warfare is military action involving the use of electromagnetic and directed energy to control the electromagnetic spectrum or to attack the enemy (JP 3-85). The Army continues to build EW competencies to support decisive action and win across all levels of combat operations and throughout all domains. Leaders must consider EA, EP, and ES capabilities when planning for counter-small unmanned aircraft system operations.

Divisions of Cyberspace Operations	Electromagnetic Attack (EA)	Electromagnetic Protection (EP)	Electromagnetic Support (ES)
Types of Operations	Attack personnel, facilities, or equipment	Protect friendly Electromagnetic Spectrum (EMS)-dependent capabilities	• Intercept • Identify • Locate • Evaluate
Types of Enabling Operations	• Reconnaissance • Enemy Attack	Preemptive Protection	• Situational Understanding • Combat Information • Targeting • Intelligence Preparation of Battlespace (IPB) Development
Common Tactical Mission Tasks	• Employing Directed Energy Weaponry • Electromagnetic Pulse • Reactive Countermeasures • Deception Measures • Electromagnetic Intrusion • Electromagnetic Jamming • Electromagnetic Probing • Meaconing	• Deconflict Electromagnetic Environmental Effects • Ensure Electromagnetic Compatibility • Electromagnetic Hardening • Emission Control • Electromagnetic Masking • Preemptive Countermeasures • Electromagnetic Security • Conduct Wartime Reserve Modes	• Conduct Electromagnetic Reconnaissance • Threat Warning • Direction Finding
Common Effects	• Disrupt • Degrade • Neutralize • Destroy • Deceive	• Deception • Denial • Disrupt • Neutralize	• Exploit • Detect

Ref: FM 3-12 (Aug '21), fig. 2-3. Electromagnetic warfare taxonomy.

Electromagnetic Attack (EA)

Electromagnetic attack is the division of electromagnetic warfare involving the use of electromagnetic energy, directed energy, or antiradiation weapons to attack personnel, facilities, or equipment with the intent of degrading, neutralizing, or destroying enemy combat capability and is considered a form of fires (JP 3-85). EA has the unique potential to affect enemy use of the EMS and attack the enemy through the EMS. Various capabilities can be used to destroy, degrade, disrupt, and deceive the threat's military EMS actions. Once the commander's intent is understood, the CEMA section, in coordination with the staff, should carefully consider the following-

- Interference with friendly communications.
- Intelligence gain or loss.
- Spectrum use by neutral actors.
- The persistence of effects.
- Electromagnetic signatures.
- Personnel and equipment capabilities and limitations.
- Persistency of effect: Effects of jamming only persist as long as jammer itself is emitting and is in range to affect the target.

Planning & Preparing

Electromagnetic Protection (EP)

Electromagnetic protection is the division of electromagnetic warfare involving actions taken to protect personnel, facilities, and equipment from any effects of friendly or enemy use of the electromagnetic spectrum that degrade, neutralize, or destroy friendly combat capability (JP 3-85). EP includes actions taken to ensure friendly use of the EMS, such as frequency agility in a radio or variable pulse repetition frequency in radar. EP is essential to prevent the adversary from learning behavior and intentions within the EMS. EP techniques that should be considered include:

- Spectrum management.
- Electromagnetic hardening.
- Electromagnetic masking.
- Emission control.
- Electromagnetic compatibility.
- Terrain masking.
- Wartime reserve modes.

Note. Refer to ATP 3-12.3 for more information on EP.

Electromagnetic Support (ES)

Electromagnetic support is the division of electromagnetic warfare involving actions tasked by, or under direct control of, an operational commander to search for, intercept, identify, and locate or localize sources of intentional and unintentional radiated electromagnetic energy for the purpose of immediate threat recognition, targeting, planning and conduct of future operations (JP 3-85). Electromagnetic support gathers information about friendly and threat electromagnetic emissions to support the commander's situational awareness and decision making. ES is accomplished through the following tasks:

- Electromagnetic reconnaissance.
- Threat warning.
- Direction finding.

Note. Refer to ATP 3-12.3 for more information on ES.

Electromagnetic interference is any electromagnetic disturbance, induced intentionally or unintentionally, that interrupts, obstructs, or otherwise degrades or limits the effective performance of electromagnetic spectrum-dependent systems and electrical equipment (JP 3-85). Electromagnetic interference prevents successful EM transmissions; units must recognize and mitigate electromagnetic interference to create the conditions required to use the EMS to communicate.

V. C-UAS Reporting (SALUTE Format)

Ref: ATP 3-01.81, Counter-Unmanned Aircraft System (C-UAS) Operations (Mar '25), pp. 24 to 25. (See also pp. 5-24 to 5-25.)

Units must set conditions allowing forces at all echelons, to recognize and report airspace violations and confirmed enemy actions that occur within their AO. Planners must include current threat sets and standardized reporting guidance. See table 4-1 for an example of a standardized spot report for reporting initial contact. Theater specific reports may vary based on command directive, mission and environmental requirements. This information should include its description and flight profile.

A unit's standard operating procedures may also include model and capability profiles for known threat UAS. These profiles will provide C-UAS planners the suite of capabilities associated with the threat. Flight profiles associated with a threat UAS can assist unit C-UAS planners in planning for a specific UAS threat. This will assist in conducting unit-specific C-UAS tactics, techniques, and procedures (TTP).

Detecting threat UAS by organic sensors or through observation can assist in determining its model and flight path. Calculating speed and heading and knowing the UAS's endurance, based on flight profiles, improves a unit's chance of mitigating the UAS ability to conduct ISR and execute attacks on friendly forces. While sensors may detect many of the larger UAS, it is likely that unit air guard may be the primary detection means for UAS. Air guard must be familiar with UAS capabilities and be able to recognize behaviors and patterns. Air guard must provide continuous updates, through reports to unit operations personnel, to develop an evolving defense against threat UAS.

Soldiers must train to report contact as soon as possible following detection of a UAS operating in the area. The size, activity, location, unit, time, and equipment (known as SALUTE) acronym/format can be used to immediately generate and deliver a spot report to the chain of command. The chain of command will then determine what immediate actions need to be taken and will generate required follow-on contact reports.

Note. The SALUTE format is intended to be a starting point in the reporting structure. It is not meant to take the place of any local/command mandated report format. This is a baseline format for personnel to report first contact; upon receipt this report should trigger follow-on information collection and security actions as required.

Line	Information	Example
1	Size	Report the number of UAS, or size of the formation (e.g., single, multiple, or swarm).
2	Activity	Report detailed account of actions: What is the direction of movement? Was there any hostile action? Is the threat UAS loitering in one spot (possibly has spotted reporting unit)? Is it flying straight (enroute to loitering location)? Was the threat UAS approach observed or was it spotted overhead?
3	Location	Report where you saw the activity. Include 6 to 8 digit grid coordinate of reporting element, grid or distance and direction from reporting element location (known point).
4	Unit (Description of UAS)	Report fixed-wing or multi-rotor, wingspan, height, color, tail configuration, rotor configuration, lights, other distinguishable markings.
5	Time	Report the time the activity was observed, not the time you report it. Annotate it in either local or Zulu time. (Example: Date-time group: 091024ZMAR18)
6	Equipment	Report all equipment associated with the activity, such as payload, weapons, and camera.
UAS Unmanned Aircraft System		

Ref: ATP 3-01.81 (Mar '25), table 4-1. Standardized Counter- Unmanned Aircraft System Detection Report Example (SALUTE Format).

Planning & Preparing

I. C-UAS Roles & Responsibilities

Ref: ATP 3-01.81, Counter-Unmanned Aircraft System (C-UAS) Operations (Mar '25), chap. 2.

This section provides guidelines for the roles and the responsibilities of commanders and specific staff to assist in developing C-UAS techniques. It discusses actions needed for preparing and to take when anticipating or encountering possible UAS threats while on the battlefield. This information complements C-UAS planning considerations, commander's intent and guidance, threat environment, and standard operating procedures.

I. Roles and Responsibilities

Commanders direct the continuous integration of C-UAS operations whether in a tactical environment or at home station. By leveraging cyberspace operations and EW as part of the combined arms approach, commanders can sense, understand, decide, act, and assess faster than the adversary assesses and achieve a decisional advantage in multiple domains during operations.

Commanders should-

- Include C-UAS within the operations process.
- Continually enforce cybersecurity standards and configuration management.
- Understand, anticipate, and account for C-UAS capabilities, constraints, and limitations, including second and third order effects.
- Understand the legal and operational authorities to affect threat portions of C-UAS.
- Understand the implications of C-UAS operations on the mission and scheme of maneuver.
- Understand how the selected course of action affects the prioritization of resources.
- Leverage effects in and through cyberspace and the EMS to support the concept of operations.
- Develop and provide intent and guidance for actions for C-UAS.
- Identify critical terrain relevant to C-UAS that may affect critical tasks or missions by phase.
- Ensure active collaboration across the staff, subordinate units, higher headquarters, and unified action partners that enable a shared understanding of C-UAS.
- Approve collection priorities and risk mitigation measures.
- Ensure the synchronization of C-UAS with other lethal and nonlethal fires to support the concept of operations.

II. Staff Roles and Responsibilities

The staff assists the commander in understanding, visualizing, and describing the OE, making decisions, and directing military operations. Staffs must conduct analysis and make recommendations based on their specific roles and areas of expertise. C-UAS operations require extensive coordination and cooperation across a wide range of staff elements. Staffs synthesize information coordinated and synchronized across multiple disciplines and areas of focus and provide it to commanders in the form of running estimates to help commanders build and maintain their situational understanding. Staff elements with critical areas of input in planning and executing C-UAS operations are outlined in this chapter.

S-2

The S-2 advises the commander and staff on intelligence aspects of C-UAS operations. Their role is to provide, timely, relevant, accurate and predictive intelligence support to enable commanders to make timely and relevant decisions. The S-2 must coordinate closely and constantly with the Division G-2 for updated threat information regarding C-UAS operations.

S-3

The S-3 is responsible for the overall planning, coordination, and supervision of C-UAS activities. Additionally, the S-3 is responsible for all matters concerning plans, operations, and training at brigade and below. The S-3-

- Plans for and incorporates C-UAS into operation plans and orders, within the fire support plan and the information operations plan (in joint operations).
- Tasks C-UAS activities to assigned and attached units.
- Exercises control over C-UAS, including electromagnetic deception plans. • Directs EP measures based on recommendations from the S-6 staff, the cyber electromagnetic warfare officer, and the CEMA working group.
- Coordinates EW and C-UAS training requirements. • Issues C-UAS support tasks within the information collection plan. These tasks are developed, according to the collection plan and the requirements tools, by the S-2 and the requirements manager.
- Ensures through the CEMA working group, that EW and C-UAS activities support the plan.
- Integrates EA within the targeting process. • Integrates C-UAS into the overall protection plan via the protection cell.

Planning & Preparing

S-6

The network defense technician, network management technician, information services technician, spectrum manager, and information security manager participate in planning EW techniques for C-UAS operations. The S-6-

- Assists the cyber electromagnetic warfare officer with the preparation of the EP policy.
- Reports enemy electromagnetic attack activity detected by friendly communications and electronic elements to the CEMA working group for counteraction.
- Assists the unit cyber electromagnetic warfare officer with resolving EW systems maintenance.
- Identifies and deconflicts electromagnetic interference.
- Issues the signal operating instructions.
- Ensures network connectivity for all EW computer systems.
- Provides EMS resources to the unit or task force (refer to ATP 6-02.70).
- Coordinates for EMS usage with higher echelon G-6 (S-6), communications system directorate of a joint staff, and applicable host-nation and international agencies as necessary.
- Prepares the restricted frequency list and issuance of emissions control guidance.
- Coordinates frequency allotment, assignment, and use.
- Supports the CEMA working group by assisting in the development of electromagnetic deception plans and activities that include EMS resources.
- Coordinates with higher echelon spectrum managers for EMS interference resolution. • Assists the cyber electromagnetic warfare officer in issuing guidance to the unit, including subordinate elements, regarding deconfliction and resolution of interference problems and processes involving EW systems.
- Participates in the CEMA working group to deconflict friendly EMS requirements with EW C-UAS activities and information collection efforts.

C-UAS Responsibilites in the Operations Process (by Warfighting Function)

Ref: ATP 3-01.81, Counter-Unmanned Aircraft System (C-UAS) Operations (Mar '25), p. 8.

Warfighting functions are the physical means that tactical commanders use to execute operations and accomplish missions assigned by superior operational and tactical level commanders. The purpose of warfighting functions is to provide an intellectual organization for common critical capabilities available to commanders and staffs at all echelons and levels of warfare (ADP 3-0). Each warfighting function has requirements throughout the C-UAS planning, preparing, executing, and assessing process.

Warfighting Functions	***PLAN***	***PREPARE***	***EXECUTE***	***ASSESS***
Movement & Maneuver	-Coordinate vulnerability assessment -Conduct Commander's risk assessment -Determine CCIR -Identify passive/active defense measures -Identify air guards	-Rehearse battle drill for UAS event	-Detect/Identify UAS platform -Conduct host nation coordination -Defeat	-Review TOCSOP/battle drill execution
Intel	-Conduct IPOE -Coordinate coverage beyond organic sensors -Provide input to CCIR/essential element of friendly information -Identify (ID) NAIs/TAIs	-Conduct sensor/ information collection rehearsal synchronization drill	-Detect/Identify UAS platform	-Track UAS exploitation -Report exploitation findings -Identify trends and TTP
Fires	-Disseminate weapons control status -Establish level of control	-Rehearse clearance of fires	-Deconflict air space	
Protection	-Refine protection priorities -Update protection priority list -Establish general & local air defense warnings	-Rehearse deconfliction	-Initiate air defense warning -Deconflict air space	-Update TTP based off exploitation and trends
Sustainment	-Plan for sustainment of C-UAS systems (CL III, V, IX) -Maintenance (organic or FSR support)	-Check battery/cable status -PCC/PCI CL III, V, IX -Systems PMCS	-Sustain C-UAS systems	-Review -Issue Resupply -Order(s)
Command and Control	-Define UAS ROE -Determine identification authority -Determine engagement authority -Establish notification procedures	-Review reporting procedures -Validate PACE plan	-Execute reporting procedures along PACE	-Review reporting procedures -Review ROE -Review engagement authority

CCIR commander's critical information requirement PCI precombat inspections
FSR field service representative ROE rules of engagement
IPOE intelligence preparation of the operational needs statement, operational needs statement, TAI target area of interest
NAI named area of interest TOCSOP tactical operations center standard operating procedure
PACE primary, alternate, contingency, and emergency TTP tactics, techniques, and procedures
PCC precombat checks UAS unmanned aircraft system

Ref: ATP 3-01.81 (Mar '25), table 2-1. Counter-Unmanned Aircraft System Warfighting Functions.

See p. 5-7 for an overview and discussion of countering the threat (by warfighting function).

- Supports all subordinate unit software updates and communications security requirements.
- Compiles and distributes the joint restricted frequency list (spectrum manager).
- Assists the EW section with computer maintenance and troubleshooting.

Air Defense Airspace Management/Brigade Aviation Element Cell

The air defense coordination/management officer in the brigade combat team level ADAM/BAE cell is responsible for ensuring that all organic elements (staff, organic units, and attached units and personnel) understand and employ the necessary passive defense measures to mitigate the effectiveness of enemy UAS. Passive air defense measures are the first line of defense against UAS. Based on the unit's assigned mission and rules of engagement, the commander will have discretion as to engage any approaching confirmed hostile aerial platforms. The Air Defense Airspace Management/Brigade Aviation Element (ADAM/BAE) cell will also assist in the clearance for lethal UAS engagements.

Staff Judge Advocate or Representative

The staff judge advocate is responsible for all legal advice. The staff judge advocate or representative reviews all C-UAS operations to ensure they comply with existing Department of Defense directives and instructions, rules of engagement (ROE), and applicable domestic and international laws, including the law of war. They can also assist the command in requesting additional required authorities for employment of C-UAS systems.

Cyber Electromagnetic Activities Section

Cyber electromagnetic activities sections are assigned to the G-3 or S-3 within corps, divisions, brigade combat teams, and combat aviation brigades. Commanders are responsible for ensuring that CEMA sections integrate cyberspace operations and EW into their concept of operations. The CEMA section plans, coordinates, and integrates cyberspace operations and EW in support of the commander's intent. It is responsible for the coordination to deconflict spectrum management efficiency. The CEMA section collaborates with numerous staff sections to ensure unity of effort in meeting the commander's total operational objectives. The CEMA section is responsible for synchronizing and integrating cyberspace operations and EW with the operations process and through other integrating processes. Personnel assigned to the CEMA section are the-

- Cyber electromagnetic warfare officer.
- Cyber warfare officer.
- EW technician.
- EW sergeant major (corps) or EW noncommissioned officer in charge (division).
- EW noncommissioned officer.
- CEMA spectrum manager.

The CEMA section leads the CEMA working group. The CEMA working group is not a formal working group that requires dedicated staff members from other sections. When needed, the cyber electromagnetic warfare officer uses a CEMA working group to assist in synchronizing and integrating cyberspace operations and EW into the concept of operations. The CEMA section normally collaborates with key stakeholders during staff meetings established as part of the unit's battle rhythm and throughout the operations process. Membership in the CEMA working group will vary based on mission requirements.

If scheduled, the CEMA working group will integrate into the staff's battle rhythm. The CEMA working group is responsible for coordinating horizontally and vertically to support operations and assisting the fires support element throughout the execution of an operation. Generally, the CEMA working group is comprised of staff representatives with equities in CEMA, and typically includes-

III. C-UAS Running Estimate

Ref: ATP 3-01.81, Counter-Unmanned Aircraft System (C-UAS) Operations (Mar '25), pp. 10 to 12.

A running estimate is the continuous assessment of the current situation used to determine if the current operation is proceeding according to the commander's intent and if planned future operations are supportable (ADP 5-0). Information which may appear in an S-6 specific C-UAS running estimate would include committed and reserved assets, maintenance status of C-UAS equipment, and training proficiency of EW personnel. Resources that are useful in developing a running estimate are the maintenance report and the commanders' training assessments. Threat information is available from online databases, unit intelligence assets, and national intelligence sources.

The purpose of the running estimate is to provide a consolidated list of information about cyberspace and the EMS to assist other staff sections in planning, preparing, and executing operations. The information serves as a foundation and is dependent on information requirements with other staff such as operations, fires, intelligence, and signal as sources of information. Some of this information will be redundant with other staff section planning products. Figure 2-1 is an example of a C-UAS running estimate that an S-6 section might produce. All staff sections have responsibilities in conducting running estimates and other aspects of the military decision-making process during the planning for and conduct of C-UAS operations. Paramount considerations should include-

- Survivability of personnel and equipment.
- The time required to build or improve the unit's EP posture and position EA and electromagnetic support capabilities.
- Ability of C-UAS resources to achieve the desired effects.
- Reprogramming of C-UAS and EW assets.
- Capabilities, limitations, advantages, and disadvantages of available EW and signals intelligence assets equipped with electromagnetic support capability.
- Intelligence available for targeting.
- Current and projected maintenance C-UAS status.
- Facts.
- Assumptions.
- Friendly Forces status.
- Enemy Activity/capabilities.
- Civil Considerations.
- Readiness of equipment, personnel, and factors of mission accomplishments.
- Relative to your staff sections/functional areas.
- Conclusions and recommendations.

Refer to ATP 3-01.81 (Mar '25), fig. 2-1 for an example of Counter-Unmanned Aircraft System running estimate.

Planning & Preparing

- The G-2/S-2.
- The staff weather officer.
- The G-6/S-6.
- The information operation officer or representative.
- Spectrum manager.
- The fire support officer or a fires support element representative.
- The staff judge advocate.
- The protection officer.

Protection Cell *(See p. 4-30.)*

The protection cell advises, visualizes, and outlines protection requirements to the commander; prioritizes and synchronizes mitigating strategies, and coordinates protection. It recommends measures to deny enemy stand-off, reducing the degradation, disruption, and denial of friendly force operations. Primary members of the protection cell typically include the chief of protection and representatives from air and missile defense, personnel recovery, provost marshal, CBRN, EOD, engineer, medical, and antiterrorism/force protection sections.

The protection cell (supported by the protection working group) develops the scheme of protection after receiving guidance and considering the principles of protection in relation to mission variables and the systems requiring protection. The scheme of protection is based on the mission variables thus, it includes protection priorities by area, unit, activity, or resource. It addresses how protection is applied and derived during all phases of an operation. For example, the security for routes, bases/base camps, and critical infrastructure is accomplished by applying protection assets in dedicated, fixed, or local security roles; or it may be derived from economy-of-force protection measures, such as area security techniques. It also identifies areas and conditions where forces may become fixed or static and unable to derive protection from their ability to maneuver. The protection cell considers the following items, at a minimum, as it develops the scheme of protection:

- Protection priorities (critical capabilities, areas, and information).
- Air defense positioning guidance.
- Specific terrain and weather factors.
- Areas or events where risk is acceptable.
- Protected targets and areas.
- Vehicle and equipment safety or security constraints.
- Personnel recovery actions and control measures.
- Force protection condition status.
- Mission-oriented protective posture guidance.
- Environmental guidance.
- Explosive ordnance and hazard guidance.
- Ordnance order of battle.
- Operations security risk tolerance.
- Fratricide avoidance measures.
- Rules of engagement, standing rules for the use of force, and rules of interaction.
- Escalation of force and nonlethal weapons guidance for C-UAS equipment.
- Operational scheme of maneuver.
- Military deception.
- Obscuration.
- Electromagnetic spectrum status.
- Availability of personnel recovery capable assets, gaps in recovery coverage, preparation of individuals.

Chap 4

II. Air & Missile Defense Planning (Overview)

Ref: FM 3-01, U.S. Army Air and Missile Defense Operations (Aug '25), pp. 90 to 96.

Understanding the integration of **air defense artillery (ADA) assets** is critical when planning and conducting counter–unmanned aircraft system (C-UAS) operations because ADA provides the structural foundation and doctrinal framework for layered, defense-in-depth protection against aerial threats. *(See chap. 3.)*

I. Air & Missile Defense (AMD) Planning

ADA commanders and staffs develop AMD plans based on the guidance from their higher headquarters and in accordance with their supporting or supported relationships. Planning considers all AMD operational elements, as appropriate, across the levels of war. Planning considerations at the operational level can be more extensive than those at the tactical level as the integration of joint and multinational AMD capabilities normally occurs at the operational level. Planning at the tactical level is nevertheless necessary to develop appropriate defenses against air platforms that may threaten maneuver units. **Plans must consider the entirety of the aerial threat, as small unmanned aircraft systems (sUASs), fixed-and rotary-wing aircraft, cruise missiles, and ballistic missiles may threaten a division and corps and theater rear areas.**

At the operational level, the army air and missile defense command (AAMDC) normally lead the joint and multinational AMD planning effort for the area air defense commander (AADC). The AAMDC commanding general, serving the theater army air and missile defense coordinator, directs ADA planning priorities and force apportionment. The AAMDC, in coordination with the AADC's staff and joint and multinational partners, supports the development and coordination of the theater area air defense plan (AADP), and provides planning guidance to deployed ADA units. The result of this planning process is an integrated defense plan that leverages all available joint and multinational AMD capabilities to effectively and efficiently defeat aerial threats in accordance with the AADC's concept of operations and directives.

At the tactical level, much of the planning resides with the divisional air defense (DIVAD) battalion and is executed in coordination with the division's AMD section and the BCT's ADSE. The AMD section and ADSE identify priority assets which may require active AMD. The DIVAD battalion commander, serving as the division's air defense coordinator, leads the planning effort to formulate a divisional AMD plan to meet the air threat. The divisional AMD plan should account for any higher echelon ADA capabilities while leveraging all available dedicated and nondedicated, SHORAD systems to defend maneuver forces. ADA personnel in ADSEs assist the brigade commanders and their staffs in integrating any supporting ADA units into the BCT's scheme of maneuver. The supporting ADA unit commander is ultimately responsible to formulate an AMD plan to meet the BCT commander's needs.

AMD planning involves joint, multinational, and Army units including Service or functional component commands, AAMDCs, ADA brigades, and ADA battalions. Defense planning and decision support tools are available in an array of planners, ranging from Command and Control Battle Management and Communications planner at the strategic level to the air and missile defense workstation to the Terminal High Altitude Area Defense (THAAD) portable planner and Patriot tactical planning workstation at the operational and tactical levels.

II. CAL & DAL Development (AMD Planning)

Ref: ATP 3-01.8, Techniques for Combined Arms for Air Defense (Jul '16), pp. A-4 to A-6.

Criticality Assessment *(See p. 4-23.)*

A criticality assessment identifies key assets that are required to accomplish commander's mission. It addresses the impact of temporary or permanent loss of key assets or a unit's ability to perform its mission. It examines costs of recovery and reconstitution including time, dollars, capability, and infrastructure support. The staff must gauge how quickly a lost capability can be replaced before giving an accurate status to the commander. The general sequence for a criticality assessment follows:

- List the key assets and capabilities.
- Determine whether critical functions or combat power can be substantially duplicated with other elements of the command or an external resource.
- Determine the time required to substantially duplicate key assets and capabilities in the event of temporary or permanent loss.
- Prioritize responses threats to personnel, physical assets, and information.

DAL development is an interactive process that involves subordinate commands. After reviewing the initial DAL, subordinate commanders and their staffs may nominate additional assets for inclusion in the DAL. The joint force commander and his staff may incorporate one or more nominees and issue an updated (re-prioritized) DAL, which then becomes the basis for AMD planning and defense design.

Other critical planning guidance provided by joint force planners includes the ACO and the ATO. The ACO implements the airspace control plan, and provides the details of the approved request for airspace control measures. The ATO also provides specific instructions for tasking forces/capabilities/sorties to specific missions and targets. The ACO is part of the ATO, although it may be transmitted separately. Both are provided to all subordinate echelons of command. All components of the ACO and the ATO should be included in the planning process to give commanders and staff a complete understanding of the air battle.

The protection cell staff continuously updates the hazard assessment during the operations process. As the staff identifies and modifies a friendly COA, reconnaissance confirms or denies information requirements. As the threat changes, the risk to the force changes. New risks may require different mitigating measures. The protection cell analyzes changes that may require modifications to protection priorities.

Vulnerability Assessment *(See p. 4-26.)*

A vulnerability assessment is a command, or unit-level evaluation (assessment) to determine the vulnerability of an air terrorist attack against an installation, unit, exercise, residence, facility, or other site. It identifies areas of improvement to withstand, mitigate, or deter air attack acts of violence or terrorism. The staff addresses the questions of who or what is vulnerable, and how it is vulnerable. The vulnerability assessment determines the susceptibility of the commander's assets to an air attack by the previously assessed threats. The product of the vulnerability assessment is the identification of physical characteristics or procedures that render critical assets, areas, or special events vulnerable to the range of known or potential threats. The assessment provides a basis for developing controls to eliminate or mitigate vulnerabilities. Vulnerability is the component of risk over which the commander has the most control and greatest influence. The general sequence of a vulnerability assessment is below:

- List assets, capabilities, and the threats against them.
- Determine common criteria for assessing vulnerabilities.
- Evaluate the assets and capabilities for their vulnerability

After determining what assets to protect and comparing them to the risk, the protection cell develops risk mitigation measures and makes recommendations to the commander. The CAL is the primary tool used by the protection cell to mitigate risk.

Critical Asset List (CAL) *(See pp. 4-28 to 4-29.)*

The CAL is a prioritized list of assets, normally identified by phase of the operation and approved by commander that should be defended. Once the staff has completed the hazard, criticality and vulnerability assessments, the prioritized list of critical assets is presented to the commander for approval. Commanders typically operate in a resource-constrained environment and have a finite amount of combat power to use for protecting assets. Therefore, the staff must determine which assets are critical for mission success and recommend priorities with available resources. The list will vary depending on the factors of METT-TC.

Defended Asset List (DAL) *(See pp. 4-28 to 4-29.)*

The vulnerability and criticality assessments, when compared to the assessed hazards, provide the commander with information to make decisions of which assets are the most critical and must have combat power dedicated to their protection and which ones to he can accept risk. For instance, one of five critical assets considered in a given criticality assessment cannot be replaced during an operation, while the other four can. This lack of replacement may then cause that critical asset to become the first priority for protection.

Not all assets listed on the CAL will receive protection from continuously applied combat power. Critical assets with some protection from applied combat power become part of the DAL. The DAL is a listing of those assets from the CAL prioritized by the commander to be defended with the resources available. The combat power applied may be a weapons system, electronic sensor, obstacle, or some combination of them all. Both the CAL and DAL are dynamic lists.

Figure A-2 illustrates how the protection cell determines the CAL during planning. This list is dynamic; it changes depending on the mission, phase of the operation, and situation. The protection cell helps commanders to integrate the risk management process with the decision-making process.

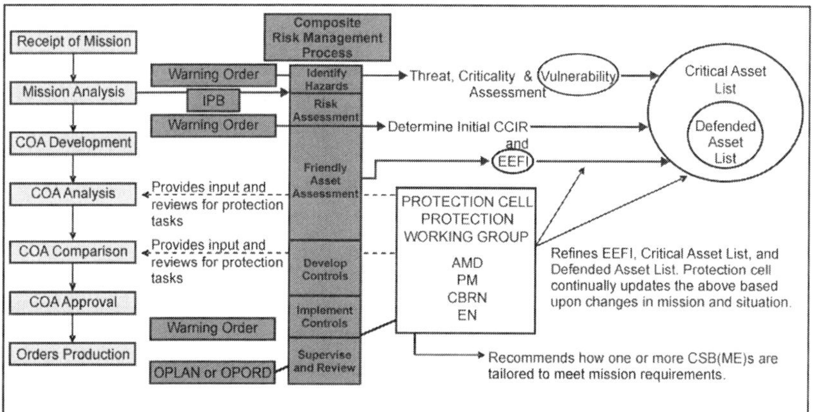

AMD	air and missile defense	EN	engineer
CBRN	chemical, biological, radiological, and nuclear	IPB	intelligence preparation of the battlefield
CCIR	commander's critical information requirements	OPLAN	operation plan
COA	course of action	OPORD	operation order
CSB(ME)	combat support brigade (maneuver enhancement)	PM	provost marshal
EEFI	essential elements of friendly information		

Ref: ATP 3-01.8 (Jul '16), fig. A-2. Example CAL and DAL development.

III. AMD Planning (the UAS Threat)

Ref: FM 3-01, U.S. Army Air and Missile Defense Operations (Aug '25), pp. 169 to 172.

An unmanned aircraft system is that system whose components include the necessary equipment, network, and personnel to control an unmanned aircraft (JP 3-30). Threat UASs are categorized in groups (table C-2), typically based on weight, operating altitude, and speed. Groups 1 through 3 are further described as small systems.

	Speed / Altitude / Weight	***Characteristics***	
Group 1 Micro / Mini UAS	• Normally operates below 1,200 feet AGL at speeds less than 100 knots. • 0 to 20 pounds with payload.	• Generally hand-launched. • Real time video and control. • Small payloads; focus on reconnaissance, surveillance, and intelligence gathering.	• Operates within LOS of user (limited range) unless flight is preprogrammed or autonomous.
Group 2 Small Tactical	• Normally operates below 3,500 feet AGL at speeds less than 250 knots. • 21 to 55 pounds with payload.	• Launched in unimproved areas by a small number of personnel. • Medium range and endurance. • Payload focus: reconnaissance, surveillance, and intelligence gathering; may add weapons. • Requires LOS to ground control station unless flight is preprogrammed or autonomous.	
Group 3 Tactical	• Normally operates below 18,000 feet MSL at speeds less than 250 knots. • 56 to 1,320 pounds with payload.	• Launched in unimproved areas by a small number of personnel. • Range and endurance vary significantly. • Payload focus: reconnaissance, surveillance, and intelligence gathering; may add weapons. • Requires larger logistics footprint than Groups 1 and 2.	
Group 4 Persistent	• Normally operates below 18,000 feet MSL at any speed. • Greater than 1,320 pounds	• Can be used both strategically and tactically. • Requires a runway for launch and recovery. • Extended range and endurance. • Payloads: reconnaissance, surveillance, intelligence, gathering, and ASM weapons. • Operates at medium-to-high altitudes.	
Group 5 Penetrating	• Normally operates higher than 18,000 feet MSL at any speed. • Greater than 1,320 pounds.	• Strategic-level asset. • Requires an improved runway for launch and recovery. • Greatest range, endurance, and airspeed. • Payloads: suite of optics for targeting and weaponry for engagements. • Operates at medium-to-high altitudes. • Logistical footprint similar to that of a manned aircraft.	

AGL above ground level
ASM air-to-surface munition
LOS line-of-sight
MSL mean sea level
UAS unmanned aircraft system

Ref: FM 3-01 (Aug '25), fig. C-2. UAS groups.

UASs include drones and remotely piloted vehicles. Drones fly preprogrammed flight paths and patterns; remotely piloted vehicles are controlled by ground-based operators. Each can perform a variety of missions, ranging from reconnaissance and battlefield surveillance to attack and electromagnetic warfare. UASs have extensive commercial applications and, as a result, are readily available and arguably the most developed system in many armies around the world. Due to the prolific use of unmanned systems, commercial off-the-shelf UASs are readily available and can be weaponized with relative ease. There are over 1,000 current and developing UAS programs worldwide. UASs are typically comprised of a control element, communication systems, support elements, an unmanned aircraft, a payload or pod, and a human or trucks, and on ships; it could even be placed in orbit around the Earth. (One-way attack UASs are normally pre-programmed to fly to a target; once programmed and launched, they are not controlled or in communication with the control element.)

UASs serve as intelligence, surveillance, and reconnaissance platforms for target detection, identification, and location; weapon targeting; target designation; and battle damage assessment. UASs can assist commanders with early warning to develop friendly courses of action. State-of-the-art sensors and data links provide real-time targeting for fire support systems, maneuver forces, and aircraft. UASs equipped with laser designators provide immediate targeting assistance and terminal guidance of munitions. UAS platforms also can serve as a means for weapons delivery or may themselves serve as the weapon, and have been used extensively to attack high payoff targets without endangering pilots. Threat actors consider UASs expendable.

UASs have relatively low radar cross sections, low speed, and low thermal signatures, thus making them difficult to detect, track, and engage. Mission-dictated flight profiles take full advantage of terrain, increasing system survivability and optimizing coverage. Flight altitudes for UASs vary by their size and mission and can range from meters to thousands of kilometers. UASs conducting intelligence, surveillance, and reconnaissance missions operate at altitudes consistent with their sensor systems. They normally fly at altitudes safe from small arms fire and can stand off and detect from up to 25 kilometers.

UAS optical payloads consist of daylight television, infrared video, and recording cameras (for reconnaissance missions). Other major payload categories include electromagnetic warfare (jammers), electronic intelligence, radar, communications relay and attack warheads. Several nations are developing and fielding anti-radiation homing UAS capabilities with the primary mission of attacking battlefield radio frequency emitters (radars and communications). These platforms have a variety of launch options and are usually fire-and-forget systems.

UASs, especially in groups 1-3, have been extensively used in the war in Ukraine. UASs have changed war as we know it. Both Russia and Ukraine are using inexpensive, sUASs for battlefield surveillance, many of which were designed for commercial use. Russia is developing the capability to build low-cost sUASs closer to the front to minimize the time to get them to the fight. Russia employs one-way attack drones (expendable ammunition with fire-and-forget missions) in massed multi-axis attacks with decoy UASs to increase chances of successful attacks against critical Ukrainian assets. UASs have inflicted great damage to Ukraine facilities as well as military equipment. Ukraine is also using UASs for short-range missions and for long-range missions deep into Russia, as far as Moscow.

Questions regarding UASs that need to be answered by ADA analysts include:

- Organizations?
- Type (attack, surveillance, other)?
- Numbers available?
- Performance (speed, altitude, and launch restrictions)?
- Endurance and range?
- Contour flying or terrain limiting factors?
- Target acquisition and standoff range?
- Loiter time (how long can the UAS stay on station)?
- Sensor package and payload (maximum weight, type, and load mixture)?
- Visibility effects on acquisition?
- Modes of recovery and turnaround time?
- Real-time data link capability?
- Guidance modes (ground controlled or preprogrammed)?
- Crew proficiency.

(See chap. 2 for discussion of UAS Threat.)

Planning & Preparing

IV. Planning Integrated Air Defenses

Ref: FM 3-01, U.S. Army Air and Missile Defense Operations (Aug '25), p. 95.

Defenses are created to optimize the protection of designated assets in compliance with the initial AMD plan. Defense plans are developed to maximize available ADA capabilities against the expected threat, thereby allowing for the greatest probability of engagement success. Ideally, defenses should incorporate dedicated and non-dedicated requisite capabilities to overmatch threat capabilities and deny an enemy surveillance and attack options. The threat's weapons, speed, altitude, and delivery techniques must be determined or estimated. Information collected by ADA and Army G-2/S-2 personnel is used to build intelligence estimates. The defense is then implemented so as to provide balanced effectiveness against those combinations of weapons and techniques that are known to be available to the enemy in meaningful quantities and are evaluated as likely to be employed.

Integrated Defenses

Integrated defenses are developed for the joint force commander's or Army forces commander's designated assets and planned in detail to defeat all aerial threats before enemy ordnance can impact a defended asset. AMD systems can be mixed with other ADA systems or sensors, such as Avenger with Patriot, or paired with the AMD systems of another Service or multinational partner (for example, the U.S. Navy's Aegis and the United Kingdom's Sky Sabre) to achieve layered defenses. Integrated active AMD measures are established and executed concurrently with attack operations and passive defense capabilities across echelons using a combined arms approach.

In implementing defense plans, commanders apply the AMD principles and employment tenets and consider the technical capabilities of the systems. (For more information on the principles and tenets, see the definitions and descriptions beginning on page 6; also, see additional discussions throughout this chapter). Defenses that have two or more types of ADA systems, with different weapons and sensors, are developed by emplacing the longest-range system first, based on its employment guidelines. Other ADA systems are deployed according to their guidelines, with emphasis on filling gaps in the coverage of the longer-range systems. Such defenses are typical of those activated for fixed sites.

Supported Commander's Scheme of Maneuver

In support of maneuver forces, the defense plan must be synchronized with the supported commander's scheme of maneuver. ADA commanders should participate in and use the same types of planning processes used by the supported force to integrate their capabilities into the maneuver operations. Coordination with the supported commander and staff sections must be continuous throughout the operations process to facilitate understanding and integration requirements of the supporting AMD system. AMD principles and employment tenets are adapted to the types of operations conducted by the supported maneuver force and changed as required to mitigate operational risk to the friendly forces and adjust priorities. Considerations for the defense planning should also include the quantity, capabilities, and potential locations of non-dedicated air defense teams. Interspersing these teams with ADA systems, such as M-SHORAD and Avenger, will further enhance defenses.

III. Protection Planning (Asset Protection)

Ref: ADP 3-37, Protection (Jan '24), chap. 2. See also p. 7-6.

Planning occurs as a holistic analysis of all threats, hazards, vulnerabilities, and criticalities. Commanders and staff must consider-

- Most likely threats and hazards.
- Protection priorities.
- Which personnel, physical assets, and information to protect.

Commanders and staff must address C-UAS relevant areas within the protection planning process. Initial planning requires various assessments to establish protection priorities including-

- Criticality assessments.
- Vulnerability assessments.
- Threat assessments.
- Hazard assessments.

Protection Planning

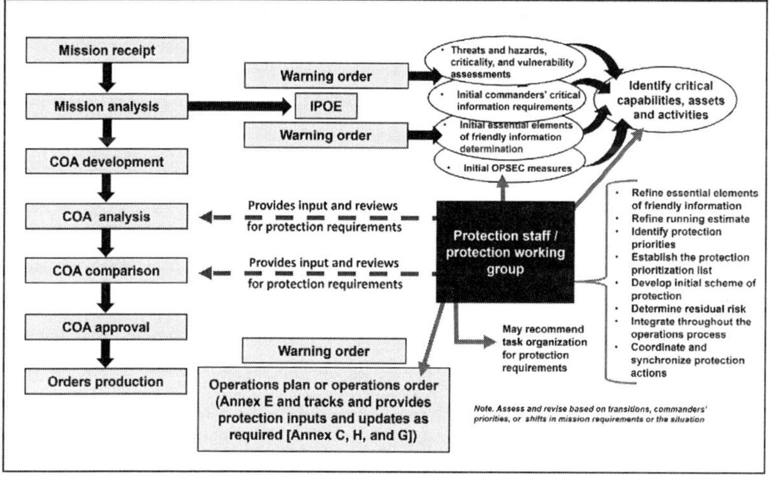

Ref: ADP 3-37 (Jan '24), fig. 3-2. Protection planning.

The keys to protection planning are establishing a protection working group; identifying threats and hazards; assessing the threats and hazards to determine the risks; developing preventive measures; directing, coordinating, and synchronizing protection capabilities; and enabling operations and protection capabilities into a comprehensive scheme of protection that includes mitigating measures (see figure 3-2). Integrating planning considerations from all warfighting functions into the scheme of protection and the integration of protection into the planning of all other warfighting functions provide common critical capabilities available to commanders and staffs at all echelons. Commanders integrate and synchronize capabilities of one warfighting function with other warfighting functions to achieve objectives and accomplish missions.

I. Initial Assessments

Initial protection planning requires various assessments to establish protection priorities. Assessments include threats, hazards, vulnerability, and criticality. These assessments determine which assets must be protected given no constraints and which assets can be protected with available resources. There are seldom sufficient resources to simultaneously provide all critical assets the same level of protection. For this reason, commanders make decisions on acceptable risks and provide guidance to the staff so they can coordinate and synchronize protection capabilities based on the protection priorities.

Initial Assessments (Protection)

- **A** Threat and Hazard Assessment
- **B** Criticality Assessment
- **C** Vulnerability Assessement

Protection planning is a continuous process that must include an understanding of the threats and hazards that may affect operations from the deep area back to the strategic support area. Protection capabilities are aligned to protect critical capabilities, areas, and information and mitigate effects from threats and hazards. The protection cell and protection working group must prioritize protection during competition below armed conflict, crisis, and armed conflict.

During Competition Below Armed Conflict

During competition below armed conflict, the focus is on cooperation, prevention, and deterrence. Within planning, the protection cell conducts continuous assessments to understand the environment; cooperate with and support partners to build a network of protection; and safeguard the force through active and passive measures, training, and exercises.

During Crisis

During crisis, the main purpose is to deter adversary actions. The protection cell estimates the protection assets necessary for future operations and the increased threat according to the commander's priorities, force availability, and adversary assessment. Units prioritize protection capabilities and align them to defend critical capabilities, areas, and information. It is imperative to conduct information advantage activities to deny adversaries the ability to obtain information.

During Armed Conflict

During armed conflict, planning prioritization considerations includes efforts to—

- Preserve combat power.
- Protect critical facilities, nodes, lines of communication networks, routes, and areas.
- Counter enemy fires and maneuver by placing personnel, systems, and units in locations or positions that are difficult to locate, strike, and destroy.

- Gain air, space, and electromagnetic spectrum superiority.
- Proactively target enemy systems.
- Use defensive information advantage activities.
- Recover and reintegrate isolated personnel.

Success in consolidating gains is obtained through setting the conditions for a stable environment. Staffs should weigh the prioritization of protection capabilities throughout defensive, offensive, and stability operations required to support the consolidation of gains.

A. Threat and Hazard Assessment

Personnel from all staff sections and warfighting functions help conduct threat and hazard analysis starting during mission analysis. This analysis comprises a thorough, in-depth compilation and examination of information and intelligence that address potential threats and hazards in the area of operations. The integrating processes (intelligence preparation of the operational environment, information collection, targeting, risk management, and knowledge management [see ADP 5-0]) provide an avenue to review and refine knowledge of threats and hazards. Threat and hazard assessments are continuously reviewed and updated as the operational environment changes.

The threat and hazard assessment results in a comprehensive list of threats and hazards and determines the likelihood or probability of occurrence of each threat and hazard. Protection considerations for the threat and hazard assessment include—

- Hostile actions.
- Environmental conditions.
- Nonhostile activities.
- Criminal activities.
- Civil unrest.

See following pages (pp. 4-24 to 4-25) for an overview of threats and hazards.

B. Criticality Assessment

The criticality assessment provides the commander a list of key assets and infrastructure based on the necessity for mission success. An asset's criticality is determined by the impact of its loss on the mission. A criticality assessment addresses the impact of a temporary or permanent loss of key assets on the unit's ability to conduct the mission. For installation, garrison, area support, and similar commands, a criticality assessment should include high-density population facilities (recreational centers, theaters, sports venues) that may not be mission-essential. Echelons above corps should consider urban population centers in their criticality assessments. They examine the costs of recovery and reconstitution, including time, expense, capability, and infrastructure support. The staff gauges how quickly a lost capability can be replaced before providing recommendations to the commander. The general sequence for a criticality assessment is—

Step 1. List the key assets and capabilities.

Step 2. Determine if critical functions or combat power can be substantially duplicated with other elements of the command or an external resource.

Step 3. Determine the time required to substantially duplicate key assets and capabilities in the event of temporary or permanent loss.

Step 4. Set priorities for the response to threats toward personnel, physical assets, and information.

Threats and Hazards

Ref: ADP 3-37, Protection (Jan '24), pp. 1-3 to 1-4.

The protection warfighting function prevents or mitigates detection, threat effects, and hazards to preserve combat power and enable freedom of action and survivability of the force by providing protection from threats and hazards. Threats and hazards have the potential to cause personal injury, illness, or death; equipment or property damage or loss; or mission degradation. Hazards lead to risk when people interact with equipment or their environment. Hazards exist in all types of environments and activities, including combat, stability, base support, training, garrison activities, and off-duty activities. Commanders and protection cells analyze the following potential threats and hazards:

- **Hostile actions.** Threats from hostile actions include any capability that enemy forces or criminal elements can use to inflict damage on personnel, physical assets, or information. These threats may include direct and indirect fires; unmanned aircraft systems; explosive hazards; suicide bombings; network attacks; asset theft; air attacks; cyberspace networks and other forms of electromagnetic attacks; or chemical, biological, radiological, and nuclear (CBRN) weapons.
- **Nonhostile activities.** Nonhostile activities include hazards associated with Soldier duties within their occupational specialty, Soldier activities while off duty, and unintentional actions that cause harm. Examples include on-and off-duty accidents, operations security (OPSEC) violations, network compromises, equipment malfunctions, unexploded ordnance, or CBRN hazards.
- **Environmental conditions.** Environmental hazards associated with the surrounding environment could degrade readiness or mission accomplishment. Common examples include weather, natural disasters, complex terrain, and diseases. Weather effect knowledge is critical to the commander's situational understanding and decision making. By exploiting this knowledge, commanders can minimize the impact of environmental threats to friendly forces while simultaneously capitalizing on environmental conditions that maximize their advantage while operating at the limits of their capabilities. The staff also considers how military operations may affect noncombatants in the area of operations. Such considerations prevent unnecessary collateral damage and consider how civilians may affect the mission.

Commanders use the METT-TC (I) mission variables to describe the characteristics of the area of operations, including threat-and accident-based hazards that may impact protection. In most cases, they can obtain relevant information from an ongoing analysis of the operational environment by using the PMESII-PT operational variables.

Threats

The various actors in any area of operations can qualify as a threat, enemy, adversary, neutral, or friendly. Land operations often prove complex because actors intermix, often with no easy means to distinguish one from another.

- A **threat** is any combination of actors, entities, or forces that have the capability and intent to harm United States forces, United States national interests, or the homeland (ADP 3-0). Threats are, by nature, hybrid. Threats may include individuals, groups of individuals (organized or not organized), paramilitary or military forces, nation-states, or national alliances. When threats execute their capability to do harm to the United States, they become enemies.
- A **peer threat** is an adversary or enemy with the capabilities and capacity to oppose U.S. forces across multiple domains worldwide or in a specific region where they have a position of relative advantage. Peer threats possess roughly equal combat power in geographical proximity to a conflict area with U.S. forces
- An **enemy** is a party identified as hostile against which the use of force is authorized (ADP 3-0). An enemy is also called a **combatant** and is treated as such under the law of war.

- An **adversary** is a party acknowledged as potentially hostile to a friendly party and against which the use of force may be envisaged (JP 3-0)
- An **insider threat** is a person with placement and access who intentionally causes loss or degradation of resources or capabilities or compromises the ability of an organization to accomplish its mission through espionage, providing support to international terrorism, or the unauthorized release or disclosure of information about the plans and intentions of United States military forces (AR 381-12).
- A **neutral** is an impartial or unbiased country or person that neither helps nor supports either side in a conflict or disagreement.
- A **hybrid threat** is the diverse and dynamic combination of regular forces, irregular forces, terrorist forces, and/or criminal elements unified to achieve mutually benefitting effects (TC 7-100).

Hazards

A hazard is a condition with the potential to cause injury, illness, or death of personnel; damage to or loss of equipment or property; or mission degradation (JP 3-33). Hazards are usually predictable and preventable and can be reduced through effective risk management efforts. Commanders differentiate hazards from threats and develop focused schemes of protection and priorities that match protection capabilities with the corresponding threat or hazard. However, hazards can be enabled by the tempo or friction or by the further complacency that sometimes develops during extended military operations.

Refer to ATP 3-34.20 and TM 3-11.91 for additional information on hazards.

Levels of Threat

There are three levels of threat: Level I, Level, II, and Level III. These different levels provide commanders with general descriptions and categorizations of threat activities and identify protection requirements to preserve combat power; enable the freedom of action; and identify, prevent, or mitigate the effects of threats and hazards.

Commanders and staffs should consider the sizes and types of potential threats to friendly forces when determining and describing levels of threat. Threat levels should be based on the activity, capability, and intent of enemy agents or forces. They can be further described by looking at mission impact. A Level I threat may require only a routine response by assembly area, base camp, or base cluster security forces and have a negligible impact on the mission, or a Level I threat may have a catastrophic impact.

- **Level I threats**. Level I threats include enemy agents, terrorists, and criminals whose primary missions include espionage, sabotage, assassination, and subversion. These include a potential for insider attacks by elements or individuals of host-nation partners and security forces.
- **Level II threats**. Level II threats include small-scale forces that can cause serious harm to military forces and civilians. Attacks by Level II threats can cause significant disruptions to military operations and the orderly conduct of local governments and services. Forces constituting Level II threats are capable of conducting well-coordinated, but small-scale, hit-and-run attacks; improvised weapons attacks with roadside or vehicle-borne improvised explosive devices; raids; and ambushes. Level II threats may also include special operations forces.
- **Level III threats**. Level III threats have the capability of projecting combat power by air, land, or sea or anywhere into the area of operations. Specific examples include airborne, heliborne, and amphibious operations; large, combined-arms, ground-force operations, or penetrations; and infiltration operations involving large numbers of individuals or small groups. Level III threats are beyond the capability of support and rear area forces and can only be effectively defeated by a tactical combat force or other significant forces.

See also pp. 1-23 to 1-27 from FM 3-0.

The protection cell and working group continuously update criticality assessments during the operations process. As the staff develops or modifies a friendly COA, information collection efforts confirm or deny information requirements. As the mission or threat changes, initial criticality assessments may also change, increasing or decreasing the subsequent force vulnerability. The protection cell members monitor and evaluate these changes and begin coordination among the staff to implement modifications to the protection concept or recommend new protection priorities. Priority intelligence requirements, running estimates, measures of effectiveness (MOEs), and measures of performance (MOPs) are continually updated and adjusted to reflect the current and anticipated risks associated with the operational environment.

C. Vulnerability Assessment

A vulnerability assessment is an evaluation (assessment) to determine the magnitude of a threat's or hazard's effect on a critical capability, asset, or activity by phase of an operation. It identifies areas of improvement required to withstand, mitigate, or deter acts of an enemy attack. Commanders must understand their vulnerabilities to being attacked in each domain and develop mitigations to protect the force. The protection cell continually assesses the risk to friendly forces. When assessing friendly force vulnerabilities and weaknesses, the protection cell views the target that the force presents through the eyes of an enemy targeting analyst. The cell addresses who or what is vulnerable and how it is vulnerable against threats. They consider ways to reduce vulnerability and determine the appropriate protection capabilities to prevent and mitigate detection and the effects of threats and hazards.

The vulnerability assessment identifies physical characteristics or procedures that render critical capabilities, areas, and information vulnerable to known or potential threats and hazards. Vulnerability evaluation criteria may include the degree to which an asset may be disrupted, the quantity of the asset available (if replacement is required due to loss), dispersion (geographic proximity), and key physical characteristics. When a vulnerability is identified and assessed, commanders ensure that capabilities, assets, and activities critical for operations are aligned with the protection principles and priorities to accomplish the mission.

The general sequence of a vulnerability assessment is—

Step 1. List assets and capabilities and the threats against them

Step 2. Determine the common criteria for assessing vulnerabilities

Step 3. Evaluate the vulnerability of assets and capabilities

DOD has created several decision support tools to perform criticality assessments in support of the vulnerability assessment process, including—

MSHARPP (Mission, Symbolism, History, Accessibility, Recognizability, Population, and Proximity)

MSHARPP is a targeting analysis tool that is geared toward assessing personnel vulnerabilities, but it also has application in conducting a broader analysis. The purpose of the MSHARPP matrix is to analyze likely terrorist targets and to assess their vulnerabilities from the inside out.

CARVER (Criticality, Accessibility, Recuperability, Vulnerability, Effect, and Recognizability)

The CARVER matrix is a valuable tool in determining criticality and vulnerability. The CARVER targeting matrix assesses a potential target from a terrorist perspective to identify what the enemy might perceive as a good (soft or valuable) target.

Refer to ATP 3-37.2 for more information on MSHARPP and CARVER.

II. Scheme of Protection Development

Ref: ADP 3-37, Protection (Jan '24), pp. 3-12 to 3-13.

The scheme of protection describes how protection tasks and systems support the commander's intent and concept of operations, and it uses the commander's guidance to establish the priorities of support to units for each phase of the operation. A commander's initial protection guidance may include protection priorities, civil considerations, protection task considerations, potential decision points, high-risk considerations, and acceptable risk.

The protection cell (supported by the protection working group) develops the scheme of protection after receiving guidance and considering the principles of protection in relation to mission variables and the systems requiring protection. The scheme of protection is based on the mission variables; thus, it includes protection priorities by area, unit, activity, or resource and should support the scheme of maneuver. It addresses how protection is applied and derived during all phases of an operation.

The protection cell develops the scheme of protection by considering the following items, at a minimum, to create a secure operational environment:

- Protection priorities (critical capabilities, areas, and information).
- Work priorities for survivability assets.
- Coordination of air and missile defense support.
- Specific terrain and weather factors.
- Information focus and limitations for security efforts.
- Areas or events where risk is acceptable.
- Friendly forces information requirements.
- CCIRs.
- Civilians and noncombatants in the area of operations.
- Vehicle and equipment safety or security constraints.
- Personnel recovery actions and control measures.
- Force protection condition (FPCON) status.
- Force health protection measures.
- Mission-oriented protective posture guidance.
- Environmental guidance.
- Scheme of information.
- Explosive ordnance and hazard guidance.
- Ordnance order of battle.
- OPSEC risk tolerance.
- Fratricide avoidance measures.
- Rules of engagement, standing rules for the use of force, and rules of interaction.
- Escalation of force and nonlethal weapons guidance.
- Operational scheme of maneuver.
- Military deception.
- Obscuration.
- Identified threats and hazards.
- Radiation exposure status or operational exposure guidance.
- Contractors in the area of operations.
- Electromagnetic spectrum status.
- Availability of personnel-recovery-capable assets, gaps in recovery coverage, preparation of individuals.

III. Protection Prioritization List (See p. 5-6.)

Ref: ADP 3-37, Protection (Jan '24), pp. 3-8 to 3-11.

Protection Priorities

Criticality, vulnerability, and recoverability are some of the most significant considerations in determining protection priorities *(previous page)* that become the subject of commander guidance and the focus of area security operations. The scheme of protection is based on the mission variables and should include protection priorities by area, unit, activity, or resource.

Although all military assets are important and all resources have value, the capabilities they represent are not equal in their contribution to decisive operations or overall mission accomplishment. Determining and directing protection priorities may involve the most important decisions that commanders make and their staffs support. There are seldom sufficient resources to simultaneously provide the same level of protection to all assets.

Most prioritization methodologies assist in differentiating what is important from what is urgent. In protection planning, the challenge is to differentiate between critical assets and important assets and to further determine what protection is possible with available protection capabilities.

Protection prioritization lists are organized through the proper alignment of critical assets and activities. The commander's priorities and intent determine critical assets and activity or operation for protection by the assets available. Critical assets can be people, property, equipment, activities, operations, information, facilities, or materials. For example, bridging companies might be identified as critical to the execution of military operations and, therefore, receive additional protection. The lack of a replacement may cause a critical asset to become a top priority for protection. A critical activity or operation could be a gap crossing or command post displacement.

The protection prioritization list is a key protection product developed during mission analysis, but it cannot be completed until the COA development step during MDMP. The protection cell and working group assess criticality, vulnerability, and probability of surveillance or attack to prioritize identified critical assets. Once the protection working group determines which assets are critical for mission success, it recommends protection priorities and establishes a protection prioritization list. It is continuously assessed and revised throughout each phase or major activity of an operation.

Criticality

Criticality is the degree to which an asset is essential to accomplish the mission. It is determined by assessing the impact that damage to, or destruction of, the asset will have on the success of the operation. Damage to an asset may prevent, delay, or have no impact on the success of the plan.

- **Catastrophic**. Complete mission failure or the inability to accomplish the mission, death or total disability, the loss of major or mission-critical systems or equipment, major property or facility damage, mission-critical security failure, or unacceptable collateral damage.
- **Critical**. Severely degraded mission capability or unit readiness; total disability, partial disability, or temporary disability; extensive damage to equipment or systems; significant damage environment; security failure; or significant collateral damage.
- **Marginal**. Degraded mission capability or unit readiness; minor damage to equipment or systems, property, or the environment; lost days due to injury or illness; or minor damage to property or the environment.
- **Negligible**. Little or no adverse impact on mission capability, first aid or minor medical treatment, slight equipment or systems damage (remaining fully functional or serviceable), or little or no property or environmental damage.

Threat Vulnerability

Threat vulnerability measures the ability for a threat to damage the target (asset) using available systems (people and material). An asset's vulnerability is greater if a lower-level threat (Level I) can create damage or destruction that would result in mission failure or severely degrade its mission capability. If an asset can withstand a Level I or Level II threat, its vulnerability ability is less and may not require additional protection assets, depending on the asset's criticality. The following mitigating factors must be considered when assessing the vulnerability of a target: survivability (the ability of the critical asset to avoid or withstand hostile actions by using camouflage, cover [hardening], concealment, and deception), the ability to adequately defend against threats and hazards, mobility and dispersion, and recoverability (which measures the time required for the asset to be restored, considering the availability of resources, parts, expertise, manpower, and redundancies).

- **Level I threat**. Enemy agents, saboteurs, sympathizers, terrorists, civil disturbances.
- **Level II threat**. Small tactical units. Irregular forces may include significant standoff weapons threats.
- **Level III threat**. Large tactical force operations, including airborne, heliborne, amphibious, infiltration, and major air operations.

Threat Probability

Probability assesses the frequency an asset will be targeted for surveillance or attack by a capable threat. Determinations of the intent and capability of the threat are key in assessing the probability of attack.

- **Frequent**. Occurs very often; known to happen regularly.
- **Likely**. Occurs several times; a common occurrence.
- **Occasional**. Occurs sporadically but is not uncommon.
- **Seldom**. Remotely possible; could occur at some time.
- **Unlikely**. Rarely occur, but not impossible.

The protection prioritization list helps Army commanders to identify or assess assets that require protection prioritization within their assigned areas. Not all assets listed on the protection prioritization list receive continuous protection. Some critical assets only receive protection assets based on available resources. It is the responsibility of the protection cell and working group to provide the assessment and recommended protection prioritization list to the commander for approval.

Shifts in Protection Priorities

Transitions mark a change of focus in an operation. Leaders plan transitions as part of the initial plan or parts of a branch or sequel. They can be unplanned and cause the protection cell and working group to react to unforeseen circumstances and reassess protection priorities. Changes to protection priorities should be anticipated and assets reassessed as transitions occur throughout an operation or as commanders shift their priorities.

Protection cells and working groups must account for the time required to plan, prepare, execute, and assess shifts in protection priorities in support of transitions. They must develop information requirements, including CCIR for decisions, to support shifts in protection priority and transitions They should also account for likely friction due to the environment, degraded communications, and enemy action. Effective transitions require protection cells and working groups to plan and prepare for changes in protection priorities well before their execution so the force can maintain the momentum and tempo of operations. Risk increases during transitions, so commanders and staffs establish protection requirements and priorities and direct, coordinate, and synchronize protection efforts and capabilities to support the transition.

IV. Protection Cell and Working Group

A. Protection Cell *(See p. 4-14.)*

The protection cell forms the core membership of the protection working group, which includes other agencies, as required. Protection cell and protection working group members differ in that additional staff officers are brought into the working group. These additional officers meet operational requirements for threat and hazard, criticality, and vulnerability assessments and protection priority recommendations. The protection working group calls upon existing resources from the staff as required. The echelon chief of protection leads the protection working group.

B. Protection Working Group

Protection working group meetings have similar purposes, regardless of the echelon. Protection functions at different echelons of command differ mostly in the size of the area of operations and the number of available protection capabilities. Commanders augment the team with other unit specialties and unified action partners, depending on the operational environment and unit mission. The chief of protection determines the working group agenda, meeting frequency, composition, input, and expected output. Each protection working group meeting should have a unique purpose. For example, protection working group meetings during the military decision-making process (MDMP) may focus on threat, hazard, and risk analysis while conducting mission analysis; on criticality and vulnerability analysis and initial scheme of protection development during course of action (COA) development; or on refining the scheme of protection and mitigating additionally identified risk during COA analysis.

The purpose of a protection working group includes:

- Determines likely threats and hazards and provides updates on enemy tactics, the environment, and accidents.
- Determines vulnerabilities as assessed by the vulnerability assessment team.
- Establishes and recommends protection priorities.
- Reviews and provides recommendations for the protection prioritization list.
- Reviews, coordinates, and synchronizes unit protection measures.
- Recommends force protection conditions and random antiterrorism measures.
- Makes recommendations to commanders on protection issues that require a decision.
- Performs tasks required for a force protection working group and a threat protection working group.
- Assesses assets and infrastructure that are designated as critical by higher headquarters.
- Analyzes and provides recommendations for the protection of civilians in the area of operations.
- Considers incorporating host-nation and constabulary forces into the scheme of protection.
- Develops and refines the protection running estimate.
- Develops a scheme of protection, ensuring that it nests with the operational concept.
- Establishes the personnel recovery coordination center.
- Determines organic personnel recovery capabilities at each echelon, tracks open personnel recovery events, and disseminates isolated personnel guidance.
- Provides input and recommendations for defensive cyber operations.
- Identifies risks to the mission.
- Provides input and recommendations on protection-related training.

Chap 4

IV. Preparing (for C-UAS Operations/SHORAD)

Ref: FM 3-01.44, Short-Range Air Defense Operations (Jul '22), chap. 5.

Understanding the preparation activities of a SHORAD (Short Range Air Defense) unit helps a C-UAS planner effectively integrate counter-UAS capabilities into the broader air defense scheme. SHORAD preparation processes—such as coordination and liaison, security operations, network setup, sustainment, rehearsals, and confirmation briefs—mirror many of the same tasks required for C-UAS employment. By studying how SHORAD units organize sensors, plan coverage, establish communication links, and manage terrain for detection and engagement, a C-UAS planner can align system placement, power and data requirements, and airspace coordination to create seamless layered defense. This understanding ensures unity of effort, reduces redundancy, and enables synchronized detect-track-identify-defeat operations that complement SHORAD's mission of protecting forces and critical assets from aerial threats. *(See chap. 3, Air & Missile Defense.)*

Preparation consists of those activities performed by units and Soldiers to improve their ability to execute an operation (ADP 5-0). A SHORAD battery may operate as part of an ADA task force or as a separate unit supporting a designated asset, with its platoons assigned to defend elements within that asset. This chapter focuses on the SHORAD battery and platoons as separate units and their preparatory considerations and actions in support of a BCT and its subordinate maneuver battalions.

The time that is left between finalizing the planning (with a commander's decision including the commander's intent and concept of operations) and the actual start of the execution of the operation determines how extensively and thorough the preparation can be performed.

All activities during the preparation phase are to validate the planning results by selecting and assessing positions for sensors, weapon systems, headquarters and C2 nodes, and routes and areas for such sites and their supporting elements (for example, maintenance areas). In addition, a visual inspection of all terrain aspects, in the actual and predicted weather conditions, will contribute to validating an adversary's projected COAs and the commander's own developed COAs. All validating results from this preparation phase are used to review, refine, and adjust, as necessary, the initial plans and develop more detailed taskings for the subordinate and supporting units.

The need for continuous coordination between the SHORAD battery and its supported unit or asset, the SHORAD battalion, adjacent ADA units, and other air control elements cannot be overemphasized.

*Refer to BSS7-1: The Battle Staff SMARTbook, 7th Ed. (w/ SMARTupdate 1), pp. 1-35 to 1-40 for detailed discussion of **preparation** as part of the operations process from ADP 5-0, The Operations Process (Jul '19). . Preparation creates conditions that improve friendly forces' opportunities for success and include activities such as rehearsals, training, and inspections. It requires commander, staff, unit, and Soldier actions to ensure the force is ready to execute operations.*

Planning & Preparing

I. C-UAS Preparation Activities

Ref: Adapted from ADP 5-0, The Operations Process (Jul '19), pp. 3-4 to 3-9.

Preparation activities help commanders, staffs, and Soldiers understand a situation and their roles in upcoming operations. Commanders, units, and Soldiers conduct the activities listed below to help ensure the force is prepared for execution.

Coordinate and Establish Liaison

Units and organizations establish liaison early in planning and preparation. For C-UAS, this includes coordination with higher, adjacent, and supported headquarters for air defense coverage, spectrum management, airspace control measures, and C-UAS engagement authorities. Liaison ensures mutual understanding of C-UAS roles, handoff procedures, and reporting channels during operations.

Initiate Information Collection

Commanders continuously improve situational understanding through information collection. C-UAS integration includes early employment of sensors—radar, EO/IR, and RF detection systems—to establish baseline air activity and identify potential UAS threats. Coordination with intelligence and air defense assets ensures overlapping coverage and timely threat reporting.

Initiate Security Operations

Security operations protect the force during vulnerable preparation periods. C-UAS measures enhance security by monitoring airspace above assembly areas, command posts, and supply routes. Units conducting local security must include aerial surveillance in their security plans and ensure quick-reaction C-UAS capabilities are on alert.

Initiate Troop Movements

Units reposition C-UAS sensors and defeat systems to cover critical movement routes and assembly areas. Commanders ensure detection and defeat assets are emplaced early to provide overhead protection as forces move. Movement plans must account for power, communication, and line-of-sight requirements of mobile C-UAS systems.

Complete Task Organization

Commanders organize C-UAS assets across echelons to ensure layered protection and responsiveness. This may involve assigning mobile C-UAS teams to maneuver elements or integrating fixed systems into base defense plans. Task organization ensures the right mix of sensors, effectors, and command links are available at each echelon.

Integrate New Units and Soldiers

When new Soldiers or units arrive, commanders and staffs ensure they understand C-UAS procedures, reporting formats, and air defense integration. All Soldiers—especially those in security and sustainment roles—must know basic C-UAS detection, reporting, and protective actions against UAS threats.

Train

Training during preparation includes familiarization with C-UAS detection and defeat equipment, air threat recognition, and reporting. Rehearsals and training events should practice C-UAS engagement drills and airspace deconfliction procedures. Training reinforces discipline in radio use and electromagnetic emissions to avoid interference with friendly systems.

Conduct Pre-Operations Checks and Inspections

Units conduct pre-operations checks to ensure C-UAS sensors, power systems, and communications are functional. Crews verify calibration, connectivity, and ROE for C-

UAS engagements. These checks ensure all systems, Soldiers, and networks are ready to detect, track, and defeat aerial threats before mission execution.

Initiate Sustainment Preparation

C-UAS readiness depends on maintaining power, spare parts, and consumables such as batteries and cooling supplies. Sustainment elements ensure resupply of electronic components and replacement systems to maintain continuous coverage. Repositioning sustainment assets supports both ground and air defense elements.

Initiate Network Preparations

During preparation, units tailor networks to handle C-UAS data and sensor feeds. Commanders ensure integration of C-UAS systems into the tactical network and COP. Frequency management, encryption, and bandwidth prioritization must support uninterrupted command and control of detection and defeat systems.

Manage Terrain

Terrain management includes identifying high-ground or clear-field locations for C-UAS sensors and antennas. Commanders allocate terrain to optimize sensor line of sight and minimize interference. Coordination ensures radar and jammer placement does not disrupt friendly aviation or communications.

Prepare Terrain

Units improve positions to support C-UAS coverage—clearing vegetation for radar fields, emplacing masts, or reinforcing sensor platforms. Terrain preparation also includes constructing protective shelters for electronic systems and camouflage against UAS observation.

Conduct Confirmation Briefs

Subordinate leaders conduct confirmation briefs to confirm understanding of the commander's intent, mission, and C-UAS responsibilities. These briefs should cover escalation procedures, ROE, and coordination between maneuver and C-UAS units. Understanding the relationship between C-UAS missions and adjacent unit operations ensures integrated defense.

Conduct Rehearsals

Rehearsals validate how C-UAS assets integrate into the detect–track–identify–defeat process. Units should rehearse sensor cueing, reporting procedures, airspace coordination, and engagement approval. Rehearsals enhance coordination between C-UAS teams, air defense artillery, and maneuver elements.

Conduct Plans-to-Operations Transition

As plans shift to execution, C-UAS responsibilities transition from planners to operators. Current operations cells monitor real-time C-UAS tracks and threat reports, ensuring rapid response to UAS incursions.

Revise and Refine the Plan

Commanders adjust C-UAS plans as intelligence and reconnaissance refine threat patterns. New UAS activity may require repositioning sensors, adjusting defeat authorities, or issuing FRAGORDs to reallocate C-UAS coverage. Refinement ensures continued alignment between operational plans and C-UAS readiness.

Supervise

Leaders supervise all preparation activities to ensure readiness across C-UAS systems, crews, and supporting networks. Commanders inspect equipment, monitor rehearsals, and confirm that Soldiers understand detection and engagement procedures. Effective supervision ensures full integration of C-UAS capabilities into mission execution.

II. Coordination

During planning, coordination of efforts between the SHORAD unit and the supported asset is established by the integration of the SHORAD battery commander into the BCT's operational planning process. The battery commander and platoon leaders continue interacting with their supported counterparts to advise them of the air defense capabilities they will bring in support, the current and projected air threat situation, and the local air defense posture. Commanders share and synchronize plans, commanders' intents, and operational concepts. SHORAD leaders must be cognizant of the friendly and enemy ground situation, while the supported commanders must be informed of the air environment and pertinent air defense directives. Situational information must identify who is operating where, when, and in what strength. SHORAD commanders must advise the supported commanders of any logistical support requirements, other than SHORAD-system specific items, and potential needs for force protection beyond the inherent capabilities in the SHORAD unit. The more frequent and thorough the interplay between the supported and supporting commanders, the more likely the potential for mission success.

The battery commander must be in constant communications with the supported unit's air defense airspace management cell, in the case of a maneuver BCT, maneuver enhancement brigade, field artillery brigade, combat aviation brigade, division artillery, or with other ADA personnel (if present) in a fixed site's operations center, such as a base defense operations center. The air defense airspace management cell or its counterpart is the conduit for air defense information within the supported asset. Cell ADA personnel keep the battery commander informed of current and future brigade operations. The battery commander, in turn, advises the supported asset of the SHORAD unit's on-going operations, capabilities, and limitations. The battery commander develops the air defense plan for the supported unit and coordinates with the air defense airspace management cell to ensure that the plan is nested with supported unit's operation plan.

The ADAM cell or operations center personnel may also be aware of another AMD unit or units operating within the asset's area and will so advise the SHORAD battery commander. The battery commander should coordinate with that commander to synchronize and ensure mutually supporting operations.

III. Reconnaissance

Reconnaissance provides information about the AO. Preparation activities begin with a reconnaissance of the route to the area. The available preparation time and resources will determine to a large extent how (extensive and detailed) the reconnaissance activities can be performed and therefore influence the choice of reconnaissance type. There are three types of reconnaissance: map, aerial, and ground.

See facing page for an overview and further discussion.

IV. Defense Design

Defense design for ADA is a strategy for defense based on a compiled list of tasks required to protect against air and missile threats in support of operations. SHORAD units are positioned to best protect the designated assets against the projected air threats in accordance with detailed defense designs. Sensor and weapon system placements are selected to maximize surveillance, tracking, and engagements. In designing a defense, sensors are positioned to provide surveillance and fire control tracking capability sufficient to protect assigned assets and prevent gaps in the coverage. Weapon systems and teams are positioned to optimize the defense of the defended assets, enabling lethal coverage over the assigned assets and extending firepower through as much of the defense coverage area as possible.

See chap. 5.

Reconnaissance Types

Ref: FM 3-01.44, Short-Range Air Defense Operations (Jul '22), pp. 5-2 to 5-3.

Map Reconnaissance

Map reconnaissance is the fastest way to conduct a reconnaissance and will always precede any other type of reconnaissance. When geographical limitations hamper physical reconnaissance, map reconnaissance is the only way. The positions of SHORAD sensors and weapon systems are plotted on a map and represent the best locations for providing air defense. Detailed physical and digital maps will provide specific information on terrain, buildings, roads, rivers, forest, and other terrestrial and man-made features. The most current and detailed digital and physical maps can be requested from the unit's space element. Overlapping current imagery and terrain data, space elements can provide the most accurate maps possible even if the ground topography has changed due to a recent battle or geographic event. The platoon leader must consider the AMD employment guidelines and tenets, weather, and mission variables in this initial reconnaissance.

Alternate positions should be identified at this time. After the weapon positions are plotted on a map, check points and primary and alternate routes are selected and plotted. The platoon leader identifies a reconnaissance route by conducting a map reconnaissance to the proposed area. Positions selected by map reconnaissance must be confirmed by additional reconnaissance if the tactical situation allows.

Aerial Reconnaissance

If time is limited and an aircraft is available, the battery commander or the reconnaissance officer (generally a platoon leader) can see the terrain. Aerial reconnaissance provides the fastest way to see the terrain, but is likely unavailable to SHORAD leaders. Even when available, aerial reconnaissance has significant limitations. For example, fields of fire, ground conditions, and local threat cannot be determined from the air. In addition, aircraft activity, to include that of UASs, may expose friendly interest in a particular area.

Ground Reconnaissance

Ground reconnaissance is an on-site examination of the terrain. A ground reconnaissance is the most accurate and desirable type of reconnaissance, though the slowest method. The route can be evaluated for road worthiness, obstacles, choke points, and key terrain. Firing positions, fields of fire, and air avenues of approach can be analyzed. However, ground reconnaissance is time-consuming and dangerous. Covering the distances over potential routes and checking alternate positions takes time. The small reconnaissance party could be subjected to threat observation en route or at the objective area.

Map reconnaissance is an important part of the preparation of the actual ground reconnaissance. The reconnaissance party will benefit from a detailed and thorough map reconnaissance.

Ground reconnaissance is performed to select the best SHORAD positions, march routes, start points, release points, checkpoints, sensor positions, and communications sites. The SHORAD battery commander or platoon leader should coordinate with the local maneuver commander and the supported element S-3 to determine what areas maneuver units plan to occupy, or, when in support of a fixed asset, with the on-site commander and base defense operations center to identify the best locations for SHORAD weapon systems and sensors. Mutual agreement must be established to make the best use of the available terrain.

The AMD employment tenets and the latest intelligence assessments inform the defense design. Considerations for the defense design should also include the quantity, capabilities, and potential locations of combined arms for self-defense teams. Interspersing these teams with SHORAD systems will further enhance defenses.

A. Selecting Positions

Site selection begins upon receipt of a movement order. Careful consideration must be given to ensure that the selected positions for the sensors, launchers, and C2 nodes are oriented in the direction of the expectant air threats, are capable of maintaining required communications with the battery and supported asset's C2 elements, and, above all, can collectively provide the required defense. The need for dispersion of equipment, camouflage throughout the site, appropriate deception measures, and operational security must also be addressed.

Sensors must be carefully placed to ensure optimal surveillance of the area. Their positioning is generally the first consideration in establishing a defense. Sentinels have a 360-degree capability to detect and classify aerial threats. Lightweight Counter Mortar Radars in Land-based Phalanx Weapon System units provide 360-degree detection of RAM threats. See ATPs 3-01.48 and 3-01.60, respectively, for more information on positioning Sentinel and the Lightweight Counter Mortar Radar.

Considerations in selecting launcher positions are dependent upon the type of SHORAD weapon to be emplaced, asset to be defended, and the threat types. Positions should have clear observation and primary and secondary fields of fire, communications with the platoon command post and early warning sources, and available cover and concealment to facilitate survivability. Launchers positioned to counter potential rotary-and fixed-wing threats are generally near projected air avenues of approach. These positions may require adjustments if UASs are also a threat, as UASs can surveil or attack from any direction.

- **M-SHORAD and Avenger systems and Stinger man-portable air defense** (hereafter referred to simply as Stinger) teams require frequent movements once they commence engagements, as the firing signature of their missiles discloses their positions. Units should have primary, alternate, and supplementary positions with planned routes between them. A primary position is one from which the M-SHORAD, Avenger, or Stinger weapon system intends to accomplish its tactical mission. An alternate position is one to which the system moves when the primary position becomes untenable or unsuitable for carrying out the assigned mission. The primary positions should generally be near the plotted positions in the original defense design, at distances determined by clear line-of-sight, security, and survivability considerations; if not near the plotted positions, the defense may need to be redesigned. Communications must also be maintained with the other squads in the platoon.
- The **Land-based Phalanx Weapon System's** limited mobility and emplacement requirements may restrict its positioning options. Its position should be close to, but set back from, the perimeter of the base or other static asset which it defends. Its communication requirements include a linkage to the base defense operations center.

The primary consideration in selecting the headquarters or C2 node site is the ability to communicate with the supported asset and higher-and lower-echelon SHORAD units. Dependent upon terrain and the locations of the other SHORAD and supported asset's elements, the headquarters or C2 node should be positioned where it can best control the SHORAD units and communicate with the supported element. Other considerations for the site selection include available cover and concealment, alternate entrance and exit routes, and defensibility against ground attacks.

After selecting the positions, the platoon leader ensures the ground guides know exactly where all the vehicles and equipment should be emplaced. Preparation should include marking the location of each major piece of equipment. Everyone at the new position is given the new challenge and password, information on any known enemy activity in the area, and the approximate arrival time, location, and order of vehicles for the main party.

If time is limited and the unit must move before the reconnaissance party returns, road guide positions are selected, and Soldiers necessary for these tasks are included in the reconnaissance team. Actual inspection of the chosen routes and positions on the ground is desirable to confirm selections made from the map, or to make necessary adjustments in plans.

Key leaders provide the results of the reconnaissance and relay appropriate instructions for the movement of the main body. The platoon sergeant organizes the vehicles as ordered by the platoon leader. Platoon vehicles use movement techniques dictated by the factors of METT-TC; tactics, techniques, and procedures; and the unit SOP.

B. Occupying Positions

The occupation of positions should be coordinated with the supported unit to avoid mutual interference. When the mission-configured platoon arrives at the position, all vehicles move off the road into the position, without halting and without closing the interval between vehicles. The unit SOP should state the requirements and the priorities for occupying positions.

The first consideration is emplacing the sensor or sensors. Sensors are emplaced on terrain that provides the best, longest range line-of-sight in all directions. Sentinels should be positioned within the supported asset's AO along likely enemy air avenues of approach to surveil, detect, and provide early warning. Sentinel should not be collocated with any command post due to an enemy's ability to target emission signatures. The Sentinel's rotating antenna must remain clear of obstructions while operating to optimize the radar tracking range and air picture reliability. The Lightweight Counter Mortar Radar should be emplaced on a hill, rooftop, or flat plain. Any radar's performance is degraded when placed near obstructions, such as buildings, trees, vehicles, or power lines. The radars should be emplaced no closer than 1,000 meters from one another, and operating frequencies should be separated as much as possible.

The SHORAD platoon leader uses the ADA employment tenets to position launchers in accordance with METT-TC. The most important tenets for the M-SHORAD platoon leader to consider are normally early engagement, weighted coverage, and defense in depth. The Avenger platoon leader stresses mutual support, or overlapping coverage when possible, to enhance engagement opportunities. If fixed-wing aircraft are considered a major threat, the platoon may weight coverage to the most likely air avenue(s) of approach. The Land-based Phalanx Weapon System platoon leader positions the system launchers in pairs with mutual support if possible, or overlapping fires between pairs as a minimum, for RAM engagements given the relatively short range of the Land-based Phalanx Weapon System. The Land-based Phalanx Weapon Systems should be emplaced inside the perimeter, but no closer than 500 meters from the perimeter when possible. Gun system emplacement requires relatively flat (±5-percent slope) on hard-packed ground or pad. During emplacement, the unit will perform gun system mapping to ensure that no coverage areas are in the "no fire zone." Stinger teams are generally positioned along the most likely air avenues of approach. Care must be taken in positioning M-SHORAD, Avenger, Stinger systems to avoid placing them near terrain features easily recognized from the air. Positions are more vulnerable to enemy fires once spotted near an identifiable object.

When all squads have reported that they are in position and capable of providing coverage of their assigned sectors, the platoon leader will report the platoon "ready for action" to the supported force commander. Once in place, SHORAD units should implement passive defense measures, such as camouflage and protective works, and continuously improve them as long they remain at a location. Crew chiefs begin improving positions as dictated by the SOP or additional orders. Position improvements include:

- Using camouflage nets and supplementing them where possible with indigenous materials (for example, branches, leaves, and snow). These materials should be from the vicinity of the squad so they will blend naturally into the surroundings.
- Preparing deliberate fighting positions.
- Providing cover for M-SHORAD and Avenger systems by having the vehicles dug in, hull down, for protection from blast, fragmentation, and small arms. Engineer support is required and should be requested through the supported unit. If such support is not available, the vehicles should be placed in defilade positions.
- Developing alternate and supplementary positions. Work is initiated as early as possible.

A total security plan is then placed in effect. Positioning SHORAD units with maneuver forces or on installations allows the units to take advantage of security measures and the security forces of those assets. However, regardless of those added forces, the platoon must provide its own local security. This includes establishing observation points, positioning automatic weapons overlooking the main avenue of approach and alternate routes into the position, and initiating other security positions around the area to preclude gaps in the perimeter. A sector of fire, with a primary target line, is assigned to each squad, and communications are established with all the positions. Range cards should be prepared. At a minimum, range cards should indicate ranges to critical points on all likely avenues of approach, dead spaces, and likely targets (air and ground). Obstacles enhance security by blocking, disrupting, or canalizing enemy attacks. All obstacles must be coordinated with the supported asset's S-3 and the engineer officer. Other measures taken should also be integrated with the ground defense plan of the supported asset. Emission control measures should be developed and initiated, when directed, to reduce the electromagnetic signatures of the sensors and command posts.

V. Rehearsals

Units conduct combat rehearsals to gain agility, ensure synchronization, increase individual and team initiative, and improve the proficiency of a force. Rehearsals are a critical part of the preparation and supervision steps in troop-leading procedures and occur throughout troop-leading steps. Commanders and leaders can determine the validity and suitability of their plan's concept of the operation, in addition to identifying any shortcomings in the concept not previously recognized, while performing their rehearsals.

Rehearsals allow leaders and their Soldiers to practice key aspects of the concept of operation. These actions help Soldiers orient themselves to their environment and other units before executing the operation and build a lasting mental picture of the sequence of key actions within the operation. Rehearsals are the commander's tool to ensure subordinate leaders understand the commander's intent and the concept of operations.

SHORAD units and leaders participate in the supported asset's rehearsals as well as conduct their own rehearsals.

V. Training for C-UAS Operations

Ref: ATP 3-01.81, Counter-Unmanned Aircraft System (C-UAS) Operations (Mar '25), chap. 7.

The threat posed from UAS employed by our adversaries has, and likely will continue, to increase at a rapid pace. Technology advances and capabilities in commercial-off-the-shelf UAS make it extremely easy for threat actors to employ these commercial-off-the-shelf systems to accomplish their goals. These platforms can easily be modified to carry payloads creating impacts that are even more devastating. In addition, nation states are developing military grade UAS that have even greater threat capabilities. Continual improvement of C-UAS training is paramount to enable U.S. forces to organize, plan, and conduct C-UAS activities; properly employ C-UAS equipment; and better understand and UAS threats.

Commanders and staffs must continually identify problems or issues related to current and future doctrine, organization, training, materiel, leadership and education, personnel, facilities, and policy that affect mission accomplishment. As commanders and staffs identify issues; analysis is required to determine what actions or coordination should take place to resolve the issue. Identified issues and requirements that cannot be resolved internally should be passed on for assistance outside the command.

Rapid development, acquisition, and fielding of C-UAS solutions, synchronized across doctrine, organization, training, materiel, leadership and education, personnel, facilities, and policy to address operational requirements in the field, requires a deliberate effort. The Army proactively identifies and analyzes C-UAS capability gaps to develop, train, and employ solutions to counter the threat. The Joint C-UAS Office has been established to execute C-UAS Executive Agent responsibilities for the Army to synchronize C-UAS efforts across all services within Department of Defense. The Joint C-UAS Office operates as a central hub for collecting identified requirements and coordinating, synchronizing and integrating C-UAS training, doctrine, materiel and policy solutions.

To accomplish the task of synchronizing and executing development and implementation of C-UAS requirements, several key areas should be addressed:

- Identifying operational needs.
- Collecting observations, insights, and lessons leading to development and assimilation of lessons learned.
- Provide relevant C-UAS training across all three training domains (self-development, institutional, and operational).

I. Training

The Army trains to fight and win-it is what we do in order to preserve the peace and to decisively defeat our enemies. We do this through challenging, relevant, and realistic training performed to the highest standards. The Army trains the way it operates. It does this by replicating in training how we expect to fight, closing the gap between operations and training (FM 7-0). Leaders, Soldiers, and units achieve tactical and technical competence while building confidence and agility. U.S. forces train using doctrine that sustains their tactical and operational capabilities. Throughout this training, commanders treat UAS threats as a condition on the battlefield with an understanding that the UAS threat is inherent across the range of military operations and may be present in any location where U.S. forces operate.

A. Providing Relevant C-UAS Training

The training and development of leaders and Soldiers in any organization requires implementation of informed and proven training strategies to achieve success. Commanders provide vision, focus, direction, and an action plan in the development of training strategies. A commander ensures that the organization is trained and mission-ready by actively engaging in all aspects of training and integrating C-UAS capabilities, resources and activities into training.

Training time is always a premium. Relevant C-UAS training will likely never be conducted as a stand-alone training event within the context of collective and combined arms training. Rather C-UAS will largely be trained in terms of reacting to a threat or condition that is threaded throughout planned training.

It should be adaptive if it is to meet training requirements while balancing resource and time constraints. The training of combined arms C-UAS tasks requires planning and execution that begins at the lowest possible level. Commanders may face time constraints, resource limitations, and schedule conflicts in their efforts to achieve optimal training results. The ability to incorporate and sustain organizational capabilities depends on the quality of modernization and sustainment training. Commanders should tailor training and education with relevant information based on lessons learned, enemy trends, and friendly C-UAS TTP to ensure that the training meets the commander's intent and results in the unit being mission ready.

B. Leaders and Individual Training

The Army develops leaders to solve problems. Adaptive leaders are those who have a feel for the fight, who possess the ability to read the situation, and who can analyze the cause-and-effect relationships from the beginning to the end of an operation. An adaptive leader can translate operational and mission variables into situational understanding, is willing to accept prudent risk, and is able to make rapid adjustments based on knowledge assessments. In the modern fight, expanded access to information has compressed response times, resulting in squad level tactical actions that have immediate and lasting strategic consequences. The result is the need for adaptive leaders at the lowest echelon of U.S. forces.

The adaptive-leader skills that a commander develops and brings to the fight are the critical determinants in effectively applying combat power. Leaders improve their ability to adapt through knowledge and experience that is gained from multiple, complex, and unexpected situations in challenging, unfamiliar, and uncomfortable conditions. Commanders should address small-unit problem solving as a centerpiece of instruction based on relevant operational challenges. Leaders should gain the knowledge and experience necessary to think adaptively and critically. Outcome-based training developed by commanders should stress the philosophy of mission command for leaders (provide intent) and their subordinate leaders (determine how to achieve that intent).

Unit leaders must be able to understand the OE and apply C-UAS capabilities and measures within an overall protection framework. This includes employing and enforcing survivability measures and layered defense techniques in the employment of C-UAS capabilities. They must understand and adopt TTP for supporting friendly networks, influencing neural networks, and neutralizing threat networks. This requires training in nontraditional areas. Commanders should also integrate C-UAS activities and capabilities during home station training and focus on integrating collective tasks that plan, coordinate, and control those capabilities.

There are several resources for C-UAS training. These resources support commanders in training individual level C-UAS skills and enable the development of C-UAS crew training. Most C-UAS training resources can be found on the Joint Counter-small UAS Office Community page housed on the Joint Knowledge Online portal. A list of C-UAS training resources that are available can be found in table 7-1.

(See facing page.)

II. C-UAS Training Resources

Ref: ATP 3-01.81, Counter-Unmanned Aircraft System (C-UAS) Operations (Mar '25), table 7-1. Counter-Unmanned Aircraft System Training Resources.

C-UAS Training Resources	
Source	**Description**
Joint Knowledge Online C-UAS Community Page (https://jkodirect.jten.mil/Atlas2/page/desktop/DesktopHome.jsf)	A repository of material and links to online training developed by the Joint C-UAS Office. To access: go to the JKO Online portal at: https://jkodirect.jten.mil/ Login using government issued common access card (CAC). Go to "Community" on the gray banner at the top of the page. Select "Joint C-UAS Office"
JCO Policy Training Module https://intelshare.intelink.gov/sites/JCO/Shared%20Documents/1_JKO%20Policy%20Training%20Module	The JCO Policy Training Module provides instructions for DOD planners and leaders on policies, authorities, and authorized actions that affect C-UAS and UAS procedures, planning, and operations.
Joint C-UAS University (JCU) at Fort Sill, OK	The JCU currently offers three distinct courses: 1. Installation C-UAS Protection Course 2. Joint C-UAS Planner Course 3. Joint C-UAS Operator Course
Center for Army Lessons Learned (CALL) (https://call2.army.mil/)	Repository for Army lessons learned as well as observations, insights, after action reports, etc.
Joint Lessons Learned Information System (JLLIS) (https://www.jllis.mil/apps)	Repository for joint lessons learned as well as observations, insights, after action reports, etc.
Joint Knowledge Online C-UAS Training Modules	Training modules developed by the Joint C-UAS Office to basic C-UAS training for individuals.
Basic counter-small unmanned aircraft systems (C-UAS) Awareness Training. CAC controlled access. (https://jkodirect.jten.mil/html/COI.xhtml?course_prefix=JCUAS&course_number=A-US1401)	Basic counter-unmanned aircraft systems (C-UAS) Awareness Training provides a basic awareness of general UAS threats, types and categories of UAS, and basic reporting requirements. This module is a prerequisite for all other C-UAS modules on JKO.
Basic counter-small unmanned aircraft systems (C-UAS) Familiarization Training. CAC controlled access. (https://jkodirect.jten.mil/html/COI.xhtml?course_prefix=JCSUAS&course_number=A-US1402)	Basic counter-unmanned aircraft systems (C-UAS) Familiarization Training provides a review of the basic UAS awareness and an overview of select C-UAS systems and capabilities.
Basic counter-small unmanned aircraft systems (C-UAS) Installation Training. CAC controlled access. (https://jkodirect.jten.mil/html/COI.xhtml?course_prefix=JCUAS&course_number=-US1403)	Basic counter-unmanned aircraft systems (C-UAS) Installation Training provides a threat overview, information on policies and authorities associated with UAS and counter UAS operations at installations and fixed sites, planning considerations, and reporting procedures.
Basic counter-small unmanned aircraft systems (C-UAS) Tactics, Techniques, and Procedures (TTP). CAC controlled access. (https://jkodirect.jten.mil/html/COI.xhtml?course_prefix=JCUAS&course_number=-US1404)	Basic counter-unmanned aircraft systems (C-UAS) Tactics, Techniques, and Procedures (TTP) provides an overview of the threat and operational environment, general TTP for C-UAS, reporting procedures, and select UAS case studies.
For Counter-Small Unmanned Aircraft Systems (C-sUAS) Policy Training. CAC controlled access. (https://jkodirect.jten.mil/html/COI.xhtml?course_prefix=JCSUAS&course_number=-US1405)	The Policy Training Module is a C-sUAS and UAS Policy training tool and reference developed for all Services' C-sUAS leaders, planners, and operations personnel.
https://www.faa.gov/UAS/commercial_operators/become_a_drone_pilot	To fly a drone under the FAA's Small UAS Rule (Part 107), you must obtain a Remote Pilot Certificate from the FAA. This certificate is needed if running a C-UAS range off military installation.

Planning & Preparing

C. C-UAS Crew Training

There is a significant amount of C-UAS employed within operational theaters and being developed to become programs of record. Many of these systems are complex system-of-systems capabilities. Commanders of units being equipped with these systems must assess manning requirements for the capability and ensure all operators are adequately trained on the systems. In some cases, crews must be identified, trained and certified to effectively employ these capabilities. Commanders must—

- Identify crew size/composition.
- Develop/train crew drills.
- Certify crews to employ the C-UAS capabilities.

Certified crews should be incorporated into unit collective training whenever possible to maintain crew proficiency and ensure the C-UAS mission is fully coordinated, integrated, and synchronized into operations and all relevant staff elements such as targeting cells, ADAM/BAE, intelligence, protection, and CEMA.

D. Combined Arms Training

Commanders focus on critical skills in combined arms training that enable rapid adaptation to the challenges of specific missions and conditions. Those conditions include incorporating realistic battlefield noise, obscuration, loss of key leaders, casualties, limited visibility (night), force-on-force training with an uncooperative enemy force (civilian and media role players), and a variety of terrain. C-UAS will rarely if ever be a stand-alone mission for commanders across the force. Rather, the C-UAS effort must be thought of and trained as a condition of the battlefield. This enables commanders and trainers to thread C-UAS events into existing training. This lessons the overall impact of obtaining limited training time for C-UAS at the collective level. Commanders, subordinate leaders, and any Soldiers or crews assigned C-UAS missions should experience the stressful, near-combat training environment provided at home station and combat training centers.

Where possible, leaders leverage specifically designed training aids, devices, simulators, and simulations to enhance collective task training in the defeat and mitigation of C-UAS threats. Simulations enable adjustments to battlefield C-UAS conditions, ranging from experienced to less experienced training audiences. These training aids, devices, simulators, and simulations devices can be used to train individuals and crews. Training aids, devices, simulators, and simulations devices can be incorporated into collective training to train commanders and staffs during large-scale, highly compressed training events, thus forcing attention on decision making.

Commanders should leverage training opportunities at the various combat training centers. Combat training centers may also be a source for C-UAS TTPs, and lessons learned and facilitate the development, fielding, and testing of new equipment, training, and concepts to enable C-UAS efforts. Commanders are encouraged to contact the combat training centers before home station training events to aid in integrating joint capabilities and new technologies.

Chap 5

Layered Defense (Overview)

Ref: ATP 3-01.81, Counter-Unmanned Aircraft System (C-UAS) Operations (Mar '25), pp. 28 to 39.

I. Layered Defense

Commanders and staff must plan for defense in depth, employing successive layers of C-UAS capability to overlap area and local defense asset coverage areas and mitigate dead spaces. The unit should develop and refine C-UAS training tailored to the expected threat environment. Individual and collective tasks need to be identified, planned, and rehearsed through live, virtual, and constructive training methods and evaluated during training exercises.

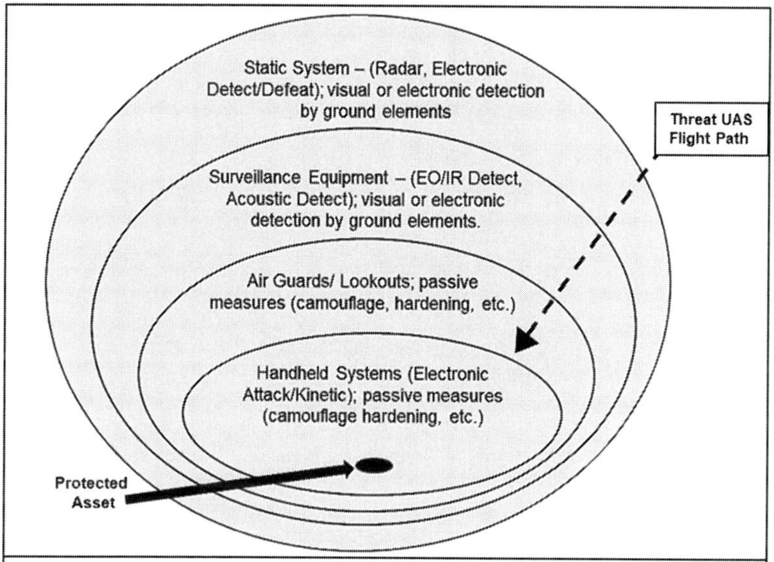

EO	electro-optical	IR	infrared
UAS	unmanned aircraft system		

Ref: ATP 3-01.81 (Mar '25), fig. 4-2. Layered defense unmanned aircraft system example.

A. Principles of Protection

Commanders, staffs, and Soldiers execute C-UAS activities as part of an overall protection plan employing the five principles of protection: comprehensive, integrated, layered, redundant, and enduring.

- **Comprehensive-Protection** is an all-inclusive utilization of complementary and reinforcing protection tasks and systems available to commanders and incorporated into the plan to preserve the force.
- **Integrated-Protection** is unified with other activities, systems, efforts, and capabilities associated with the conduct of land operations to shape security environments, prevent conflict, prevail in ground combat, and consolidate gains across the range of military operations. Integration must occur vertically and

Layered Defense

horizontally with unified action partners throughout the operations process.

- **Layered-Protection** capabilities are deliberately sequenced across multiple domains to eliminate, mitigate, or assume the risk of threat effects.
- **Redundant-Protection** efforts for identified critical vulnerabilities require dedicated primary and alternate protection capabilities.
- **Enduring-Protection** is a continuous activity. Commanders and leaders preserve combat power and reduce the risk of loss, damage, or injury to their formations.

See pp. 4-21 to 4-30 for related discussion of protection planning.

B. ADA Employment Tenets *(See pp. 3-10 to 3-11.)*

Seven ADA employment tenets tie directly into layered defense and optimum protection from UAS. They are mutual support, overlapping fires and overlapping coverage, balanced fires, weighted coverage, early engagement, defense in depth, and resilience.

- **Mutual Support.** Weapons are positioned so that the fires of one weapon can engage targets within the dead zone of the adjacent weapon. For guns, this dead zone is usually small. For missiles, the dead zone may be large, and mutual support is a critical element. Mutual support can also cover nonoperational weapons or weapons at lower states of readiness. Mutual support, when applied to sensors has the same connotation; that is, sensors are deployed to cover the dead zone of adjacent sensors.
- **Overlapping Fires and Overlapping Coverage.** Weapons are positioned so that their engagement envelopes overlap. Because of the many altitudes, directions, and ranges from which the enemy can attack or conduct surveillance operations, defense planners must apply mutual supporting and overlapping fires vertically and horizontally. Overlapping coverage is the positioning of sensors such that their coverage does not leave any seam in the defense that might be used by incoming threats. Overlapping fires and overlapping coverage should be planned during defense design.
- **Balanced Fires.** Weapons are positioned to deliver an equal volume of fires in all directions. This is necessary for AMD in an area where the terrain does not canalize the threat or when the avenues of approach are unpredictable. Balanced fires are a desired characteristic of defense design against cruise missiles and other air threats.
- **Weighted Coverage.** Weapons coverage is combined and concentrated toward the most likely threat air avenues of approach or directions of attack. Based on the tactical situation, a commander may risk leaving one direction of attack unprotected or lightly protected to weight coverage toward another direction. Weighted coverage and balanced fires are not mutually achievable, requiring the defense designer to give up most aspects of one to achieve the other.
- **Early Engagement.** Early engagement generally requires extending the defense away from the defended asset. Sensors and weapons are positioned so they can engage the threat before ordnance release or detection of friendly forces. Early engagements enable destruction of enemy platforms over enemy forces and unoccupied areas, thereby reducing the possibility of friendly collateral damage and fratricide. As with weighted coverage, early engagement is achieved at the expense of balanced fires.
- **Defense in Depth.** Sensors and weapons are positioned so that the threat is exposed to a continuously increasing volume of fire as it approaches the friendly protected asset or force. Defense in depth decreases the probability that attacking missiles, aircraft, or RAM will reach the defended asset or force.

- **Resilience.** Resilience is the quality of the defense to maintain continuity of operations regardless of changes in or unanticipated tactics by enemy air or losses of critical air and missile defense components (FM 3-01). ADA planners must plan for defense design adjustment based on ADA system attrition. Resilience is a key determinant when considering which tenet (or tenets) to use in maintaining the defense.

See pp. 3-10 to 3-11 for an overview and further discussion.

C. Layering Protection Capabilities

Applying these principles increases capability density as a threat approaches the inner ring of defense. A UAS is faced with layered, collocated or integrated systems; increased sensor coverage; and no seams in the area or local defense (all dead spaces are covered).

Layering protection capabilities enables commanders and staffs to deliberately sequence capabilities and mitigation measures across multiple domains to eliminate, mitigate, or assume the risk of threat effects.

- Layer area and local defense assets.
- Coverage increases as the UAS approaches.
- Ensure dead space coverage.
- Co-locate or integrate systems.

II. Layered Defense Approach

A layered defense approach provides strength and depth. Units use their available capabilities to defend personnel and critical assets through layering of capabilities to prevent threat UAS attacks. When the threat cannot be stopped, techniques are applied to reduce the destructive effect of threats and hazards. Individuals are protected at the lowest level by awareness, personal protective equipment, an understanding of the ROE, and fratricide avoidance measures. Additional protection measures are implemented in the surrounding area such as fighting positions, observation posts, and equipment with specific protection capabilities to achieve layered collective protection. Techniques employed by all Soldiers, application of well-planned and executed survivability measures, coupled with application of C-UAS capabilities employed in depth and mutually supporting, are critical to the overall C-UAS effort. *(See Figure 4-2 on p. 5-1.)*

Passive and Active Measures

Commanders, leaders and planners must include the simultaneous application of both passive and active measures during operational planning and employment as part of a layered defense strategy against UAS threats. Leaders must understand active measures help deny the initiative to the enemy or adversary, while the execution of passive defense measures prepares the force against threat and hazard effects and accelerates the mitigation of those effects.

See pp. 5-15 to 5-26 for discussion of passive measures and pp. 5-27 to 5-30 and chap. 6 for active measures in a layered defense.

Employing a successful fixed-site layered defense, or defense in depth, against the UAS threat requires layering area and local defense assets to ensure redundancy, mutual support, and complementary capabilities in depth. A layered defense ensures that coverage increases as the threat UAS approaches protected assets/areas (number and density of C-UAS assets increases as range to protected assets reduces). Leaders must ensure that dead space is covered through mutual support of systems and capabilities; this is accomplished through co-location and integration of complementary supporting systems and capabilities.

See following pages (pp. 5-4 to 5-5) for an overview and discussion of key tasks.

A. Layered Defense (Key Tasks)

Ref: ATP 3-01.81, Counter-Unmanned Aircraft System (C-UAS) Operations (Mar '25), pp. 30 to 31.

Commanders, leaders and planners must include the simultaneous application of both passive and active measures during operational planning and employment as part of a layered defense strategy against UAS threats. Leaders must understand active measures help deny the initiative to the enemy or adversary, while the execution of passive defense measures prepares the force against threat and hazard effects and accelerates the mitigation of those effects.

Key tasks to integrate into plans and combined arms unit training strategies addressing UAS threats are:

- Identify critical assets to be protected.
- Perform cover and concealment.
- Harden unit positions.
- Employ dedicated air guards.
- Perform visual aircraft recognition training.
- Conduct air threat avoidance techniques.
- Conduct deception/decoys and dispersion techniques.
- Establish a security force and quick reaction force.
- Establish an organic early warning sensor network.
- Conduct UAS reporting procedures.
- Conduct downed UAS recovery/battle damage assessment.
- Select UAS defeat mechanisms.
- Perform unit recovery and reconstitution.
- Disseminate UAS threat warning and response option readiness levels.
- Coordinate threat representative UAS red team flights for system operational readiness and training purposes.
- Understand and employ C-UAS systems to maximize range and effectiveness.
- Employ multiple capabilities to provide redundancy, mutual support, and complementary capabilities in depth.
- Determine likely launch sites and controller locations relative to friendly operations in time and space.

For C-UAS systems site surveys, system validation, and employment, identify and consider leveraging-

- ADA assets within the organic unit, higher echelons, and adjacent units.
- EW assets within the organic unit, higher echelons, and adjacent units.
- Available field service representatives.
- Mobile training team support (when available).

Integration of C-UAS Equipment/Systems

Units may be equipped with systems designed specifically to defeat UAS threats. These systems may be employed by an individual, may be employed by a small team of two or three personnel, may be carried to provide coverage over a dismounted element, may be mounted on vehicles, or they may be complex system-of-system with capabilities requiring specialized training and crew certification. When employing these systems commanders and leaders should consider—

Threat

- Identify and understand all potential threat UAS for the area of operations.
- Identify and understand common enemy TTP.

Site Selection

The following is a list of criteria involved in site selection:

- Line of Sight (LOS). Many systems require LOS to a target. Analyze terrain to identify dead spaces and determine coverage strategies. LOS verification can be accomplished through use of 3D mapping tools and visual LOS analysis.
- Multiple systems should be employed to ensure no dead zones.
- Systems are emplaced to maximize LOS.

Background Electromagnetic Environment (EME)

System placement in areas of high electromagnetic density can create a masking effect, preventing UAS detection. Analyze terrain to ensure optimum placement of sensors to reduce signal disruption. Ensure that all C-UAS EW plans and operations are integrated. Consider adjacent equipment compatibility and effects of background EME for the following:

- Radars.
- Communications equipment.
- Counter improvised explosive device/EW equipment.
- Wi-Fi networks.
- Other wireless systems.

Priorities for Protection

- Identify commander's critical assets for protection.
- Focus defense based on priorities. Emplace assets to secure high priority assets first.

Clearance of Fires (for Lethal Capabilities)

- System limitations.
- Range of the system (lethal and non-lethal).
- Atmospheric and terrain effects.
- Equipment optimization.
- Jamming frequencies.
- System library limitations.
- Other radio frequency (RF) transmitting systems (vehicle Wi-Fi, remote control doors, etc.) that may appear as unmanned aircraft signal detections.

B. Commander's Protection Priority List (See pp. 4-28 to 4-29.)

To plan and execute a successful layered defense strategy, commanders and staffs must first identify the critical assets being protected. It's accomplished through criticality assessments, vulnerability assessments, and threat and hazard assessments. The results of these assessments are the commander's protection priority list. This provides the prioritized list of assets to be defended.

C. The First Layer of Defense

Countering the threat is not focused on solely defeating the threat UAS; much of the activities to counter the threat are centered on mitigation efforts to lessen or eliminate the impact of a threat UAS mission against U.S. forces. These efforts are conducted as part of a unit's overall protection plan; they are not stand-alone counter UAS plans and activities. Countering the threat requires individual Soldier actions and survivability activities, blended with employment of equipment and systems ranging from organic weapons, to specialized handheld EW and lethal devices, to complex systems designed specifically to detect, identify, and defeat UAS. Not all units will be equipped with equipment designed specifically to counter the threat.

The first layer of defense includes all the individual Soldier skills and tasks required for personal protection and defense in the immediate area. Next, survivability tasks are employed to harden critical assets to mitigate the effects of an attack by UAS or other enemy capability. These tasks and activities protect against many risks present in the operational area not just UAS threats. These tasks and activities are conducted as part of a holistic protection plan. Considerations and requirements include-

- Individual Soldier task proficiency.
- Trained proficiency in attack and emergency response battle drills.
- Survivability measures applied to critical assets.
  - Camouflage.
  - Cover and hardening.
  - Concealment
- Use of decoys and deception to fool adversaries.
- Compartmentalization to mitigate threat impacts.
- Emplacing observation posts.
- Bunkers and fighting positions.
- Conducting combined arms for air defense.
- Conducting direct fire attacks against airborne assets.

Observation Positions (OPs)

Commanders and leaders must identify where observation positions combined with multiple C-UAS capabilities and systems are emplaced in a layered and integrated manner to provide maximum protection from a threat UAS. A thorough terrain analysis should be conducted to determine lines of sight and associated dead space that must be accounted for in the overall plan.

When C-UAS systems with EW capability are present, a thorough analysis and survey of the area of operation should be conducted prior to system emplacement. Special emphasis should be placed on understanding the background EME of the AO and optimizing equipment for operation, ensuring systems have clear line of sight and are not placed in defilade or other areas that may affect RF propagation.

III. Countering the Threat (by Warfighting Function) *(See also p. 5-7.)*

Ref: ATP 3-01.81, Counter-Unmanned Aircraft System (C-UAS) Operations (Mar '25), pp. 19 to 20.

Countering the UAS threat requires a combined arms approach that leverages capabilities across many disciplines. **Commanders must consider a wide range of capabilities to counter the UAS threat including active measures from direct fires, EW, and aviation as well as passive survivability and protection measures.** Examples of contributions for countering UAS threats across the combined arms force include:

Fires forces-

- Integrate surveillance capabilities.
- Identify targets.
- Facilitate and conduct engagements.
- Employ EA and/or kinetic capabilities to defeat or destroy UAS threats.
- Support direct action against threat elements to prevent UAS threats prior to launch.

Protection elements assist commanders to-

- Identify critical capabilities, areas, information and prioritize protection resources.
- Integrate cover, concealment, and deception (CCD) and C-UAS capabilities to build a layered defense.
- Employ electromagnetic protection (EP) capabilities to protect critical networks and systems.

Intelligence forces-

- Identify UAS threat capabilities and associated tactics.
- Collect and analyze threat sustainment networks to enable direct action against UAS systems, logistics and manufacturing/test facilities prior to launch.
- Conduct exploitation activities to gain critical intelligence from captured threat UAS.
- Conduct electromagnetic support to exploit UAS information gathered from the electromagnetic spectrum (EMS) and employ EA capabilities to defeat or destroy UAS threats.

Maneuver forces-

- Contribute to surveillance and identification of UAS threats.
- Conduct direct action engagements of threat systems employing ground and aviation assets.
- Employ EA capabilities to defeat or destroy UAS threats.
- Conduct established recovery procedures to safely recover threat UAS for exploitation.
- Employ survivability and protection measures across their formations.

Command and control used to-

- Integrate capabilities to enhance situational awareness and understanding.
- Synchronize and converge all elements of combat power required to combat UAS threats at echelon.
- Enable decision-making.

IV. Building the Defense for Fixed Sites

Ref: ATP 3-01.81, Counter-Unmanned Aircraft System (C-UAS) Operations (Mar '25), pp. 33-35.

When building a defense in depth leaders must first identify and protect critical assets. This requires employing survivability techniques to protect these assets within the perimeter of the base of operation (see figure 4-3). Additional survivability measures will be required if observation points and fighting positions are placed outside the base perimeter. Blast mitigation and stand-off should always be considered when employing survivability measures.

BDOC base defense operations center ECP entry control point

Ref: ATP 3-01.81 (Mar '25), fig. 4-3. Fixed site.

Once the maximum detect range is identified, the maximum defeat range must be established. The maximum defeat range will vary by system but must be identified in order to prevent threat UAS from completing its assigned task. Security forces positions (air guards/observation posts) should be identified and prepared to visually detect threat UAS during hours of day light and limited visibility. These positions should be placed to mitigate any potential dead space where detection equipment capabilities are limited (see figure 4-4).

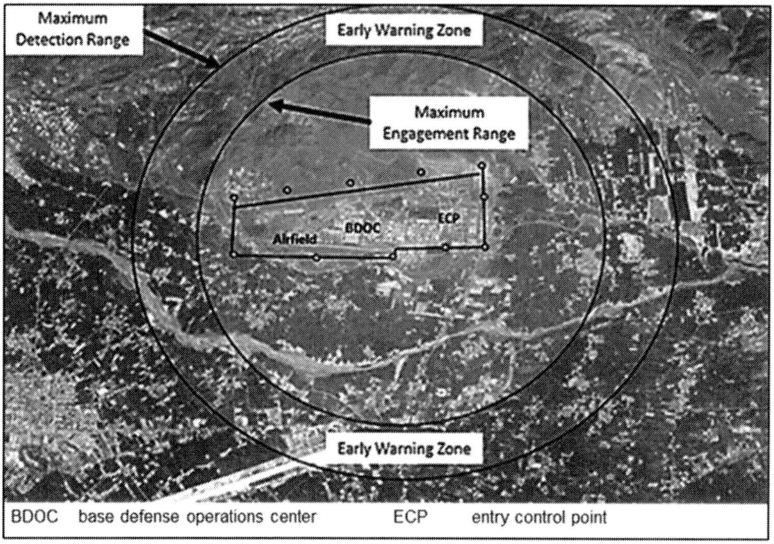

Ref: ATP 3-01.81 (Mar '25), fig. 4-4. Detection and engagement ranges.

If handheld or man-packed C-UAS systems are available, these can be distributed to observation posts to cover dead space or on critical avenues of approach (see figure 4-5). Leaders should consider emplacing these systems on elevated positions to prevent masking from terrain and structures. Commanders and leaders should consider emplacing some portable systems at static posts to defend critical assets. Others should be assigned to mobile elements to allow for response to any location on the site. Units may also be equipped with optical persistent surveillance systems. These systems are omnidirectional and can be slewed to give visual confirmation to an incoming UAS.

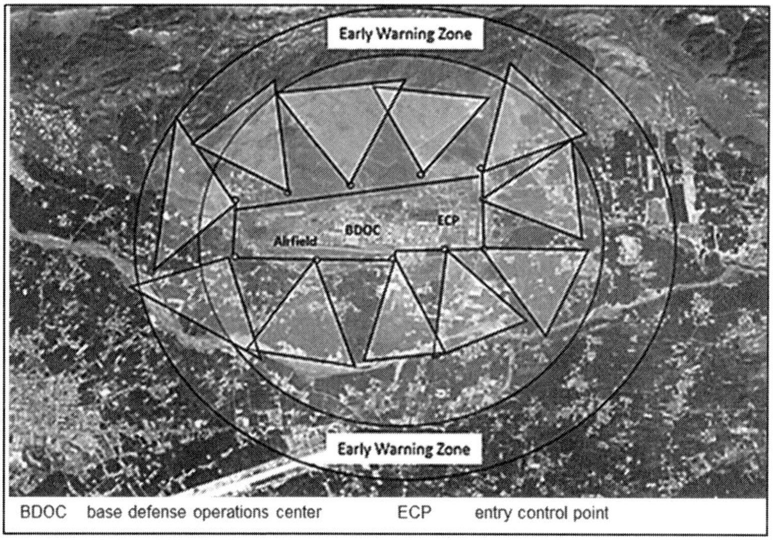

Ref: ATP 3-01.81 (Mar '25), fig. 4-5. Observation points.

Layered Defense

V. React to Threat UAS

Understanding the capabilities and limitations of UAS can increase a Soldier's ability to react and defeat a threat. Reacting to a threat UAS should include reporting to the operations center and security personnel if the UAS does not pose an immediate threat. Defeat does not equate lethal means; however, it is an option. Other defeat solutions could be limiting a surveillance threat from gaining information or following the air path of the UAS to the operator.

Editor's note: In addition to "React to Threat UAS" (pp. 5-10 to 5-14), ATP 3-018.81 also outlines steps in terms of "Active Defense Measures" (see pp. 5-27 to 5-30).

React to Threat UAS

A Detect

B Track and Identify

C Defeat

D Report

Techniques for reacting to UAS include:

- Treat unidentified UAS as enemy and/or with hostile intent.
- Use visual, electromagnetic, or procedural means to detect, track, and identify threat UAS.
- Initial spot reports must be sent to higher headquarters to facilitate the identification process.
- Once a UAS is identified, simultaneous processes begin thereafter.
- Commanders have discretion to initiate a quick reaction force and begin the target defeat decision.
- All target defeat decisions must be given by an engagement authority and receive airspace deconfliction/clearance.
- Dependent on the mission and the best method of engagement, lethal or non-lethal engagement responses come from fixed-site or mobile systems.
- After engagements have ceased, the battlespace owner conducts a cordon and search to locate the down UAS. The battlespace owner then submits a 9-line report to request explosive ordnance disposal (EOD), and support to render safe the downed UAS.
- Once EOD renders safe any explosive hazards, weapons intelligence or other analysis can commence on the downed UAS.
- All reports must be sent within the respective theater's reporting timeframe.

A. Detect *(See also p. 5-28.)*

Detection is priority in counter-UAS operations. Smaller platforms are used for supportive functions such as detection, targeting assistance, or surveillance prior to an actual attack. There are three methods of identification: visual, procedural, and electromagnetic.

React to Threat UAS (Flow Chart)

Ref: ATP 3-01.81, Counter-Unmanned Aircraft System (C-UAS) Operations (Mar '25), p. 36.

Understanding the capabilities and limitations of UAS can increase a Soldier's ability to react and defeat a threat. Reacting to a threat UAS should include reporting to the operations center and security personnel if the UAS does not pose an immediate threat. Defeat does not equate lethal means; however, it is an option. Other defeat solutions could be limiting a surveillance threat from gaining information or following the air path of the UAS to the operator.

Figure 4-6 is a simplified flow chart showing the process from UAS detection to submission of the final report to higher. Leaders must be aware that local standard operating procedures may dictate other steps and requirements.

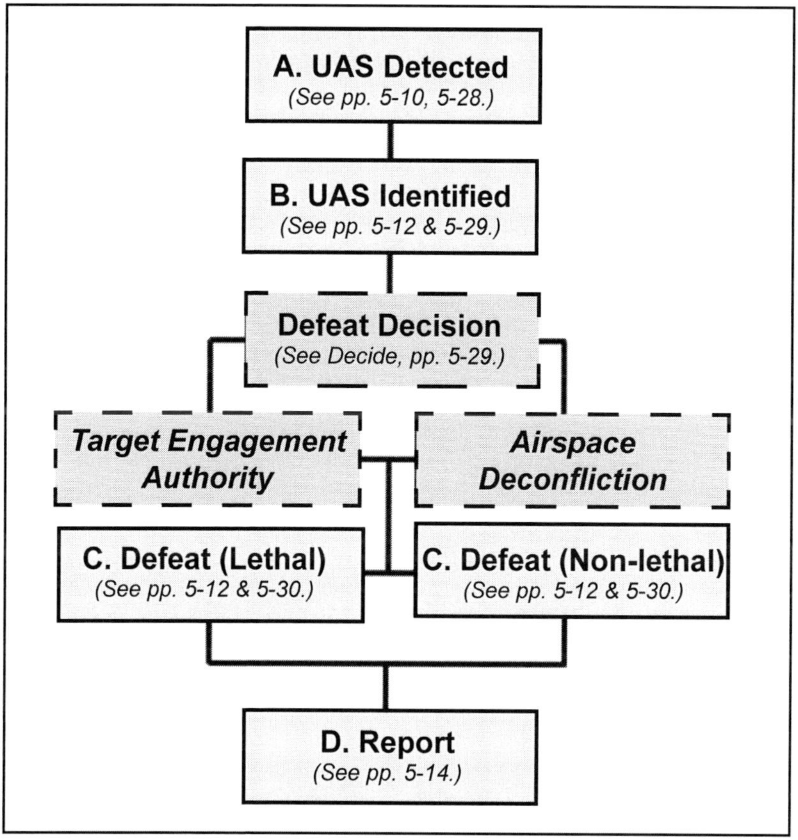

Ref: Adapted from ATP 3-01.81 (Mar '25), fig. 4-6. Unmanned aircraft system reporting flow chart. (See pp. 6-6 to 6-7 for an expanded view.)

Editor's note: In addition to "React to Threat UAS" (pp. 5-10 to 5-14), ATP 3-018.81 also outlines steps in terms of "Active Defense Measures" (see pp. 5-27 to 5-30).

Visual (See p. 5-20.)

Visual Aircraft Recognition (VACR) is an established method of visually identifying the type of aircraft in flight. Visual aircraft recognition can distinguish the UAS physical form based on its configuration. Multiple configurations exist; a few examples are rotary-wing, fixed-wing, hybrid, tiltrotor, or a combination of all.

Procedural

Procedural detection depends on UAS behavior in flight. Airspace coordinating measures that are not followed, point of origin, erratic behavior, and loitering above important locations are all behaviors that illustrate non-friendly or hostile behavior. Rotary-wing platforms can hover and collect intelligence, while Fixed-wing platforms can make several passes to achieve the same fidelity as a rotary-wing. These behaviors are indicative of intelligence-gathering platforms and can be likely be categorized as meeting hostile intent of intelligence platforms and can be detected as unordinary with hostile intent.

Electromagnetic

Electromagnetic detection is generally the fastest detection method. It requires special equipment to detect the operational controls of the UAS. Sensors can identify targets based upon radar cross-section, infrared signatures, and other electromagnetic signatures that UAS exhibit within the EMS.

B. Track and Identify (See also pp. 5-28 to 5-29.)

Tracking UAS is possible with several systems and personnel. System integration with multiple sensors is the best method to maintaining situational awareness of UAS. Tracking is achieved by multiple means of active and passive sensing measures and integrating them with current C2 infrastructure and systems. Some passive sensor systems receive RF downlink cues from the UAS while active sensor systems will use radar to track the UAS. Planning considerations for tracking include line-of-sight limitations, terrain obstruction, urban environment obstructions, and the operational RF environment. Many frequencies are shared with UAS and must be identified before classifying as hostile or using mitigation techniques.

See figure 4-7 (facing page), C-UAS Time/Distance Engagement Sequence Planning Considerations, for a linear depiction of this process.

"Wings, Engine, Fuselage, Tail" (WEFT)

After initial detection, tracking and identification will follow. There are multiple ways to identify UAS. A common method is the "wings, engine, fuselage, tail" (WEFT) identification method. The presence of wings or rotors determine the type of UAS. Most UAS are propeller driven, from either a single propeller for propulsion with wings or multiple props mounted on top of arms to provide hover capability. The fuselage will maintain and house internal components. In addition, the presence of a tail for assisting in maneuvering can identify the type of UAS. Through joint processes and procedures, identification is achieved. As mentioned before, UAS that do not adhere to airspace control measures can then be deconflicted and identified as a friendly with an emergency, an unknown with hostile intent, or hostile aircraft.

C. Defeat (See also p. 5-30.)

Defeat techniques begin once airspace deconfliction and target engagement authority has been passed down to the tactical level. To confirm the identity of the UAS and prevent fratricide, multiple processes and procedures must be initiated. Depending on the engagement method, area RF deconfliction may be required. Depending on the frequencies, there may be opportunities where operational RF spectrums overlap with the control frequency of the UAS. Key leaders must determine and mitigate frequency fratricide during defeat measures. Defeat measures can follow non-lethal or

C-UAS Time/Distance Engagement Sequence Planning Considerations

Ref: ATP 3-01.81, Counter-Unmanned Aircraft System (C-UAS) Operations (Mar '25), p. 38.

Figure 4-7 shows C-UAS time/distance engagement sequence planning considerations, for a linear depiction of this process.

- Commanders, staff, and service members execute C-UAS layered defense activities as part of an overall protection plan employing the five principles of protection: comprehensive, integrated, layered, redundant, and enduring *(See pp. 5-1 to 5-2.)*
- Using joint force employment tenets of early engagement and defense in depth will allow the maneuver greater standoff against threat UAS; however, such will shorten the timelines and distance associated with detection through defeat/engagement sequence depicted below.
  - **Early Engagement**: Sensors and weapons are positioned so that they engage the threat before ordnance release or detection of friendly forces.
  - **Defense in Depth**: Sensors and weapons are positioned so that the threat is exposed to a continuously increasing volume of fire as it approaches the friendly protected asset of force.
- Target Identification and engagement authorities, C-UAS defense weapons control statuses, and ROE are subject to theater-specific conditions and orders
- Threat UAS detection will trigger tactical SOP battle drill/reactions.

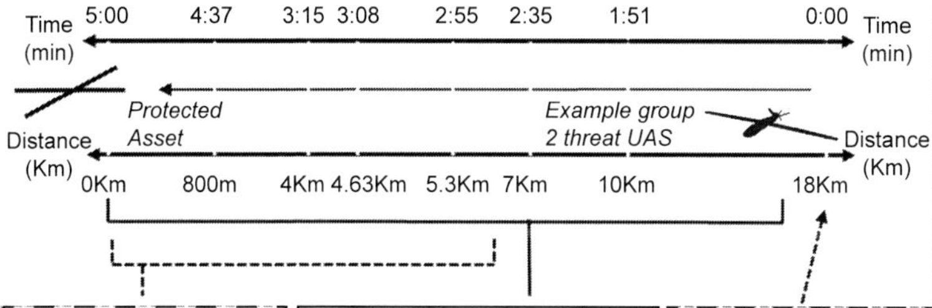

C. Defeat *(p. 5-12.)*
- Target engagement authority delegation
- Airspace/radio frequency deconfliction
- Lethal/non-lethal response
- Weapons control statuses
- ROE
- Report

B. Track & Identify *(p. 5-12.)*
- Via radar/electromagnetic, visual aircraft recognition or procedural methods
- Via "wings, engine, fuselage, and tail" WEFT method
- Via friendly procedural airspace control measures
- Target identification authority delegation

A. Detect *(p. 5-10.)*
- Via radar/electromagnetic, visual aircraft recognition or procedural detection
- Defense in depth
- Early engagement
- Layered defense

Time/distance based on example group 2 threat traveling at 4.3Km/min (160 knots) and rotational detection range of 18Km. If example group 2 threat traveled at 7.7Km/min (250 knots), the UAS would reach/strike defended asset 2:18 mins from 18Km detection.

C-UAS	counter-unmanned aircraft system	**SOP**	standard operating procedure
Km	kilometer	**UAS**	unmanned aircraft system
Min	minute	**ROE**	rules of engagement

Ref: Adapted from ATP 3-01.81 (Mar '25), fig. 4-7. Counter-Unmanned Aircraft System time/distance engagement sequence planning considerations. (See pp. 6-6 to 6-7 for an expanded view.)

lethal procedures. During non-lethal responses, it is crucial to continue jamming until the craft is inoperable. After lethal or non-lethal responses, and once the UAS has shown no control-link, EOD must render the UAS safe. Once the UAS is rendered safe, the craft may be exploited for weapons intelligence or other analysis.

D. Report *(See pp. 5-24 to 5-25.)*

Reporting requirement timeframes are based on which unit level is submitting the report. It is expected that company, battery, troop level spot reports and engagement reports are sent up according to their theater specific timelines. The initial spot report drives the UAS battle drill. Without the important information captured, the chain of command cannot make essential decisions based on UAS payload, intent, type, and possible endurance. Fixed-wing UAS have more endurance than rotary-wing aircraft. This essential piece of information can help decide what types of missions the platform can conduct. If the type of UAS payload can be identified, there may be ample time to warn friendly personnel about possible fragmentary, anti-personnel, or surveillance payloads that may affect operations.

Downed UAS

Soldiers may encounter a downed UAS. When this occurs, Soldiers will treat the UAS as an improvised explosive device with active ISR capabilities. The best practice for these situations is to conduct the five "C's" (confirm, check, clear, cordon, control) while maintaining visual contact from a safe distance of the downed UAS:

- Confirm-Use optics for stand-off, look for indicators/presence of suspicious items (does it look modified, is there a payload?).
- Check-Conduct 5/25/200-meter check for dropped payload or additional grounded UAS.
- Clear-Clear non-essential personnel from the immediate area.
- Cordon-Cordon off the area to preserve the scene for exploitation and provide protection/security from command/victim-initiated improvised explosive device.
- Control-Allow only authorized access to the site, maintain visual LOS to the UAS.
- Report status to the operations center.

Chap 5

(Layered Defense) I. Passive Defense

Ref: ATP 3-01.81, Counter-Unmanned Aircraft System (C-UAS) Operations (Mar '25), pp. 41 to 50.

Passive air defense are measures that units may employ to safeguard its personnel and equipment from air threats. Units may be in a situation where they are vulnerable to attack or taking unnecessary casualties. In cases where the enemy aerial platforms are outside the range of the unit's weapons, a unit's most lucrative option can be to disperse and seek cover. Commanders must prepare their units to actively engage air threats if attacked. The decision to engage air threats should include consideration of the unit's mission and tactical situation.

Passive Defense

Combined arms forces and small units should be ready to employ passive defense measures to protect themselves from detection, observation, and attack. Passive air defense is all measures, other than active defense, taken to minimize the effectiveness of hostile air and missile threats against friendly forces and assets (JP 3-01). Passive defense measures decrease the effectiveness of enemy attacks using UAS against friendly forces and their assets. Units must train their forces to avoid direct and indirect attacks from air threats. Two types of passive measures are damage-limiting and attack avoidance measures.

Commanders employ passive air defense measures to improve their units' survivability by increasing the likelihood of not being detected and targeted (attack avoidance) from the air and by mitigating the potential effects (damage-limiting) of an air attack (ATP 3-01.8). ADA units provide threat identification and site alarming by disseminating early warning. Units will not always have a dedicated ADA element available to prevent air attacks. Conducting passive air defense measures routinely may decrease the chance of detection or limit the damage from an air attack.

The air defense coordination/management officer in the brigade combat team ADAM/BAE is responsible for ensuring that all brigade combat team elements (staff, organic units, and attached units and personnel) understand and employ the necessary passive defense measures to mitigate the effectiveness of enemy UAS. Passive air defense measures are the first line of defense against UAS. Based on the unit's assigned mission and ROE, the commander will have to decide whether to engage any approaching hostile aerial platform.

I. Passive Defense Tasks

Ref: ATP 3-01.81, Counter-Unmanned Aircraft System (C-UAS) Operations (Mar '25), pp. 41 to 42.

When planning damage-limiting and attack avoidance measures units should consider, and subsequently train their forces, on the following passive defense tasks:

Operate at night or during limited visibility. Affects threat UAS observation but also limits the unit's operational support. Organizations must assume that the enemy has the capabilities to conduct limited visibility or night operations. Units at all levels need to ensure proper training is conducted during hours of limited visibility. Practice light restrictions and discipline.

Establish an early warning network. Establish an air defense early warning network using the radios and digital networks that are assigned to the unit command post. Ensure that supported units have trained and assigned personnel to always monitor these networks and radios.

Disseminate early warning of air threats to the lowest echelon. Early warning from the highest to the lowest echelons is essential to countering the UAS threat. Upon detection and classification of enemy UAS, the brigade must relay alert information on locations, altitudes, and time to surrounding ground forces. This information will be vital to the ground forces protection and the possible capture of the UAS and its ground platform. Automated early warning of UAS is disseminated throughout the brigade's common user network down to the lowest echelons.

Practice good operational security. Operational security is an essential part of the unit's planning process. Commanders and leaders must always enforce their units' operational security measures. A unit must implement countermeasures immediately upon detection of an enemy UAS. Practicing good operational security will safeguard critical information and reduce the unit's vulnerabilities to enemy collection efforts. Developing and employing basic countermeasures and TTP, such as limiting the number of times friendly forces are exposed to possible aerial observation or minimizing organized unit activities until the threat is neutralized or destroyed, should be considered.

Use emission control to limit electromagnetic and acoustic footprints. The controlled use will also mitigate detection by enemy sensors, mutual interference among friendly systems, and enemy interference with the ability to execute a military deception plan.

Unit dispersion. Disperse assets to minimize detection and damage if attacked. Dispersion may be the best damage-limiting measure. Proper dispersion of a unit and equipment lessens target density and reduces the lethal effects of the ordnance used against that organization. Area weapons become less effective when units disperse.

Maintain vigilance. Friendly forces must assume they are always vulnerable to enemy targeting attempts. This is especially true when conducting convoys for troop movements or performing supply actions and as they move through open areas or concentrate at choke points along the convoy route.

Survivability techniques. These are specific groups of tasks that enhance survivability. When existing terrain features offer insufficient cover and concealment, altering the physical environment to provide or improve cover and concealment enhances survivability (engineers are typically required for these tasks). *(See following pages for further discussion of survivability.)*

II. Survivability *(See pp. 1-30 to 1-31.)*

Survivability (as a Dynamic of Combat Power)

Combat power is the means for conducting warfare and consists of five dynamics -- leadership, firepower, information, mobility, and **survivability**. Each dynamic is interdependent and enabled by the protection warfighting function. Survivability is both a primary task of the protection warfighting function and a dynamic of combat power. In both uses, survivability is a quality or capability of military forces that permits them to avoid or withstand hostile actions or environmental conditions while retaining the ability to fulfill their primary mission. As a primary task of the protection warfighting function, survivability includes supporting tasks that enhance a formation's ability to avoid or withstand an enemy's actions by providing or improving camouflage, cover, and concealment; controlling electromagnetic emissions; and increasing personal protective posture (adding small arms protective insert plates, adding mission-oriented protective posture, or hardening command facilities).

- FM 3-0, Operations (Mar '25), pp. 2-3 to 2-6.

These are specific groups of tasks that enhance survivability. When existing terrain features offer insufficient cover and concealment, altering the physical environment to provide or improve cover and concealment enhances survivability (engineers are typically required for these tasks). Similarly, using natural or artificial materials such as camouflage may confuse or mislead the enemy or adversary. Together, these are called survivability operations-those protection activities that alter the physical environment by providing or improving camouflage, cover, and concealment.

See following pages (pp. 5-18 to 5-19) for an overview and further discussion.

Camouflage and Concealment Techniques

Ref: ATP 3-01.8, Techniques for Combined Arms for Air Defense (Jul '16), pp. 3-1 to 3-6.

Attack avoidance is taking the actions necessary to avoid being seen by the enemy to include concealment and, lacking concealment, camouflage. What can be seen can be hit, and if you cannot be seen, the probability of being hit diminishes to near zero. The techniques used for concealment from aerial observation are the same as used for concealment from ground observation

There are three concealment principles employed (site, discipline, and construction) to eliminate the factors of recognition. Site refers to positioning your unit in the most advantageous area in which to hide a man, equipment or an activity. Discipline success in any attack avoidance effort is the strict maintenance of concealment discipline by both the unit and by the individual Soldier. All activities should be avoided that change the appearance of an area or reveal the presence of military equipment. Laxness and carelessness will undoubtedly reveal a position. Tracks, spoil, and debris are the most common signs of military activity, which indicate concealed objects. Ensuring wheeled/track/foot movement follows existing paths, roads, fences, or natural lines in the terrain pattern. Do not end exposed routes at a position, but extend them to another logical termination. If practical, tracks should be brushed out, camouflaged, or covered. Spoil and debris must be covered or placed to blend with the surroundings. Artificial camouflage is added when the terrain and natural vegetation are such that natural concealment is not possible.

While concealing a position it is best to use natural construction materials to blend-in with the surrounding terrain, which augments this type of concealment. Concealment techniques are hiding; blending, and disguising.

See following pages (pp. 5-18 to 5-19) for an overview and further discussion.

Survivability Techniques

Ref: ATP 3-01.81, Counter-Unmanned Aircraft System (C-UAS) Operations (Mar '25), pp. 42 to 43.

Survivability (a Dynamic of Combat Power) *(See pp. 1-30 to 1-31.)*

Combat power is the means for conducting warfare and consists of five dynamics -- leadership, firepower, information, mobility, and **survivability**. Each dynamic is interdependent and enabled by the protection warfighting function. Survivability is both a primary task of the protection warfighting function and a dynamic of combat power. In both uses, survivability is a quality or capability of military forces that permits them to avoid or withstand hostile actions or environmental conditions while retaining the ability to fulfill their primary mission. As a primary task of the protection warfighting function, survivability includes supporting tasks that enhance a formation's ability to avoid or withstand an enemy's actions by providing or improving camouflage, cover, and concealment; controlling electromagnetic emissions; and increasing personal protective posture (adding small arms protective insert plates, adding mission-oriented protective posture, or hardening command facilities).

- FM 3-0, Operations (Mar '25), pp. 2-3 to 2-6.

These are specific groups of tasks that enhance survivability. When existing terrain features offer insufficient cover and concealment, altering the physical environment to provide or improve cover and concealment enhances survivability (engineers are typically required for these tasks). Similarly, using natural or artificial materials such as camouflage may confuse or mislead the enemy or adversary. Together, these are called survivability operations-those protection activities that alter the physical environment by providing or improving camouflage, cover, and concealment. See figure 5-1 for examples of survivability techniques.

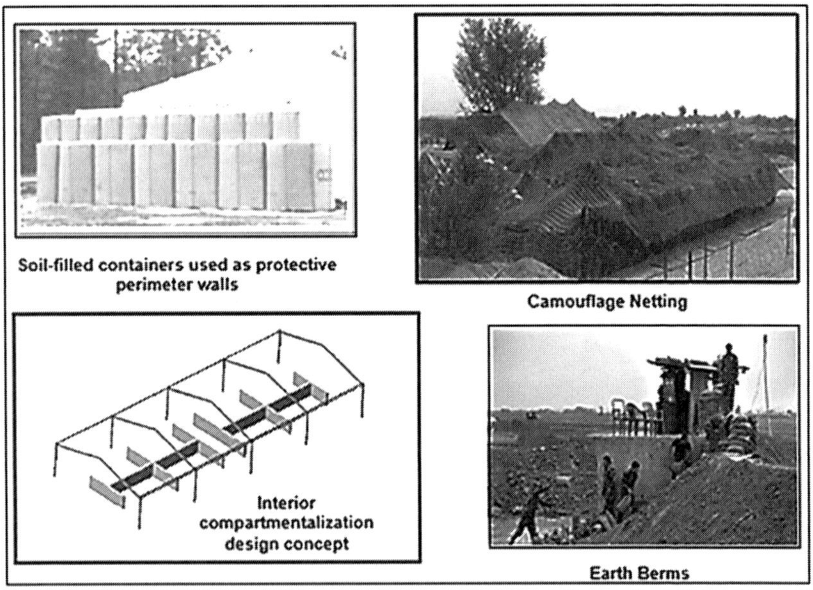

Ref: ATP 3-01.81 (Mar '25), fig. 5-1. Examples of survivability techniques.

Survivability focuses on using:

Camouflage
This is the use of natural or artificial materials on personnel, objects, and tactical positions to confuse, mislead, or evade the enemy.

Concealment
This makes use of terrain and other natural or man-made features to protect from observation and surveillance. Concealment is used to reduce the factors of recognition.

Hiding
Hiding is the concealment of an object by some form of natural or man-made screening. Hiding screens is a method to screen a target from enemy sensors. This can include the use of obscurants such as optical and noise reducing measures to limit the glare or noise of equipment. Place mud on headlights and use camouflage nets to obscure the glare of windshields. Use noise blocking barriers and foam to lower the noise of operating equipment.

Blending
Blending is the arrangement or application of camouflage materials on, over, and around the object so that it appears to be part of the background.

Disguising
This is the application of materials on a target to mislead the enemy as to its identity. Clever disguises can often mislead the enemy concerning identity, strength, and intention, and may draw fire away from real assets.

Disrupting
This includes actions or techniques that alter or eliminate regular patterns and target characteristics.

Decoys and Deception
Using such things as decoys to set up false locations, with smoke to draw attention away from an operation, or emitters and emulators to confuse collection activities can conceal unit activities from enemy detection.

Cover
Cover and hardening use protective construction and overhead cover to provide damage-limiting cover for friendly forces and equipment. The hardening and fortifying of cover will limit the threat UAS ability to visually see and limit the damaging effects of an aerial attack. Hardened overhead cover mitigates the effects of improvised munitions delivered by threat UAS on assets such as fighting positions; operation centers; petroleum, oil, and lubricant sites; ammunition supply points; critical communication nodes; and other assets deemed critical by the commander.

Compartmentalization
Compartmentalization refers to dividing a large, facility, area, or individual assets into smaller components. This can be accomplished within a smaller footprint through use of ballistic-resistant walls. It can also refer to separation of components/elements across a wider area to reduce risk from blast/impact.

III. Air Guard

Assigned Soldiers need to be vigilant with eyes on the horizon. They will perform actions such as search and scan techniques for approaching threat UAS while observing their assigned sectors (see figures 5-4 and figure 5-5 for air guard scan techniques). Early warning is key to alerting formations of any possible air threats. Sectors must cover likely avenues of approach for threat UAS. Soldiers can potentially engage threat UAS if they are equipped with dismounted C-UAS assets in accordance with ROE and weapon control status.

Visual Aircraft Recognition (VACR)

A unit's area of operation can be the deciding factor for types of UAS that will be seen in the area. Increased UAS capabilities have increased the demand for visual aircraft recognition and understanding is critical at the Soldier level. Due to their small radar cross section and low/slow/kinematic profile, it is difficult to detect and identify using radar systems. VACR helps distinguish between UAS primary physical forms (rotary-wing, fixed-wing or hybrid) to assist in combat identification. The traditional method of aircraft identification is wings/engine/fuselage/tail. Wings/engine/fuselage/tail is the accepted system of aircraft identification based on the aircraft's individual features. Wings/engine/fuselage/tail is a memory aid used in training VACR. UAS must be detected, recognized and then identified at the farthest range possible. Early detection aids leaders with making timely decisions to report and possibly engage UAS. The farther out an aircraft is positively identified, the more time is given to make engagement decisions. Estimation of search limits/visual technique (see figure 5-4), vertical scan and search (see figure 5-5), horizontal search and scan (figure 5-6) are a few techniques that can be utilized to detect and identify UAS.

See p. 5-12 for discussion of the three means of detection: visual, procedural, and electromagnetic.

Reporting threat UAS activity should include an estimate of the threat(s). This information should include reference points based on known locations, such as the air guard's position, a friendly unit's grid coordinates, or major terrain feature. Using this method can assist others in accurate locating and targeting the threat UAS. The air guard reports the approximate distance, time, duration, size, estimated elevation, and direction the UAS was heading when detected. If a low flying UAS is spotted, the air guard should immediately be on alert to identify any possible threats in close proximity, to include enemy ground troops or control teams. When spotting a UAS, compromise of the air guard's position is possible.

Standardized Reporting Formats *(See pp. 5--24 to 5-25.)*

Reporting of a threat UAS should utilize a standard reporting format. This ensures all Soldiers understand the process and information required to conduct an enemy spot report. Routine reporting allows units time to quickly adjust force protection measures and coordinate immediate response to UAS threats. Unit C-UAS TTPs must be tailored to the expected threat environment in one's AO. That means creating a TTP that focuses on setting priorities on specific lines on a spot report above other. Units can also tailor their reports to address the UAS situation by adding the reported UAS's approximate speed, heading, and altitude in line 4. When reporting UAS activities the reporting element may also add UA platform characteristics into line 3 instead of just the reported platform's size alone. Those characteristics may include if the UA is a quadcopter, fixed-wing, or single rotary platform.

See pp. 5--24 to 5-25 for example reporting formats. Table 5-1, for an example of a spot report format for threat reporting. Table 5-2 provides an example of the 16-line report that must be completed within 24 hours. Table 5-3 provides an example of the engagement report that must be completed within 48 hours of a UAS engagement.

A. Air Guard Positioning

Ref: ATP 3-01.81, Counter-Unmanned Aircraft System (C-UAS) Operations (Mar '25), pp. 44-45.

Air guards can be the first line of defense when detecting and coordinating enemy UAS location and direction. The commander of each unit typically selects them, but selection can be delegated down to other leaders. Air guards are responsible for spotting aerial threats within close proximity to the unit's location They search for aerial targets using search and scan techniques.

Position air guards in in-depth or linear positions. Air guards placed in-depth can be configured as completely vehicle-mounted air guards, dismounted air guards forward of the vehicle, or a combination of the two. In-depth air guard placement allows the company to observe its entire sector. This method works well when the company has assigned platoons sectors that contain terrain that facilitates multiple UAS avenues of approach, is heavily vegetated, or contains civilian infrastructure that provides the enemy with cover and concealment. In-depth placement allows for redundancy in observation and better interlocking coverage of the sector. Linear placement is effective when the enemy is not moving and provides optimum observation of the enemy.

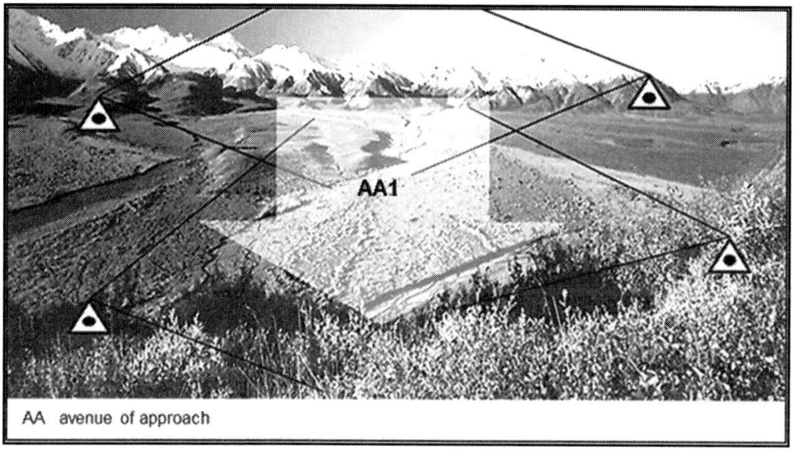

Ref: ATP 3-01.81 (Mar '25), fig. 5-2. Positioning in-depth.

Ref: ATP 3-01.81 (Mar '25), fig. 5-3. Positioning linear.

B. Air Guard Scan Techniques

Ref: ATP 3-01.81, Counter-Unmanned Aircraft System (C-UAS) Operations (Mar '25), pp. 48 to 50.

Understanding how to apply C-UAS observation techniques enhances reaction and reporting time, thus increasing survivability. Air guard proficient in applying terrain analysis and map reading techniques may be able to determine probable enemy launch locations using the UAS's direction of approach.

Commanders and leaders must conduct rehearsals to train and educate air guard personnel on their specific responsibilities and standard operating procedures for reporting the UAS threat. Rehearsals will provide company commanders the ability to make any needed changes and verify the proficiency of personnel to conduct air guard techniques.

Soldiers should have the ability to conduct all operations under all conditions: day, night, and limited visibility. They should be equipped with the necessary optical gear to perform search and scan techniques to reduce the enemy's ability to evade detection. The following figures illustrate search limits, search, and scan methods. Refer to TC 3-01.80 for details on search and scanning techniques.

When searching and scanning for UAS, do not limit the search too near the horizon and miss higher-flying aircraft or search too high above the horizon and miss lower flying targets. The correct upper limit of search is 20 degrees above the apparent horizon (see figure 5-4).

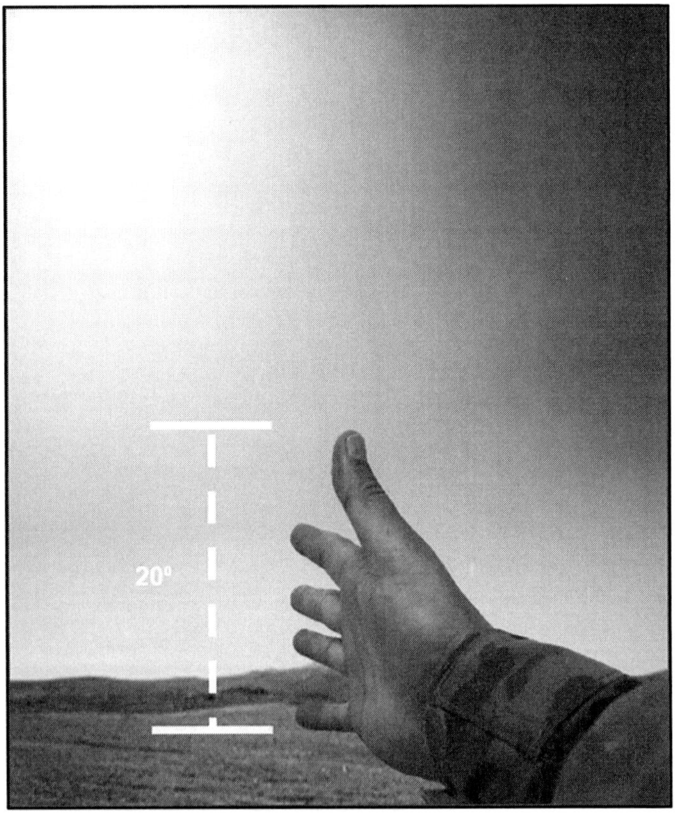

Ref: ATP 3-01.81 (Mar '25), fig. 5-4. Method of estimating 20 degrees.

Layered Defense

Vertical Scan

The vertical scan is the technique used to optimize vision for finding air threats (see figure 5-5). By using quick eye movements upward toward the sky, then downward to the horizon and continuing across the terrain, the air guard should scan using the same pattern to approximately 20 degrees below the horizon to compensate for all UAS.

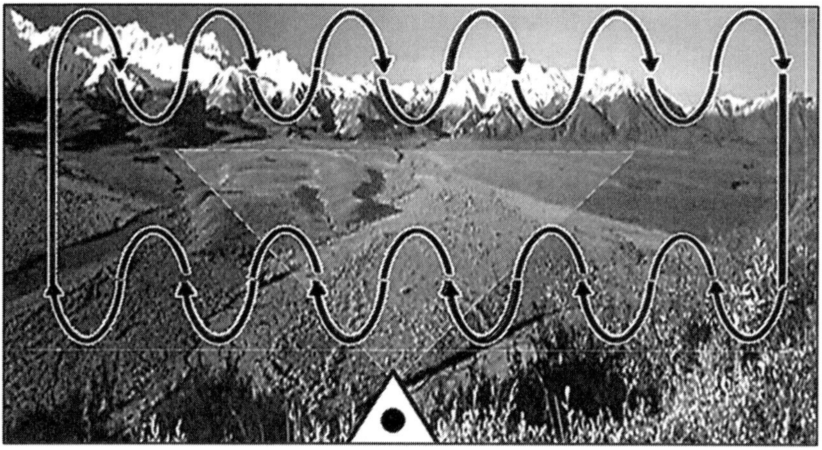

Ref: ATP 3-01.81 (Mar '25), fig. 5-5. Vertical scan technique.

Horizontal Scan

In horizontal scan viewed from the horizon (see figure 5-6 on page 50), the air guard should search using short, quick eye movements across the sky while working upward to approximately 20 degrees. As in vertical scanning, the air guard should continue the search and scan technique to 20 degrees below the horizon to detect aircraft flying nap of earth.

Ref: ATP 3-01.81 (Mar '25), fig. 5-6. Horizontal scan technique.

C. Air Guard Reports

Ref: ATP 3-01.81, Counter-Unmanned Aircraft System (C-UAS) Operations (Mar '25), pp. 46 to 48.

Based on mission tasks, air guards use the standard quick reference or pre-deployment and combat checklists developed by the brigade to focus on C-UAS tasks. The checklist should be available through standard military digital devices or in hardcopy form and include-

- Current UAS trends (type classification).
- Types and characteristics of UAS.
- Specific data on local air threats and named areas of interest.
- C-UAS equipment available.
- Secure radio operations and frequencies.
- Unit call signs to request support (quick reaction force or ADA element).
- Military map of the area.
- Binoculars and night vision devices.
- Orientation techniques (location, heading, speed, and line-of-sight).
- C-UAS spot report (see table 5-1).

C-UAS Spot Report (SALUTE Format)

Line	*Information*	*Example*
1	Size	Report the number of unmanned aircraft systems (UAS), or size of the formation.
2	Activity	Report detailed account of actions: What is the UAS's direction of movement? Was there any hostile action? Is the threat UAS loitering in one spot? Is it flying straight? Was the threat UAS approach observed or was it spotted overhead?
3	Location	Report where you saw the activity. Include 6 to 8- digit grid coordinate of reporting element and either grid or distance and direction from reporting element location (known point).
4	Unit (Description of UAS)	Include details such as: • Fixed-wing or rotor/multi-rotor. • If fixed wing. • Estimated length of wingspan. • Tail configuration. • If rotor/multi-rotor. • Number of rotors. • Height. • Payload, sensors, and weapons. • Any lights. • Other distinguishable markings.
5	Time	Report the time the activity was observed.
6	Equipment	If possible, report all equipment associated with the UAS, such as payload or weapons.

Table 5-1. Example Counter-Unmanned Aircraft System Detection Report (SALUTE Format).

Layered Defense

16-Line Report (Incident)

Line	*Information*	*Example*
1	Base or Facility name	Report the State / Country
2	Date Time Group (DTG) of the incident or incursion (Zulu) and duration.	Report using DDHHHH Zulu MMMYYYY format.
3	Geographic area	Report either USAFRICOM, USCENTCOM, USEUCOM, USINDOPACOM, USNORTHCOM, or USSOUTHCOM
3a	Reporting service / Component	Report either Army, Navy, Air Force, Marine Corps, Space Force, Coast Guard, National Guard, Air National Guard, or other.
4	Unit / Organization Contact Information	Report the Unit / Organization, Name, Rank/Grade, E-mail, and Phone Number.
5a	Geo-location of the incident or incursion	Report the Latitude-Longitude or MGRS of the observer / sensor to the UAS
5b	Location name	Report utilizing standardized naming convention.
6	Incident or incursion environment	Report either United States, Host nation, Theater of War, or International Waters.
7	Total number of UAS detected	Report the number of UAS detected in the format of (1-XXX)
8	First Detection	Report Outside installation / outside Installation-restricted area / within installation-restricted area / mobile
9	Detection method	Report if detected utilizing visual, radar, electronic, EO-IR, acoustic, or other means.
10	UAS airframe	Report the type of UAS either fixed wing, rotary wing, unknown, and other
11	UAS detailed description	Report the number of rotors, number of lights, light pattern (blinking or solid), color of lights, payload, movement pattern, and other features.
12	C-UAS equipment Operational	Report yes, no, unknown, or N/A
13	UAS engaged	Report Yes (engagement report, CJCS and CCDR CCIR), No, or N/A
14	UAS incident or incursion has a mission impact (commanders' assessment)	Report Yes (CJCS and CCDR CCIR) or No.
15	UAS incident or incursion unsafe / unprofessional (commander's assessment)	Report Yes (CJCS and CCDR CCIR) or No.
16	UAS incident or incursion narrative	Create a detailed narrative describing the UAS incident or incursion.

Engagement Report

Line	*Information*	*Example*
1	Title	Report Base or Facility name; State/country
2	Engagement start Date TIME Group (DTG) (Zulu)	Report using DDHHHH Zulu MMMYYYY format.
3	Total time of initial / follow-on engagements	Report using the HHMM format
4	Type of engagement	Report RF / Kinetic / C-PNT / other
5	C-UAS engagement Effective?	Report Yes / No / Unknown
6	Any confirmed electronic / collateral effects from C-UAS systems	Report Yes / No / Unknown (explain in narrative, line 11)
7	Was the UAS confiscated and / or captured?	Report Yes / No / Unknown
8	Attribution: Was the UAS operator located and / or interviewed?	Report Yes / No / Unknown / N/A (explain in narrative, line 11)
9	FAA and / or Host nation equivalent notified:	Report Yes / No
10	Is video and / or photo available?	Report Yes / No / N/A
11	UAS engagement narrative	Report if confiscated and / or captured explain who has it

D. Air Guard Checklist

The company commander should consider planning for redundancy in positioning guards. The need for redundancy may be dictated by the terrain, threat assessment, or overall importance of the mission. Air guard redundancy ensures sustained observation, enabling the unit to continue the mission if the enemy compromises some air guard positions.

Before assuming duty as an air guard, leaders perform precombat inspection to ensure the designated air guard has the correct equipment and has been briefed on the current threat. The air guard checklist includes, but is not limited to:

- Understanding on types and characteristics of threat UAS.
- Understanding of current UAS trends.
- Specific data on local air threats and named areas of interest.
- Detection equipment (such as binoculars, night vision devices, and thermal weapon sights).
- C-UAS equipment available.
- Secure radio operations and frequencies to send out early warning.
- Unit call signs to request support.
- Military map of area.
- Orientation techniques (location, heading, speed, and line-of-sight).
- C-UAS Range card.

(Layered Defense)
II. Active Defense

Ref: ATP 3-01.81, Counter-Unmanned Aircraft System (C-UAS) Operations (Mar '25), pp. 51 to 53.

Active air defense is a direct defensive action taken to destroy, nullify, or reduce the effectiveness of hostile air and missile threats against friendly forces and assets. It includes the use of aircraft, air defense weapons, electromagnetic warfare, and other available weapons (JP 3-01). Active defense of UAS requires early detection to permit acquisition, tracking, classification, identification, and destruction as soon as possible.

Active Defense Measures

- I **Detect**
- II **Track**
- III **Identify**
- IV **Decide**
- V **Defeat**
- V **Engage** *(See chap. 6.)*

Active air defense plans must be coordinated with airspace managers and ADA personnel. These plans should consider the use of air defense weapons, small arms, EW, and any other assets available to the brigade that can detect and mitigate hostile aerial platforms.

Active Measures for C-UAS

Active air defense planning measures taken in support of the mission must adhere to the ROE and weapons control status and consider potential collateral damage associated with the weapon system to the surrounding area.

Active defense measures are a multi-step sequence that units and service members do to detect, track, identify, decide, and potentially engage an unknown UAS. The quicker these steps can be applied, the more effective the response against threat UAS.

Editor's note: In addition to "Active Defense Measures" (pp. 5-27 to 5-30), ATP 3-018.81 also outlines steps in terms of "React to Threat UAS" (pp. 5-10 to 5-14).

I. Detect *(See also p. 5-10.)*

UAS are small, maneuverable, and quiet. Even to the trained eye they are often very difficult to observe in flight. Environmental conditions can make such systems difficult to detect without the use of special technological tracking and identification devices. Time of day, ambient light levels, meteorological conditions, and observer alertness will all impact on the ability to observe a potential hostile UAS. Moreover, an experienced operator can fly a UAS in a tactical manner as such to further enhance the ability to exploit UAS characteristics. This may include:

- Flying at low level, using terrain, obstacles with vertical extent and/or urban environment to mask an approach to a potential target location.
- Making a series of false take-offs and approaches to intended target location.
- Use of erratic, dynamic flight profiles to confuse and deceive personnel, making visual observation challenging.
- Using the sun or cloud cover to effectively mask the UAS from sight.
- Flying upwind approaches to minimize the ability to detect the sound of the UAS.
- Use of sport flying modes to increase speed and agility and minimize the time to observe the UAS.
- Use of multiple UAS to confuse, deceive, and overwhelm observers and make tracking leading to defeat more difficult.
- Flying a pre-programmed flight profile to minimize risk to the operator, allowing for the link to be disconnected in flight from the control station and re-established over the target area from another position.

The types of sensors and their placement determine the unit's detection capability. Threat UAS types, threat axis of advance, terrain, weather, time-distance analysis, friendly defended assets, desired engagement zone, surveillance requirements, and the number of available assets impact how best to place and employ sensors. For sensor placement, integrating and networking sensors to develop the enemy threat UAS situation should be applied. The use of various types of sensors is warranted as currently there is no one type of sensor that is 100% effective.

Various sensor capabilities outside of visual (observer) could include radar, radio frequency, audible, and optical devices. The goal is to form an integrated sensor network that includes various sensor types. Sensor capabilities in support of low-level air threats are planned accordingly and coordinated in advance. Leaders may have to coordinate through higher echelons for additional sensor capabilities. No matter what sensor capabilities a unit has, all service members need to be aware of the air threats and constantly look up before and during any movements. Dedicated air guards are another way units can assist in air threat detection and engagement.

II. Track *(See also p. 5-12.)*

Simultaneous with providing warning, friendly forces track the UAS and monitor its movement. Keeping track of the UAS needs to occur until a decision is made to engage or not engage the UAS and is successful. A location is a static estimated report or display of where an air threat is located at a given moment. A system track is a compilation of location reports over a set period. Depending on the system employed the system track can be reported as a heat map display, quadrant alert, or a circle to indicate estimated center and location error or line of bearing. The detection plan contributes directly to the unit's ability to track airborne objects continuously and efficiently.

III. Identify *(See also p. 5-12.)*

Identification is the process of determining the friendly or hostile characteristics of an unknown detected contact. The employment of C-UAS capabilities requires early identification of UAS to maximize engagement times and avoid fratricide. The problem of distinguishing friendly, neutral, and hostile aerial objects while employing various weapon systems against threat UAS is a highly complex task. Service members need to be familiar with the characteristics of friendly UAS operators in the area. The same UAS may be flown by both friendly and enemy forces. Accurate identification allows leaders to make engagement decisions and enhances situational awareness. Timely identification improves weapons employment options, helps conserve friendly resources, and reduces the risk of friendly fire.

There are two methods of identification, procedural and positive. Positive identification is the preferred method of operation, but generally, some combination of positive and procedural identification is used. Procedural identification separates airspace users by geography, altitude, heading, time, and maneuver. Positive identification is derived from observation and analysis of UAS characteristics including visual recognition, electronic support systems, non-cooperative UAS recognition techniques, identification friend or foe systems, or other physics-based identification techniques in accordance with JP 3-01.

Identification of a UAS should lead to a specific name or category or the exact make and model of the UAS. Identification of its payload is also important if it can be determined. The process of assigning an identification to a track likely will depend on several criteria. Characteristics that assist in identifying an unknown UAS include identifying:

- If fixed wing-
  - Estimated length of wingspan.
  - Tail configuration.
- If rotor/multi-rotor-
  - Number of rotors.
  - Height

IV. Decide

Decide consists of two decisions. First, whether there is a need to engage. Second if the decision to engage is made, the methods used to lessen or eliminate the threat posed by a UAS. These methods include physical and nonphysical. Some organizations will be equipped with cyber capabilities which include both physical and non-physical methods. The degree of delegation is consistent with the ROE, the available airspace, potential for collateral damage, and the inherent right of self-defense.

Physical methods engage with and either destroy or damage the device so that it is not operational.

Examples of physical methods include, but are not limited to:

- Explosive munitions
- Small arms
- Projectiles
- Entanglement
  - Streamers
  - Spray foam
  - Capture nets
- Directed Energy
  - Laser
  - High-power microwaves

V. Defeat *(See also p. 5-12.)*

Defeat techniques begin once airspace deconfliction and UAS engagement authority has been passed down to the tactical level. To confirm the identity of the UAS and prevent fratricide multiple processes and procedures. Depending on the engagement method, area radio frequency (RF) deconfliction may be required. Depending on the frequencies there may be opportunities where operational RF spectrums overlap with the control frequency of the UAS. Key leaders must determine and mitigate frequency fratricide during defeat measures. Defeat measures can follow non-lethal or lethal procedures.

During non-lethal responses, it is crucial to continue jamming, until the UAS is inoperable. After lethal or non-lethal responses, and once the UAS has shown no control-link, explosive ordnance disposal (EOD) must be requested to render the UAS safe. Once the UAS is safe, the UAS may be submitted for weapons intelligence and analysis.

VI. Engage *(See chap. 6.)*

Commanders have the responsibility to take whatever action is necessary to protect their forces and equipment against attack and ensure their Soldiers operate in accordance with established ROE. The air defense ROE must be adhered to when conducting engagements against UAS. Commanders at all echelons must exercise the right of self-defense to protect their forces from UAS. Hostile criteria for identification of UAS must be delegated as low as possible to facilitate C-UAS operations. Weapons control status for engagements against UAS must be as permissive as possible. The air defense staff at the brigade, division, and corps must ensure that conditions are set for a permissive fight at subordinate echelons against enemy UAS.

Once a decision is made to engage a threat UAS the capability will then be decided, the determined capability engages the threat UAS neutralizing the target or destroying it. Other C-UAS capabilities will continue to track the threat UAS in case the target was initially missed, and they can engage or re-engage as necessary.

Engagement Operations *(See pp. 6-1 to 6-10.)*

Proactive engagements are conducted in accordance with theater-level ROE and complementary directives. The best shooter-that weapon system that can best deliver the appropriate effects to achieve maximum attrition at the right time and place, and as far forward as possible, to ensure keep-out ranges and keep-out altitudes-conducts engagements.

Engagement Techniques *(See pp. 6-11 to 6-24.)*

Given the limited amount of dedicated ADA resources available within the Army, **it is incumbent on non-ADA units to contribute to their own defense against air threats using organic weapons** when ADA units are not available or to supplement any dedicated ADA units allocated to them. In addition to organic crew-served and individual weapons, divisional forces have Stinger MANPAD systems and are being fielded with C-sUAS capabilities.

Engagements can be lethal or non-lethal, and they can be executed against air and missile threats that are airborne. Engagements are conducted in accordance with the promulgated ROE and the local commander's priorities.

Maneuver forces in the close area use their organic weapons to engage surveilling or attacking rotary-or fixed-wing aircraft and UASs. Crew-served weapons are more capable against fixed-wing aircraft, rotary-wing aircraft, and sUAS; small arms are generally better against the group 1 UASs. All such efforts are considered self-defense actions. Threat systems that transit a unit's area of operation, but which are not attacking that unit, can also be engaged; however, they are more likely allowed to continue transiting given the relative short ranges and limited AMD effectiveness of maneuver force weapons. Transiting threats should be reported for early warning of and potential engagement by other units.

Layered Defense

I. Engagement Operations (Execution)

Ref: ATP 3-01.81, Counter-Unmanned Aircraft System (C-UAS) Operations (Mar '25), p. 58.

This section addresses the execution phase of the Army's operation process presented in terms of engagement operations.

I. Execution

Ground-based or aerial sensors, or combinations of both when available, are employed to perform surveillance of the airspace and provide focused early warning to at-risk forces. The sensors provide engagement authorities the time to make engagement decisions that deny threat platforms the ability to see or impact the defended assets. These engagement authorities function as a vertically and horizontally integrated kill chain.

Kill Chain (AMD)

The kill chain is the successive linkage of commanders who can authorize engagements of air and missile threats (FM 3-01). The designated engagement authority in the kill chain, providing operational control of and, ultimately, directions to the shooter, makes engagement decisions. Integrated sensors, shooters, and C2 networks will allow for greatly increased engagement options. As such, engagement authorization must be delegated to the lowest level possible.

See pp. 6-4 to 6-5 for discussion of the kill chain as a model in terms of F2T2EA.

Engagement Operations *(See following pages.)*

Proactive engagements are conducted in accordance with theater-level ROE and complementary directives. The best shooter-that weapon system that can best deliver the appropriate effects to achieve maximum attrition at the right time and place, and as far forward as possible, to ensure keep-out ranges and keep-out altitudes-conducts engagements. Keep-out range is the horizontal distance from a defended asset at which a successful engagement denies an adversary's desired weapons effects against the defended asset. It also allows C-UAS systems to counter enemy aerial surveillance and reconnaissance platforms beyond the range at which they can surveil an asset and beyond the engagement capability of a defending AMD system, contributing to the overall ability of friendly forces to see first. Keep-out altitude is the vertical distance above a defended asset at which a successful engagement denies an adversary's desired weapons effects against the defended asset. The method of fire may be stipulated in operation plans, directed by the kill chain, or influenced by the situation. Coordination is affected by the AMD C2 elements to ensure the AMD weapons, facilitating their destruction while simultaneously mitigating the potential for fratricide of friendly and neutral aircraft, overmatch the threats.

Maneuver forces using C-UAS systems can provide vital protection from UAS, thereby contributing to freedom of maneuver. The C-UAS systems enable maneuver forces to effectively detect and engage UAS at ranges and altitudes that deny them surveillance and attack options. Stinger teams organic to maneuver formations further enhance this capability. The ADAM cell in the maneuver brigade links the maneuver forces to the kill chain to implement appropriate ROEs. Intelligence enables cyber and EW elements to electromagnetically attack air targets. Intelligence elements can also provide or assist in the surveillance, identification, and classification of air targets, facilitating early warning.

II. (C-UAS) Engagement Operations (Concept of Operations)

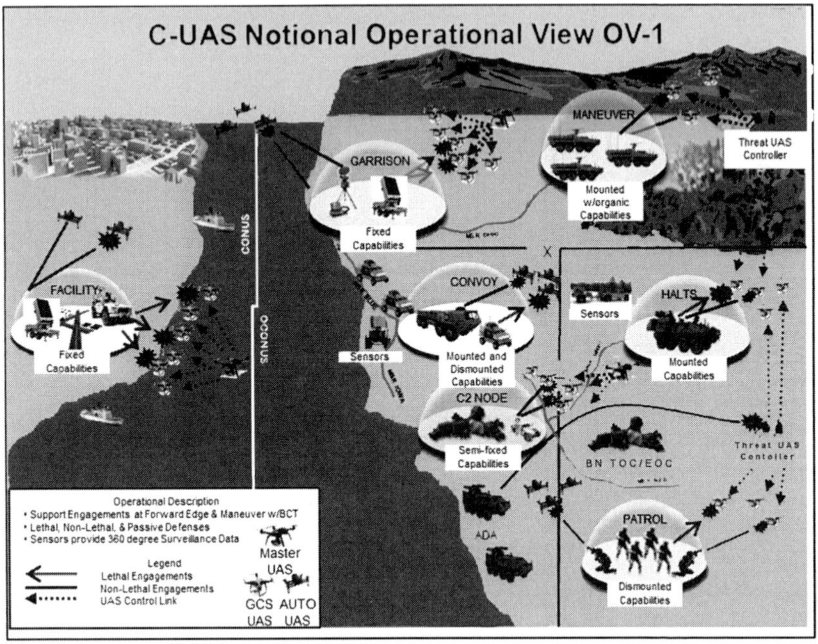

Ref: ATP 3-01.81 (Mar '25), fig. 1-1. Notional C-UAS operational view (see p. 1-20.)

A. Air Defense Artillery (ADA) *(See chap. 3.)*

The threats to be countered by ADA systems are ballistic missiles, large-caliber rockets, cruise missiles, air-to-surface missiles, hypersonic weapons, manned fixed-and rotary-wing aircraft, **unmanned aircraft systems (UASs)**, and indirect fire (rocket, artillery, and mortar) weapons. Most aerial threats will be employed in the areas of operations in which ADA units operate.

B. Non-Dedicated Air Defense

Ref: FM 3-01, U.S. Army Air and Missile Defense Operations (Aug '25), pp. 51 to 52.

Given the limited amount of dedicated ADA resources available within the Army, **it is incumbent on non-ADA units to contribute to their own defense against air threats using organic weapons** when ADA units are not available or to supplement any dedicated ADA units allocated to them. In addition to organic crew-served and individual weapons, divisional forces have Stinger MANPAD systems and are being fielded with C-sUAS capabilities.

Engagements can be lethal or non-lethal, and they can be executed against air and missile threats that are airborne. Engagements are conducted in accordance with the promulgated ROE and the local commander's priorities. Engagements are conducted by the best available system or systems positioned and capable of negating an enemy's capabilities. Key to the engagements is the detection of the threat in sufficient time and with sufficient accuracy to allow friendly forces to execute necessary responses.

See pp. 6-25 to 6-28 for discussion of air defense rules of engagement (ROE).

Figure 2-1 presents a doctrinal framework for C-UAS applicable to non-dedicated air defense and dedicated AMD Soldiers and units. It notes the non-dedicated protection task for air defense and considerations for C-UAS AMD missions. It also depicts overlaps in capabilities in engagements.

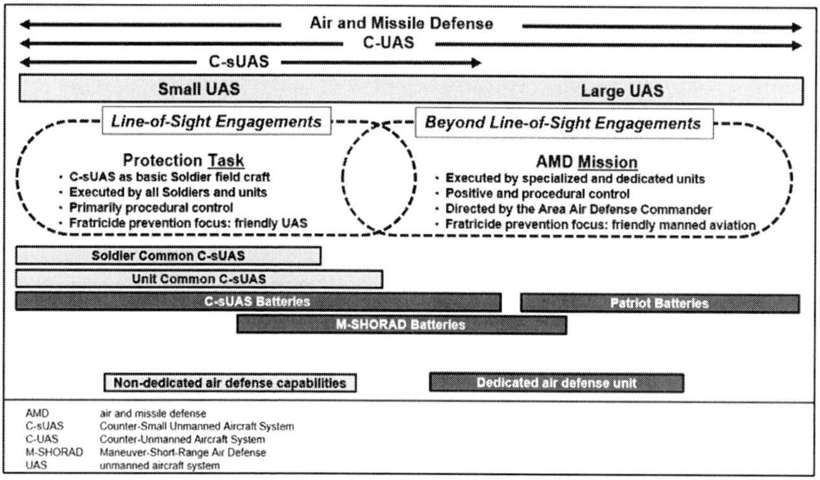

Ref: FM 3-01 (Aug '25), fig. 2-1. Doctrinal framework for counter-unmanned aircraft systems.

Dedicated C-UAS Batteries/Units *(See p. 1-35)*

Army Force Structure Transformation (2024) cites four types of units designed to provide new C-UAS capability:

- Multi-Domain Task Forces (MDTFs)
- Indirect Fire Protection Capability (IFPC) Battalions
- Counter-Small Unmanned Aerial Systems (C-sUAS) Batteries
- Maneuver Short-Range Air Defense (M-SHORAD) Battalions

Unit-Common C-UAS Systems *(See pp. 9-13 to 9-32.)*

Maneuver forces using C-UAS systems can provide vital protection from UAS, thereby contributing to freedom of maneuver. The C-UAS systems enable maneuver forces to effectively detect and engage UAS at ranges and altitudes that deny them surveillance and attack options.

Soldier-Common C-UAS Systems *(See pp.9-13 to 9-32.)*

Countering the threat requires individual Soldier actions and survivability activities, blended with employment of equipment and systems ranging from organic weapons, to specialized handheld EW and lethal devices, to complex systems designed specifically to detect, identify, and defeat UAS. Physical methods engage with and either destroy or damage the device so that it is not operational.

Maneuver Forces/Soldiers using Organic Weapons *(See pp. 6-11 to 6-28.)*

Maneuver forces in the close area use their organic weapons to engage surveilling or attacking rotary-or fixed-wing aircraft and UASs. Crew-served weapons are more capable against fixed-wing aircraft, rotary-wing aircraft, and sUAS; small arms are generally better against the group 1 UASs. All such efforts are considered self-defense actions. Transiting threats should be reported for early warning of and potential engagement by other units.

III. The Kill Chain: Find, Fix, Track, Target, Engage, Assess (F2T2EA)

A Comprehensive Approach to Countering Unmanned Aircraft Systems, Joint Air Power Competence Centre (Jan '21), pp. 104 to 107.

A UAS has three main components: the sensor, the controller, and the communication links. Each component has its own weakness. The sensor is vulnerable to Global Positioning System [US system], GLONASS [Russian system], Galileo [EU system], and BeiDou [Chinese system] signals. The controller to link protocols and the communications link are vulnerable to video and telemetry downlink. See figure 1-2, for visual representation of the communication path for a UAS.

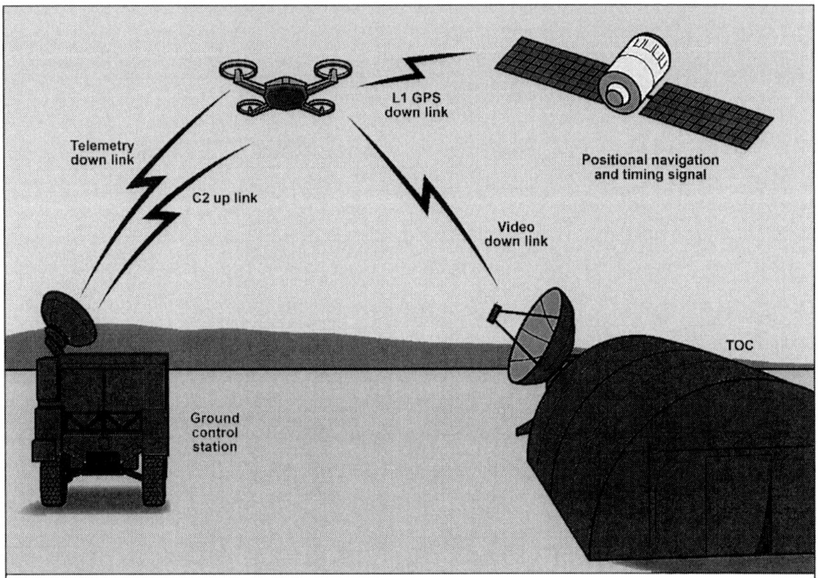

C2	command and control	L1	L-band channel 1
GPS	global positioning system	TOC	tactical operations center

Ref: ATP 3-01.81 (Mar '25), fig. 1-2. Unmanned Aircraft System communication link.

From an air defense perspective, the main objective is to prevent the delivery of effects by adversary air threats.

Additional Options to Counter Unmanned Aircraft Systems

For regular manned air threats the kill chain can be aimed at the aircraft, the payload or the logistical backbone. **When dealing with UAS, three more target options arise.** These are the control element, the human operator or the control link/mechanism.

Every individual target option requires an additional kill chain, which enables the suppression of the UAS' mission, in several cases even without physically destroying the aerial vehicle itself. Also, smaller UA might only have a very short-range and therefore will be launched well inside of any SBAMD sensor envelope, reducing available time for SBAMD systems to react. In this case, the initial act of finding an UA before it ever becomes airborne is critical.

Engagement Operations

Kill Chain Model (F2T2EA)

The United States of America developed the F2T2EA kill chain model inspired by General Ronald R. Fogleman. Analyzing the six steps defined by their keywords will shed some light on the question of how the kill chain works, and why there are additional options for UA defense.

F - Find

This is the initial element to start the entire process. Information collection in regards to Intelligence, Surveillance and Reconnaissance (ISR) to provide a proper Joint Intelligence Preparation of the Operational Environment (JIPOE) is obligatory to be sufficiently informed about the expected threat. In parts, this has to take place long before an actual C-UAS operation, but the theoretical planning process to develop defensive reactions is based on these prerequisites; on a practical note, sufficient surveillance is considered to be obligatory.. Before thinking about any active defensive actions, it is mandatory to 'find' targets. Sensors (e.g. radars, but also optical and acoustic systems) are required to detect any air targets. Sensor requirements for UA, which are similar to known air threats, should be satisfied by existing systems. Although, the smaller the UA and lower the flight altitude, the more complex it will be to reliably execute the first step of the kill chain with Surface-Based Air and Missile Defence (SBAMD) sensors. Therefore, finding UA before they are airborne could significantly enhance SBAMD operations, at least in passive defence terms.

F - Fix

The meaning of fix in the AD domain is to identify the detected air targets. This will contribute to proper situational awareness (SA) and allow valid and consequent decision-making. This becomes especially problematic, but very important in peacetime, due to the massive increase of private or recreational UA.

T - Track

Tracking the identified radar contact is mandatory to continue the dynamic decision-making process. In relation to the track history and the current track behavior, military leaders are capable of prioritizing or retaining attack options. Individual tracks may be sorted and allocated to weapon systems, which directly leads to the next step.

T - Target

Final coordination including the reassurance of the correct classification and identification is going to lead into the final approval for engagement under consideration of all applicable laws and Rules of Engagement (ROE).

E - Engage

The allocated weapon system is ready to fire and will receive the engagement order. Like with other air threats against UAS, this process will work in all modes of operation, depending on the individual situation and UA.

A - Assess

The last step in the kill chain is the assessment of engagement success. The outcome of this assessment could be an all-clear, a re-engagement or the alerting of threatened sites.

IV. React to Threat UAS (See pp. 5-10 to 5-14.)

Ref: ATP 3-01.81, Counter-Unmanned Aircraft System (C-UAS) Operations (Mar '25), pp. 35 to 39.

Understanding the capabilities and limitations of UAS can increase a Soldier's ability to react and defeat a threat. Reacting to a threat UAS should include reporting to the operations center and security personnel if the UAS does not pose an immediate threat. Defeat does not equate lethal means; however, it is an option. Other defeat solutions could be limiting a surveillance threat from gaining information or following the air path of the UAS to the operator.

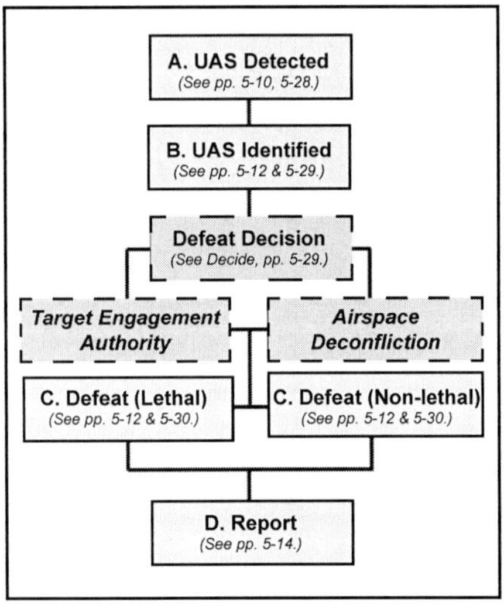

Understanding the capabilities and limitations of UAS can increase a Soldier's ability to react and defeat a threat. Reacting to a threat UAS should include reporting to the operations center and security personnel if the UAS does not pose an immediate threat. Defeat does not equate lethal means; however, it is an option. Other defeat solutions could be limiting a surveillance threat from gaining information or following the air path of the UAS to the operator.

Figure 4-6 is a simplified flow chart showing the process from UAS detection to submission of the final report to higher. Leaders must be aware that local standard operating procedures may dictate other steps and requirements. Ref: Adapted from ATP 3-01.81 (Mar '25), fig. 4-6.

Active Measures for C-UAS

Ref: ATP 3-01.81, Counter-Unmanned Aircraft System (C-UAS) Operations (Mar '25), pp. 51 to 53.

Active air defense is a direct defensive action taken to destroy, nullify, or reduce the effectiveness of hostile air and missile threats against friendly forces and assets. It includes the use of aircraft, air defense weapons, electromagnetic warfare, and other available weapons (JP 3-01). Active defense of UAS requires early detection to permit acquisition, tracking, classification, identification, and destruction as soon as possible.

Active defense measures are a multi-step sequence that units and service members do to detect, track, identify, decide, and potentially engage an unknown UAS. The quicker these steps can be applied, the more effective the response against threat UAS.

Active air defense plans must be coordinated with airspace managers and ADA personnel. These plans should consider the use of air defense weapons, small arms, EW, and any other assets available to the brigade that can detect and mitigate hostile aerial platforms.

Active air defense planning measures taken in support of the mission must adhere to the ROE and weapons control status and consider potential collateral damage associated with the weapon system to the surrounding area.

See pp. 6-25 to 6-28 for discussion of air defense rules of engagement (ROE).

C-UAS Time/Distance Engagement Sequence Planning Considerations

Ref: ATP 3-01.81, Counter-Unmanned Aircraft System (C-UAS) Operations (Mar '25), p. 38.

Figure 4-7 shows C-UAS time/distance engagement sequence planning considerations, for a linear depiction of this process.

- Commanders, staff, and service members execute C-UAS layered defense activities as part of an overall protection plan employing the five principles of protection: comprehensive, integrated, layered, redundant, and enduring *(See pp. 5-1 to 5-2.)*
- Using joint force employment tenets of early engagement and defense in depth will allow the maneuver greater standoff against threat UAS; however, such will shorten the timelines and distance associated with detection through defeat/engagement sequence depicted below.
  - **Early Engagement**: Sensors and weapons are positioned so that they engage the threat before ordnance release or detection of friendly forces.
  - **Defense in Depth**: Sensors and weapons are positioned so that the threat is exposed to a continuously increasing volume of fire as it approaches the friendly protected asset of force.
- Target Identification and engagement authorities, C-UAS defense weapons control statuses, and ROE are subject to theater-specific conditions and orders
- Threat UAS detection will trigger tactical SOP battle drill/reactions.

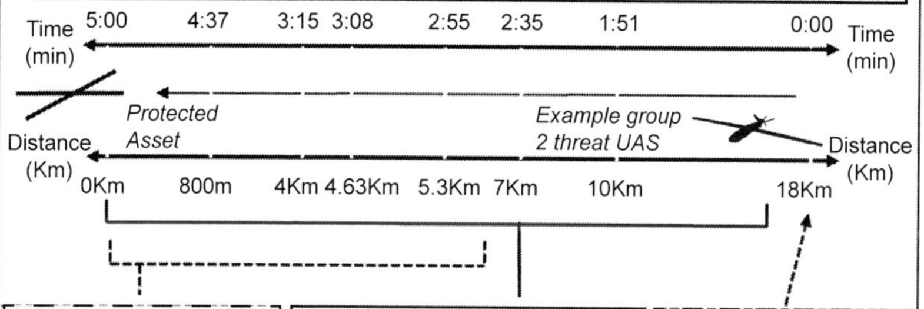

C. Defeat *(p. 5-12.)*
- Target engagement authority delegation
- Airspace/radio frequency deconfliction
- Lethal/non-lethal response
- Weapons control statuses
- ROE
- Report

B. Track & Identify *(p. 5-12.)*
- Via radar/electromagnetic, visual aircraft recognition or procedural methods
- Via "wings, engine, fuselage, and tail" WEFT method
- Via friendly procedural airspace control measures
- Target identification authority delegation

A. Detect *(p. 5-10.)*
- Via radar/ electromagnetic, visual aircraft recognition or procedural detection
- Defense in depth
- Early engagement
- Layered defense

Time/distance based on example group 2 threat traveling at 4.3Km/min (160 knots) and rotational detection range of 18Km. If example group 2 threat traveled at 7.7Km/min (250 knots), the UAS would reach/strike defended asset 2:18 mins from 18Km detection.

C-UAS	counter-unmanned aircraft system	**SOP**	standard operating procedure
Km	kilometer	**UAS**	unmanned aircraft system
Min	minute	**ROE**	rules of engagement

Ref: Adapted from ATP 3-01.81 (Mar '25), fig. 4-7. Counter-Unmanned Aircraft System time/distance engagement sequence planning considerations. See following pages (pp. 6-8 to 6-9) for further discussion of engagement operations considerations.

V. Engagement Operations (Considerations)

Ref: FM 3-01, U.S. Army Air and Missile Defense Operations (Aug '25), pp. 85 to 86.

Engagement operations are functions and activities required to execute the air, missile, and counter-surveillance battle. Engagement operations include establishing an air picture, determining the classification and identification of all tracks, evaluating the threat these tracks pose to the defended assets, and exercising engagement control over subordinate units.

AMD ground-or air-based sensors, or combinations of both when available, perform surveillance of the airspace and provide focused early warning to at-risk forces. The sensors provide engagement authorities the time to employ weapon systems that deny threat platforms the ability to see or impact the defended assets, and for early warning of a possible attack to be passed to at-risk units or assets.

AMD weapon engagements are controlled by joint and Army authorities. Joint engagements are conducted in accordance with directives issued by the area air defense commander (AADC) as the engagement authority or the AADC's representatives, a regional or sector air defense commander when so delegated. Engagement authorizations are directed by the engagement authority and typically transmitted by an air and missile defense fire coordination officer (ADAFCO) to Patriot and Terminal High Altitude Area defense (THAAD) units. ADA commanders in divisions and BCTs control engagements using ROE approved by the joint force commander. Engagements are typically decentralized to fire unit level.

Airspace Coordination and Management

Airspace control is the exercise of delegated authority over designated airspace and users through control procedures and coordination measures to maximize operational effectiveness (JP 3-52).

Competing airspace users balance the demands for and integrate their requirements for airspace. To help balance these demands, the joint force commander usually designates an airspace control authority. The airspace control authority is the commander designated to assume overall responsibility for the operation of the airspace control system (JP 3-52). An airspace control system is an arrangement of those organizations, personnel, policies, procedures, and facilities required to perform airspace control functions (JP 3-52). The airspace control authority, working with the other components of the joint force, develops policies and procedures for all airspace users. The airspace control authority uses the airspace control plan, airspace control order, and air tasking order to transmit the policies and procedures.

- The **airspace control plan** is the document approved by the joint force commander that provides specific planning guidance and procedures for the airspace control system for the joint force operational area (JP 3-52). An airspace control plan details the broad policies and procedures for airspace control within the joint force commander's operational area.
- The **airspace control order** is an order implementing the airspace control plan that provides the details of the approved requests for airspace coordinating measures (JP 3-52). An airspace control order directs the use of joint airspace and details the approved requests for airspace coordinating measures.
- **An air tasking order** is the daily operation order listing all component aviation assets directed by the joint force commander or made available to the joint force commander for joint force air component commander's taskings (FM 3-52). The order shows all missions operating in theater. The special instructions (known as SPINS) section of the air tasking order provides supplemental, corrective, or exact amplification to the general mission tasking of the specific air tasking order period. SPINS often include ROE and combat identification criteria for air defense. SPINS are published as a baseline, weekly, and daily (FM 3-52).

When delegated control authority by the airspace control authority, the Army procedurally controls assigned airspace. Division-assigned airspace, for example, is a volume of airspace in which the airspace control authority has delegated the responsibility for control of that volume, in accordance with the airspace control plan and airspace control order, to the division. Division-assigned airspace is typically between the lateral boundaries of the division's area of operations, and up to the coordinating altitude. While the Army's airspace control methodology emphasizes procedural control of airspace use, it includes the flexibility to utilize positive control or a combination of the two throughout a commander's area of operations. Positive control requires sensors to locate and identify airspace users in real-time and communications, to maintain continuous contact with the user (JP 3-52).

Note. The coordinating altitude and other airspace coordinating measures are included in the airspace control plan and distributed via airspace control orders. Army echelons should include the airspace control plan guidance and integrate the airspace control order, area air defense plan (AADP), special instructions, and air tasking order into their unit airspace plans and operation orders (FM 3-52). The unit airspace plan is the integrated set of airspace coordination measures to support Army operations submitted to the airspace control authority through the battlefield coordination detachment for integration into a future airspace control order (ATP 3-52.1).

Army elements and cells within the corps, divisions, and brigades exercise control, coordination, and/or management of their organizations' airspace. The AMD operations section and ADSE are part of an airspace control element, consisting of ADA, field artillery, aviation, and other service representatives at corps, division, and brigade levels that facilitate airspace coordination and clearance of fires. Their responsibilities for airspace management vary dependent upon echelon.

Clearance of Fires

Airspace must be cleared to allow uncontested operations by the multitude of airspace users, avoiding time and space conflicts. Airspace coordination elements, such as the air-ground integration element and ADSE at division and brigade levels, maintain contact with the joint air ground integration center for clearance. At corps and above, the coordination elements are linked to various airspace control agencies, including the Air Force control and reporting center, radar approach control, and air traffic control agencies. Responsibilities of these agencies are generally defined by the established coordinating altitude.

Most fires exceed the coordinating altitude and therefore must be cleared by the control and reporting center and radar approach control. Clearance must be obtained from the control and reporting center because it controls tactical military aircraft, and from the radar approach control because it has control of commercial and nontactical military aircraft.

Army air traffic service units control limited airspace below the coordinating altitude, specifically around airfields, forward arming and refueling points, and tactical assembly areas. All Army elements involved in airspace management must coordinate their efforts to ensure that air operations are synchronized, potential fratricide of aircraft are mitigated, and interruptions to ongoing operation are minimized. Army airspace coordinators must be cognizant of the locations, plans, and activities of other users to enable the greatest use of the airspace with minimal conflicts. The positions of land-based airspace users (such as ADA units, field artillery units, and aviation operating bases) and Army aviation transit routes must be displayed in common air pictures and updated continuously. The common air picture should also include airspace coordinating measures, such as any restrictive operating measures, in effect in the area of operations. Once appropriate agencies clear their airspace, the ADSO, fire support officer, or aviation officer can acknowledge that the airspace is cleared for the designated operation.

See *pp. 6-25 to 6-28 for related discussion of air defense warnings (ADWs) and controls.*

VI. Low-Slow-Small Unmanned Aircraft System (LIDS) Engagement Overview (See pp. 9-16 to 9-17.)

LIDS CONCEPT: The LIDS family of systems uses a range of passive and active sensors to detect, track and identify UAS and non-hostile aircraft. (Graphic by IF/RCO, U.S. Army) (See pp. 8-13 to 8-32 for discussion of LIDS and other C-UAS systems. See also pp. 6-11 to 6-24 for an overview and discussion of enagement techniques with maneuver forces using organic weapons systems.)

II. (Maneuver Forces) Engagement Techniques

Ref: ATP 3-01.81, Counter-Unmanned Aircraft System (C-UAS) Operations (Mar '25), pp. 53 to 54.

Given the limited amount of dedicated ADA resources available within the Army, **it is incumbent on non-ADA units to contribute to their own defense against air threats using organic weapons** when ADA units are not available or to supplement any dedicated ADA units allocated to them. In addition to organic crew-served and individual weapons, divisional forces have Stinger MANPAD systems and are being fielded with C-sUAS capabilities.

Engagements can be lethal or non-lethal, and they can be executed against air and missile threats that are airborne. Engagements are conducted in accordance with the promulgated ROE and the local commander's priorities. Engagements are conducted by the best available system or systems positioned and capable of negating an enemy's capabilities. Key to the engagements is the detection of the threat in sufficient time and with sufficient accuracy to allow friendly forces to execute necessary responses.

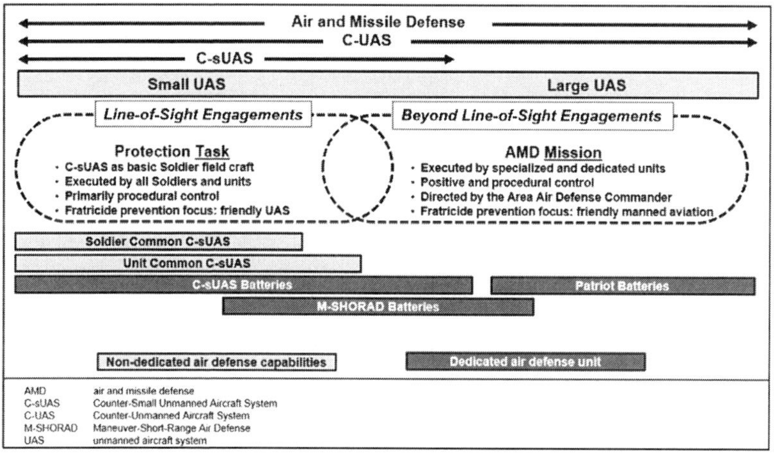

Ref: FM 3-01 (Aug '25), fig. 2-1. Doctrinal framework for counter-unmanned aircraft systems. (See pp. 6-2 to 6-3 for an overview of C-UAS "concept of operations.")

Figure 2-1 presents a doctrinal framework for C-UAS applicable to non-dedicated air defense and dedicated AMD Soldiers and units. It notes the non-dedicated protection task for air defense and considerations for C-UAS AMD missions. It also depicts overlaps in capabilities in engagements.

Maneuver forces in the close area use their organic weapons to engage surveilling or attacking rotary-or fixed-wing aircraft and UASs. Crew-served weapons are more capable against fixed-wing aircraft, rotary-wing aircraft, and sUAS; small arms are generally better against the group 1 UASs. All such efforts are considered self-defense actions. Threat systems that transit a unit's area of operation, but which are not attacking that unit, can also be engaged; however, they are more likely allowed to continue transiting given the relative short ranges and limited AMD effectiveness of maneuver force weapons. Transiting threats should be reported for early warning of and potential engagement by other units.

I. Ballistic Low Altitude Drone Engagement System (BLADE) Example

Ref: Excerpted DOW article by Timothy Rider (Sept. 3, 2025). https://www.war.gov/News/News-Stories/Article/Article/4292228/blade-cuts-down-drones-during-project-flytrap-40/ (Accessed 19 Oct 25.)

BLADE Cuts Down Drones During Project Flytrap 4.0

Soldiers assigned to the Army's 2nd Cavalry Regiment detected, tracked, engaged and defeated multiple drones at ranges between 500 and 800 meters using the Ballistic Low Altitude Drone Engagement system from a Stryker vehicle at the Bemowo Piskie Training Area in northeast Poland during a live-fire event for Project Flytrap 4.0 on Aug. 1.

Army Staff Sgt. Robert Laux, assigned to the 3rd Platoon, 2nd Cavalry Regiment, receives instruction on operating a Night Fighter portable electronic jamming device from a British soldier assigned to the 1st Battalion, Royal Yorkshire Regiment, at Bemowo Piskie Training Area, Poland, July 28, 2025.

"Some of those [drone] threats were being flown simultaneously, so the system defeated one target then quickly targeted and defeated a second target in a matter of seconds," said David Goldstein, counter-unmanned aerial systems lead for the Army Combat Capabilities Development Command Armaments Center at Picatinny Arsenal, New Jersey.

The Armaments Center ran the BLADE science and technology project from 2016 to 2019 to develop a capability to counter drone threats. BLADE leverages — and enhances — a system operated throughout existing Army formations, the Common Remotely Operated Weapons Station (CROWS).

As a remote system mounted on different types of Army vehicles, CROWS allows an operator to fire a weapon while remaining protected inside. Using an array of optics, a viewing screen and motorized gimbals and controls, the operator can effectively select and engage targets.

With BLADE, a precision radar and C-UAS fire control software are integrated with CROWS hardware and software to assist operators in identifying, tracking and pointing the weapon to a continually calculated intercept point, enabling the difficult challenge of destroying enemy drones.

Capable of functioning with numerous weapons, the BLADE/CROWS combination at Project Flytrap included an M2 .50-caliber machine gun firing multiround bursts.

Starting in late July, Project Flytrap 4.0 was the final iteration of the C-UAS training events, where soldiers from the 2nd Cavalry Regiment and the 1st Royal Yorkshire Regiment, United Kingdom, tested new C-UAS capabilities in combined exercises meant to develop and inform the future of anti-drone tactics and technologies.

"Project Flytrap is part of a series of training events designed to test and refine new counter-unmanned aerial systems technologies and tactics to respond to the evolving threat of drone warfare," said Army Col. Matt Davis, transformation chief and exercise director of V Corps, Project Flytrap, during a press engagement.

The CNALT software plug-in, also being developed by the Armaments Center, is an application that provides tactical fire control and visualization tools to reduce time in the kill chain — the cycle of finding and identifying targets, prioritizing and coordinating attacks. Time is critical in military operations with targets in constant motion, considering that potential targets threaten harm to friendly forces.

Soldiers took 15 minutes on new equipment training for CNALT operations, said Marc Sanzari, chief of the Armaments Center Networked Lethality and Collaborative Autonomy Branch. During various training scenarios, CNALT successfully ingested real-time sensor data from the BLADE precision fires radar, classified unmanned aerial vehicle tracks and generated actionable engagement recommendations for BLADE and the Stryker commander.

After the installation, Gann met with soldiers to discuss the tactics, techniques and procedures for BLADE operations within unit formation while stationary and on the move. He noted that soldier feedback from the live-fire event provided insights on future improvements for BLADE, but also, "it highlighted the possibilities of UAS engagement in a dynamic environment," Gann said.

"I was able to understand the fundamental workflow for the system and gunner actions within one hour of seeing the BLADE for the first time," said Army Maj. Jared Whitaker, deputy chief G31 from V Corps, who, as a leader, observed the entirety of the drone engagements in a seat between the gunner and radar operator.

Whitaker, who is proficient in operating CROWS, said that compared to ground engagements, there were added steps in counter-drone operations. Schooling on the radar was necessary, as was additional training to become fluid during rapid reactions while performing counter-drone scenarios.

"I found that the ability to engage UAS using a CROWS has gone from impossible to possible with the addition of BLADE," Whitaker said.

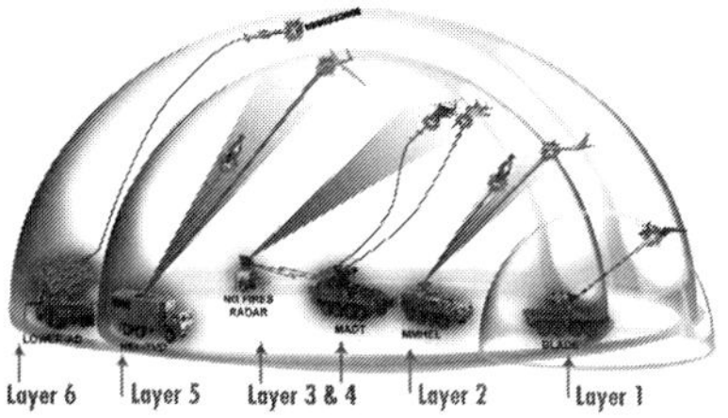

Layer 1: BLADE (smallest, most mobile dome of protection)

At the tactical edge, maneuvering Soldiers need to have local protection to shoot moving targets such as unmanned aerial systems (UAS), which range from Group 1 (very small UAS) to Group 5 (the largest UAS). The Ballistic Low Altitude Drone Engagement (BLADE) is a set of enabling technologies that are integrated with an armament system to defeat smaller unmanned aerial systems at close ranges, which includes the distance a Soldier can see a UAS without using binoculars. Its intuitive interface makes the BLADE easy for Soldiers to use.

The BLADE system works with the Common Remotely Operated Weapon Station (CROWS), and uses advanced fire control and precision targeting enablers to detect, track and defeat unmanned aerial systems. Mounted on a tactical vehicle, CROWS contains a sensor suite and fire control software that allow the warfighter to remotely engage targets. CROWS can engage targets during the day or at night, and includes a daytime video camera and a thermal camera. *(CCDC.)*

Engagement Operations

II. Soldier-Wearable C-UAS Devices

Ref: DOW article by Katie Lange, Pentagon News (Sept. 23, 2025). https://www.war.gov/News/Feature-Stories/Story/Article/4312674/drone-busting-smart-devices-work-together-to-knock-out-uas-threats/ (Accessed 20 Oct 25.)

Drone Busting: Smart Devices Work Together to Knock Out UAS Threats

What's harder to shoot down than a small drone in the sky? Students at the War Department's only joint training center for countering aerial threats will likely say it's what they practice on — balloons attached to those small drones.

Drones can fly at racecar speeds, making them difficult targets to hit by an individual on the ground. At the Joint Counter-Small Unmanned Aircraft System University, located at Fort Sill, Oklahoma, students use handheld smart devices to help them increase their probability of scoring a hit.

Known as the Dronebuster and Smart Shooter, these two handheld systems work in tandem as a layered defense that uses electronic and kinetic warfare to do the hard work for the shooter.

Students who attend Joint Counter-Small Unmanned Aircraft System University (JCU) learn about various drones and the intricate service-specific systems used to engage with them. The Dronebuster and Smart Shooter, however, are two of the few used by service members on the ground, such as patrol members or personnel guarding installation towers.

So, how do these tandem devices work?

When service members are warned of a drone presence, the Dronebuster can be held up like a weapon to scan the area. Once it identifies a hostile target, the service member simply presses a button on the device to disrupt the target using electronic warfare.

"It'll start jamming the command and controls, severing that connection between the drone and the operator," explained JCU operator's course instructor Paul Bliefernich.

The drone remains frozen in the air. That's when a second operator picks up the Smart Shooter, an optic system that attaches to a weapon to control how it fires.

"Typically, when you pull the trigger, you're used to a gun going 'bang.' [The Smart Shooter] doesn't do that," explained JCU operator's course supervisor Fred Hill. "It calculates, and then it shoots when it's ready to shoot."

Army Staff Sgt. Marcjarall Baxter uses the Dronebuster to destroy simulated opposition force drones during Operation Combined Victory at Muscatatuck Training Center, Butlerville, Ind., Feb. 14, 2025.

"When [shooters] get to the target they want, they'll hit the 'lock' button … and start squeezing the trigger," Bliefernich explained. However, the weapon doesn't immediately fire. Instead, its computer takes over and displays a set of crosshairs in the scope that the operator aligns with the target. When the computer decides there's a good probability of a hit, it sends a command to the pistol grip, allowing the operator to complete the trigger pull and fire the weapon.

Bliefernich said it can take from a few seconds up to two minutes. "It's doing calculations and taking a lot into consideration," he added.

Since shooting down several drones per class isn't cost-effective, JCU students mostly shoot balloons attached to drones. It's a more challenging target, but with practice, it can make them a better shot.

"The balloons are harder to shoot than the drone," Hill said. "Once the shooter gets toward the end of the training, we'll put up the aircraft and let them shoot at the aircraft."

"It takes a lot of time and practice," Bliefernich said.

The students are also able to train on the Smart Shooter in the virtual reality lab beforehand, which instructors said is both cost-effective and safer.

"It's getting the soldiers, airmen and sailors familiar with the equipment before they ever come and start shooting on the ground," said Army Sgt. 1st Class Alan Buhl, JCU instructor.

See p. 1-9 to 1-11 for related content and description of the Joint Counter-Small Unmanned Aircraft System University (JCU).

III. First-Person View (FPV) Drone Armed with a Claymore Mine (Kinetic Kill Example)

Ref: DoD News article by Army Capt. Jennifer French (Aug. 27, 2025). https://www.war.gov/News/News-Stories/Article/Article/4287649/army-producing-second-tranche-of-short-range-reconnaissance-drones/ (Accessed 19 Oct 25.)

Paratroopers Make History With Army's First Drone-on-Drone Strike

For Army 1st Lt. Francesco La Torre, the Army's first drone-on-drone strike wasn't exactly a Hollywood moment.

"In [the Star Wars movie], 'A New Hope,' Luke Skywalker flies an X-Wing through the Death Star trench and blows it up," La Torre said. "This wasn't like that. It was way harder. Honestly, it felt more like Darth Vader shooting down rebel pilots. He might be the villain, but he's also one of the best pilots in the galaxy."

That was the image in La Torre's mind as his team with the 173rd Airborne Brigade achieved a milestone once unimaginable for conventional Army units: destroying an aircraft in flight using a first-person-view drone carrying an explosive charge.

The strike took place earlier this month during the Army's Unmanned Aerial Systems and Launched Effects Summit, held Aug. 11-15 at Fort Rucker, Alabama. The live-fire event was led by the brigade's bayonet innovation team in partnership with the Pennsylvania Army National Guard and engineers from Army Combat Capabilities Development Command.

Army paratroopers assigned to the 173rd Airborne Brigade prepare to operate and detonate a first-person-view drone at Pabradė Training Area, Lithuania, during a joint forcible entry operation as part of Swift Response 2025, May 18, 2025. (Credit: Army Sgt. Jose Lora.)

"When I heard [Army] Chief [Warrant Officer 2] Nate Shea say, 'Arming,' when he was lined up behind the fixed wing threat UAS, I knew he'd take it down," La Torre said. "A few seconds later he detonated the claymore [mine], and I looked up from my end-user device to see the fixed-wing [aircraft] drop out of the sky."

The success came after weeks of training. Paratroopers flew for hours each day, building muscle memory while rehearsing how to engage a fast-moving aircraft. Army Chief Warrant Officer 2 Andrew Topits familiarized himself with the fixed-wing platform a month in advance, while Shea practiced with inactive claymores. Once on the ground at Fort Rucker, the operators flew daily in changing weather, preparing for a target that could appear anywhere across the range.

Technical skill proved as important as piloting.

"The ability to train and solve problems immediately afterwards was critical to mission success," La Torre said. "The team was soldering, disassembling, reassembling and performing [quality assurance] around the clock while rehearsals were underway. While tactical units and technicians may not share the same language, they do share an intellectual curiosity and desire to succeed."

La Torre emphasized what it means for the average paratrooper.

Army paratroopers assigned to the 173rd Airborne Brigade prepare to operate and detonate a first-person-view drone at Pabradė Training Area, Lithuania, during a joint forcible entry operation as part of Swift Response 2025, May 18, 2025. (Credit: Army Sgt. Jose Lora.)

"With the right kit, paratroopers will be able to utilize a low-cost system with preexisting munitions to execute a wide variety of mission tasks — and have the power to protect their unit from an endemic threat," he said.

The drone demonstration showed that innovation doesn't have to come from the top.

"It feels extremely satisfying to have this level of impact, but solving these kinds of problems should not be alien to junior officers, or any leader," La Torre said. "Every year the Army produces leaders at all levels who are better educated now than ever before. At the end of the day, rank is immaterial when it comes to problem-solving."

For La Torre, the milestone reflects how far the 173rd Airborne Brigade has come.

Army paratroopers assigned to the 173rd Airborne Brigade prepare to operate and detonate a first-person-view drone at Pabradė Training Area, Lithuania, during a joint forcible entry operation as part of Swift Response 2025, May 18, 2025. (Credit: Army Sgt. Jose Lora.)

"At this time last year, this kind of training was a pipe dream for conventional units," he said. "For the engagement itself, Chief Shea did an incredible job and so did Chief Topits. It's such a difficult mission profile for both aircraft."

Lessons from the bayonet innovation team fed directly into Fort Rucker's new lethal UAS course, directed by Army Capt. Rachel Martin and Army Maj. Wolf Amacker. The course consolidates milestones within the Army's UAS community and serves as a hub for education and experimentation.

"The more testing and experimentation we do, the less we must teach theoretically and the more we can back claims with real data," La Torre said. "Safe and effective experimentation allows commanders to understand the true nature of the risks they assume with these systems and helps make this training more commonplace."

For La Torre, the boundaries are clear: safety and scalability.

"We must safely conduct training, testing and experimentation," he said. "We also must make sure everything we do is exportable to other units so they can do the same. I'm confident what we've done is scalable, and the lethal UAS course at Fort Rucker will centralize the growth of lethal UAS engagements while also serving as a hub for subject matter experts to gather and exchange lessons learned."

IV. Small Arms Techniques for Engaging UAS

Small arms techniques used in air defense incorporate the use of volume fire and proper aiming points according to the UAS direction. If service members are trained to apply an appropriate sequence of engagement techniques for UAS based on the rules for selecting aiming points, the response will be automatic upon command. Small arms are limited to the range and destructive capability of the weapon and should be used only on low flying UAS.

The decision to fire small arms against threat UAS is the unit commander's and is based on the situation. Considerations should at a minimum include the severity of the threat versus the potential impact of the unit's effectiveness and the area of engagement (urban versus rural).

Soldiers that encounter UAS should consider them as a threat eligible for engagement unless positively identified as friendly. UAS can fly extremely low underneath traditional radar detection zones. They can fly very slowly with the ability to hover in place, preventing sensors from detecting them. They can also fly quickly, making it difficult to track them.

Volume Fire

Volume fire is an effective method to employ when using small arms fire against aerial threats. The key to success in engaging enemy aircraft with small arms is to put out a high volume of fire towards the immediate threat. The more bullets a unit can put in the sky, the greater the chance the enemy will fly into them. Even if these fires do not hit the enemy, throwing up a "wall of lead" in the sky can intimidate threat UAS pilots, ultimately breaking off their attack or distracting them from taking proper aim. One of the most important points about volume fire is that once the lead distance is estimated, you must aim at the estimated aiming point and fire at that single point until the aircraft has flown past that point. Maintain the aiming point, not the lead distance. Once you start firing, do not adjust your weapon.

When the decision is made to engage an aircraft with small arms, every weapon (M4, M240, M249, and M2) should be used with the goal of placing as many bullets as possible in the enemy's flight path. That does not mean that everyone fires in some random direction. Instead, everyone selects an aiming point in front of the UAS and fires at that point. This aiming point is determined using the football field technique. When deciding to fire upon an enemy UAS, practical considerations need to be considered before engaging. An example is engaging a UAS from a range of 3 kilometers. It is important to consider the capabilities of the available weapons. Using small arms at this range is ineffective while the best possibility may be the use of the main gun on a tank or tracked vehicle. Small arms have a low probability of kill against attacking UAS due to their size, speed, and maneuverability.

Football Field Technique

The football field technique is a simple method of estimating lead distance. The theory is that most people have played or watched football and have a concept of how long a football field is. When told to lead the UAS by one football field, everyone aims at approximately the same point in space. One person's error in making the football field estimate will be offset by another person's error. The variation in aiming points will ensure that massed fire is delivered into a volume of space in front of the UAS rather than on a small point. Also, the differing perspectives from which the service members view the UAS will act to further distribute the fire over a volume of space.

Aiming points used to engage threat UAS are different but may be used on a variety of different threats. An example of this technique would apply if enemy helicopters were detected, and the decision was made to engage it should be engaged as a group 5 rotary wing UAS. The rules for selecting aiming points are simple, easily learned, and retained. The various aiming points are summarized in table 5-4.

React to Threat UAS (Flow Chart) (See pp. 5-10 to 5-14.)

Ref: ATP 3-01.81, Counter-Unmanned Aircraft System (C-UAS) Operations (Mar '25), p. 36.

Understanding the capabilities and limitations of UAS can increase a Soldier's ability to react and defeat a threat. Reacting to a threat UAS should include reporting to the operations center and security personnel if the UAS does not pose an immediate threat. Defeat does not equate lethal means; however, it is an option. Other defeat solutions could be limiting a surveillance threat from gaining information or following the air path of the UAS to the operator.

Figure 4-6 is a simplified flow chart showing the process from UAS detection to submission of the final report to higher. Leaders must be aware that local standard operating procedures may dictate other steps and requirements.

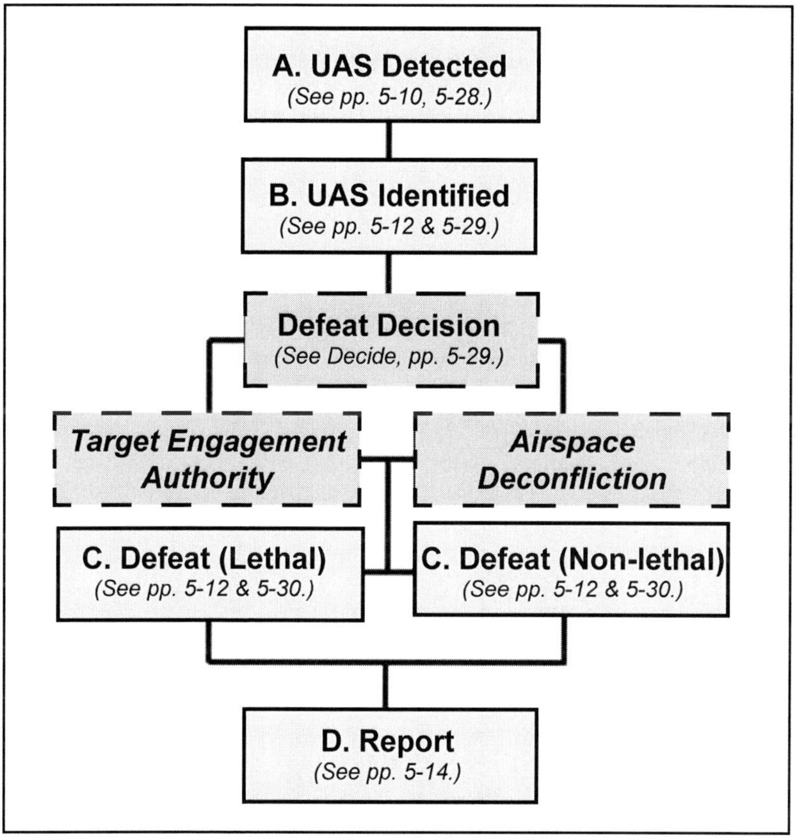

Ref: Adapted from ATP 3-01.81 (Mar '25), fig. 4-6. Unmanned aircraft system reporting flow chart. (See pp. 6-6 to 6-7 for an expanded view.)

Editor's note: In addition to "React to Threat UAS" (pp. 5-10 to 5-14), ATP 3-018.81 also outlines steps in terms of "Active Defense Measures" (see pp. 5-27 to 5-30).

UAV Type	Course	Aiming Point
Group 1 UAS	Crossing	0.5 football fields in front
Group 2 UAS	Crossing	1 football field in front
Group 3 - 5 UAS	Crossing	2 football fields in front
All UAS	Directly at you	Slightly above the UAS body
Rotary wing	Hovering	Slightly above the UAS body
Legend		
UAS unmanned aircraft system		

Ref: ATP 3-01.81 (Mar '25), table 5-4. Aiming Points

Engaging With Machine Guns

The machine gun is effective against slow-moving UAS. To sustain the volume of fire and kill a UAS, a continuous burst of 20 to 25 rounds fired using a tracer on target method, allowing the gunner to adjust rounds on target as follows:

- Vehicle's machine guns are useful weapons against UAS. They can be fired quickly unleashing a high-volume of fire. Ensure that the area is clear of friendly personnel. Use the following guidelines when using crew served weapons.
- When UAS are hovering or inbound, aim high with the machine gun and fire a continuous burst, adjusting onto the UAS by observing the travel and impact of tracers remembering, tracers may appear to be striking the target when they are going under it.
- If engaging a moving UAS, track along its flight path using a lead of 50 meters or half a football field. Fire a continuous burst, forcing the UAS to fly through the cone of fire.

When coordinated effectively, the aircraft will fly through the unit's cone of fire. Vehicle commanders must not try to track high-speed aircraft because they fly too fast. When enemy aircraft fly directly at vehicles, vehicle commanders are responsible for engaging them with controlled machine gun bursts of 50 to 100 rounds. Select an aiming point slightly above the nose to position the UAS in the cone of fire.

Due to the speed of some types of UAS, the best technique to use against them is to fire all automatic weapons in continuous bursts. If the UAS are inbound, aim slightly above the nose or fuselage and fire. If the UAS is crossing, use a lead of 200 meters (two football fields) and fire letting the UAS fly through the cone of fire from the machine guns. Do not try to track or traverse your fire for group 3 and above UAS since they fly too fast.

V. Tank Platoon Example

Ref: ATP 3-20.15, Tank Platoon (Jul '25), pp. 218 to 225.

The platoon should assume they are being observed by enemy reconnaissance and targeting systems, and not assume they are under a protective umbrella of friendly air and missile defense units. UAS is a key threat to the tank platoon's ability to maneuver. While not all hostile air threats require engagement using air defense measures from air and missile defense units, there is still a requirement to detect,identify, and be prepared to counter and defeat all classes of UASs.

Counter-Unmanned Aircraft System Planning

Platoon leaders must plan for the threat environment where the units will conduct operations. Passive air defense, combined arms for air defense and counterreconnaissance tasks training should be an integral part of the units' C-UAS practices. The unit should develop and refine C-UAS planning and tailored to the expected threat environment. The platoon and company must plan and execute operations as an integrated combined arms team employing all forms of passive air defense techniques when active air defense is limited or not available. Intelligence preparation of the operational environment provides the commander with specific threat information on known enemy locations, tactics, and threat capabilities.

Examples of key tasks to integrate into plans and combined arms unit training strategies addressing sUAS threats include the following:

- Employ dedicated observers (conducting air guard techniques).
- Perform visual aircraft recognition training.
- Conduct air threat avoidance techniques.
- Establish a security force and quick reaction force.
- Establish an early warning organic sensor network.
- Conduct UAS reporting procedures.
- Employ cover and concealment techniques.
- Select appropriate sUAS defeat mechanisms.
- Conduct hardening of unit positions.
- Disseminate the air defense warning and weapons control status (WCS).
- Employ counter tracking techniques.

Platoon leaders must understand the plan for friendly UAS employment. This assists the platoon in differentiating from friendly and threat UAS, and disseminating the information when one is launched at the platoon/company/BN.

Employing an Air Guard *(See pp. 5-20 to 5-26.)*

Platoon security operations are complemented by employing air guard techniques. An air guard may assist with mitigating the threat's use of aerial assets against the unit. Air guards need to be vigilant, focusing on the horizon. Air guards will perform actions such as search and scan techniques for approaching threat UASs while observing their assigned sectors. Air guards should position themselves where they can best observe and more importantly listen for threat UAS. When listening, OPs should exercise noise discipline, ensure all engines are off and remove their headgear to listen. Early warning is the key for air guards since it is their responsibility to alert the formation of any possible air threats. Reporting threat UAS activity should include an estimate of the threat location from the air guard position. The air guard reports the approximate distance, time, duration, size, estimated elevation, and direction the UAS was heading when detected. Reporting of a threat UAS should utilize a standard reporting format. Leaders should always have a plan for air guards to include:

- Plan for air guards while moving and at halts.
- Plan for air guards in AAs and defensive positions.
- Plan for air guards when in overwatch positions.

Based on threat activity and mission tasks relative to C-UAS observers (air guards) should consider developing quick reference or pre-deployment and combat checklists to focus the team on C-UAS. The checklist should be available through standard military digital devices or in hardcopy form and include the following:

- Current UAS trends (type classification).
- Specific data on local air threats and named areas of interest.
- Secure radio operations and frequencies.
- Unit call signs to request support (quick reaction force or reconnaissance and information collection support).
- Military map of area.
- Binoculars and night vision devices.
- Orientation techniques (location, heading, speed, and line of sight).
- C-UAS SPOTREP *(See pp. 5-24 to 5-25.)*

Battle Drill (Tank Platoon Example)

Ref: ATP 3-20.15, Tank Platoon (Jul '25), pp. 354 to 358.

Battle drills are rehearsed and well understood actions made in response to common battlefield occurrences (ADP 3-90).

Battle Drill Foxtrot: React To Unmanned Aircraft System While Mounted (17-PLT-D9516)

Conditions

The platoon is conducting operations in a training/combat environment. The designated air guard identifies a UAS which requires the platoon to take either passive or active air defense measures.

Task Steps

1. React to the threat.
2. Air guard detects the UAS.
3. Air guard alerts the platoon with a contact report containing these elements:
   a. "CONTACT."
   b. "ENEMY UAS."
   c. Announces cardinal direction. (Specify, "NORTH, SOUTH, EAST, or WEST.")
   d. Gives visual signal for air attack. *(See figure F-4.)*
4. VCs place their hatches in the open protected or closed position; other crewmen close their hatches.
5. Platoon leader/sergeant transmits contact report to higher HQ.
6. Platoon leader/sergeant gives visual or radio communication signal for the platoon to execute a herringbone. *(See figure F-5.)*
7. Platoon increases dispersion.

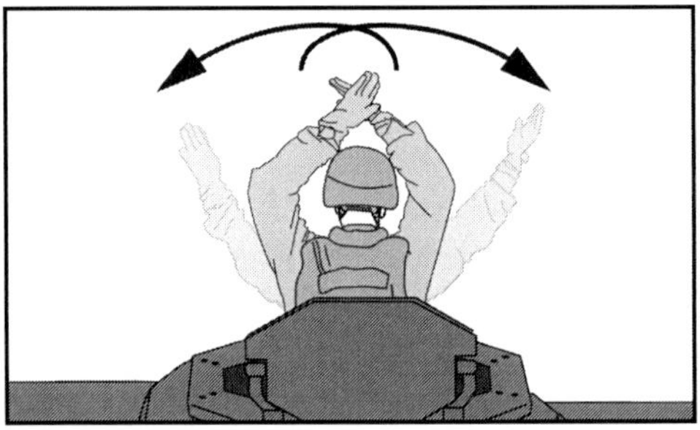

Figure F-4. Visual signal for air attack.

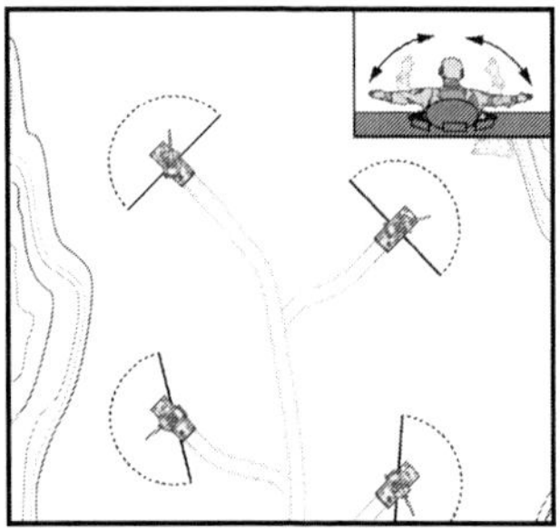

Figure F-5. Visual signal for herringbone.

Develop the Situation

1. The platoon leader assesses what action is appropriate to counter the threat based on the following: (See figure F-6.)
   a. Distance and direction to UAS.
   b. UAS activities. (Attacking, hovering, crossing.)
   c. Air defense warning, weapon control status, and engagement criteria given in OPORD.

2. Dismounted Soldiers immediately remount their vehicle or seek best available cover.

3. Vehicle crewmembers secure all hatches, doors, and ramps.

Choose a Course of Action

1. Continue mission. *(Note. Higher HQ may order the platoon to continue mission.)*

2. Seek cover and concealment. *(See figure F-7.)*

Note; Disperse and Seek cover/concealment) Vehicles can create a nonlinear target by moving as fast as possible at a 45-degree angle away from the path of flight and toward attacking aircraft. The platoon maintains an interval of at least 100 meters between vehicles, forcing aircraft to make several passes to engage the entire platoon.

3. Engage threat UAS by: *(see figure F-8).*

    a. Direct vehicles to aim based on type and course of UAS.

    b. Direct vehicles to engage with all machine guns or 120-mm canister rounds.

    c. Direct cease fire when aircraft is destroyed or no longer a threat.

4. Execute and report to higher HQ.

    a. Execute selected COA until the threat is defeated or withdraws.

    b. Platoon leader/sergeant transmits SPOTREP to higher HQ.

5. The platoon leader reports the platoon's status then continues operations according to the order and the commander's guidance.

RULES FOR SELECTING THE AIM POINT

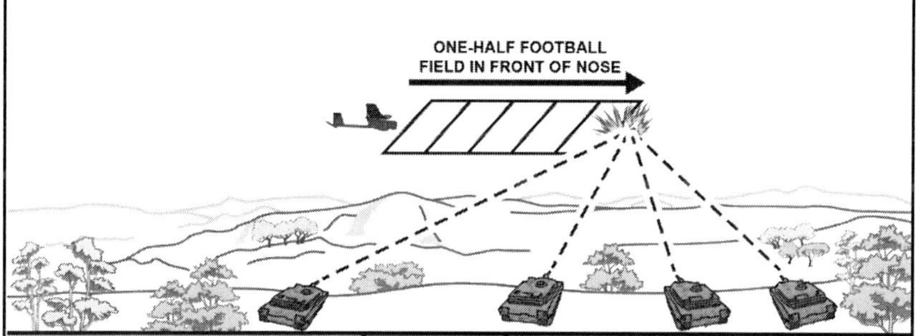

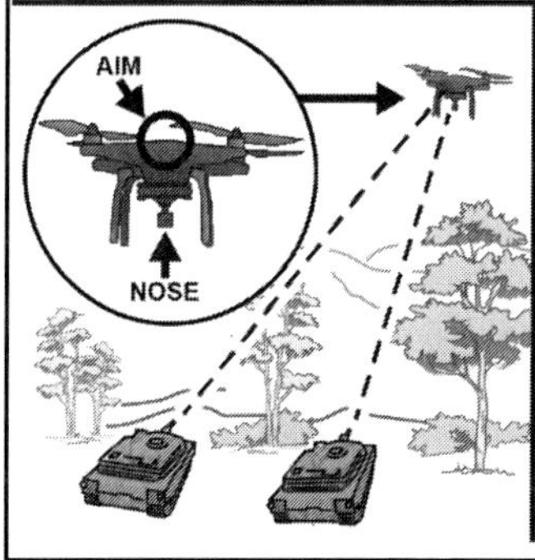

TYPES OF AERIAL PLATFORMS	COURSE	AIM POINT
FIXED WING	CROSSING	ONE-HALF FOOTBALL FIELD IN FRONT OF NOSE
BALLOON	HOVERING	SLIGHTLY ABOVE BALLOON BODY
QUAD	DIRECTLY AT YOU	SLIGHTLY ABOVE HELICOPTER BODY

Figure F-8. Point of aim.

Note. Low, slow, sUASs can fly extremely low underneath traditional radar detection zones. UAS fly very slowly and can even hover in place preventing current sensors from detecting them. UAS are very difficult to defeat using direct fire weapons.

Active Defense Measures (Tank Platoon)

Ref: ATP 3-20.15, Tank Platoon (Jul '25), pp. 224 to 225.

Platoon leaders have the responsibility to take whatever action is necessary to protect their forces and equipment against attack and ensure their Soldiers operate in accordance with established ROE. Active measures for the platoon must include basic rules that assist in the identification and defeat process for threat UASs. For example,establish SOPs for disseminating WCS and hostile criteria. The platoon leader should consider training their forces in the following active measures:

- Define characteristics for threat UAS (factors for defining the characteristics of threat UASs are speed, altitude, location, and heading).
- Develop and transmit WCS.

Weapon Control Status (WCS) *(See pp. 6-27 to 6-28.)*

WCS is a control measure designed to establish procedures for forces using surface-to-air weapons (including small arms weapons) to engage threats. WCSs can apply to weapon systems, volumes of airspace, or types of air platforms. This includes established restricted and engagement zones. Categories of WCS are as follows:

- Weapons-free
- Weapons-tight
- Weapons-hold

Using Air Guard *(See pp. 5-20 to 5-26.)*

Designate air guards for every vehicle and position to establish 360-degree security and execute the following:

- React to threat UASs by determining distance and bearing to the threat and take pictures if possible.
- Immediately report sightings of threat UAS (SPOTREP) as prescribed by the SOPs.
- If the air guards' position and personnel become threatened execute the following:
  - Respond in accordance with established unit SOP that could include moving to alternate positions.
  - Engaging the threat UAS with small arms using combined arms for air defense firing techniques.
  - Request engagement support with air and missile defense weapon systems and aviation assets.

Air guards need to be vigilant with eyes on the horizon. Air guards are responsible for spotting aerial threats within proximity to the unit's location and providing early warning by alerting the unit of possible air threats. Sector limits cover likely avenues of approach for threat aircraft. Air guards can be used during mounted and dismounted offensive and defensive operations.

When scanning for UAS, do not limit the search to near the horizon and miss higher-flying aircraft or search too high above the horizon and miss lower flying targets. When the tank is completely buttoned up, a crew member has limited ability to scan the horizon. This limited ability must be considered during the threat assessment.

Platoons should always post an air guard day and night for all operations, based on METT-TC (I). If two or more teams are posted, then assign sectors. Air guards must be trained to watch but also listen for adversary UAS. Listening and OPs should also be assigned the air guard mission. While moving or stationary all members of the platoon should be vigilant and on the watch for threat UAS.

III. Air Defense Warnings (ADW) & Controls

Ref: ATP 3-01.81, Counter-Unmanned Aircraft System (C-UAS) Operations (Mar '25), chap. 6.

This section describes how air defense warnings (ADWs), weapons control statuses, and ROE are used to mitigate the effects of potential air threats and attack.

This discussion is related to the execution phase of the Army's operation process presented in terms of engagement operations *(see pp. 6-1 to 6-10).*

I. Air Defense Warning (ADW) Conditions

Air defense C2 nodes distribute standardized air defense warnings (ADWs) to alert all forces of air and missile attacks. Commanders can issue local ADWs, which provide additional guidance alerting all members within their area of operations. Supported units may monitor the AMD early warning network that contains specific information for ADA units to include more detailed track information. However, the format of directed early warning does not necessarily follow the size, activity, location, unit, time, and equipment report format. Hostile track information and targeting data may be reported using the same systems as directed early warning. Frequencies used to monitor AMD early warning networks are located in the unit's communications plan.

ADW conditions demonstrate the degree of air attack probability based not only on current threat assessments but also on future air threats as required. The area air defense commander (AADC) issues ADWs for the joint operations area, based on current threat intelligence and disseminated through C2 channels to subordinate AMD commanders.

There are three types of ADWs: red, yellow, and white, with red being the most urgent. Any commander, after coordination and approval through air defense channels, may issue a higher level of warning for their command but may not lower the level. The chain of command must ensure that every Soldier knows the current ADW. ADWs will be disseminated through C2 systems throughout a theater. It is the responsibility of supported units at every echelon to inform their respective commanders of the current ADW. Army commanders must then inform their subordinate units of the ADW.

Air Defense Warning (ADW) Conditions

- **ADW Red (Attack imminent or in progress)**
- **ADW Yellow (Attack probable)**
- **ADW White (Attack improbable)**

Descriptions of the ADWs are as follows:

- **ADW Red.** An attack by hostile aircraft or missile is imminent or in progress
- **ADW Yellow.** An attack by hostile aircraft or missile is probable.
- **ADW White.** An attack by hostile aircraft or missile is improbable.

Refer to FM 3-01 for further discussion on ADWs.

II. Alert States and Readiness Conditions

AMD commanders use weapons alert designators, alert states and readiness conditions to ready their forces and configure their systems rapidly against an enemy air attack. An example of this is the ready for action drills conducted upon receiving a new surface to air missile tactical order, which heightens a unit's alert state. In conjunction with elevating the unit's readiness, the commander also retains the ability to stand down units for rest or maintenance.

Alert states and readiness conditions are implemented locally and coordinated with guidance from the AADC and designated controlling authorities. The combined arms force should understand ADA brigade commanders establish and control the alert state for their subordinate units operating in support of forces in the maneuver areas of operation. Actual timelines and manning requirements are determined by theater or joint operations area plans. Alert state and readiness condition information is normally found in the unit's tactical standard operating procedures which also considers the factors of mission, enemy, terrain and weather, troops and support available, time available, and civil considerations.

III. Rules of Engagement (ROE)

Rules of engagement (ROE) are directives issued by competent military authority that delineate the circumstances and limitations under which United States forces will initiate and/or continue combat engagement with other forces encountered (JP 3-84).

ROE at all echelons are positive and procedural directives that specify the circumstances and limitations under which forces will initiate or continue combat engagements with enemy forces. The joint force commander approves the theater ROE. These established ROE enable the commanders to delegate the authority to execute air defense operations.

ROE permit the senior air defense commander to retain control of an air battle by prescribing the exact conditions under which engagements may take place. ROE is a direct reflection of the centralized control with decentralized execution principles of air defense. ROE apply to all warfare participants in the theater and are disseminated to all echelons of air, land, and sea forces.

There are seven AMD ROE categories: right of self-defense, identification criteria, fire control orders, weapons control status, levels of control, autonomous operations, and modes of control. For more information on each refer to FM 3-01.

A. Right of Self-Defense

Commanders at all echelons must take whatever action is necessary to protect their forces and equipment against air attack, to include from UAS. Self-defense operations allow friendly units to defend themselves against direct attacks or threats of attack with the use of organic weapons and systems.

When under attack, the right of self-defense takes precedence over any other established rules and procedures, which normally govern engagement. The right of self-defense is inherent in all ROE and weapons control procedures (JP 3-01).

B. Identification Criteria

The employment of C-UAS systems requires early identification of friendly, neutral, or hostile UAS to maximize effective range engagement. Active defense measures protect the force, reducing delays during tactical operations, and mitigate fratricide. Positive identification of aircraft is normally the preferred method of operation. Positive identification is an identification derived from observation through visual recognition, electromagnetic support systems, identification friend or foe systems, or other physics-based identification techniques. In the absence of positive identification, procedural identification is used, which employs previously established and disseminated airspace coordinating measures. Procedural identification separates airspace users by geography, altitude, heading, time, and/or maneuver. Generally, some combination of positive and procedural identification will be used.

Proper employment of air defense weapon systems enables early detection, identification, classification, and alert of threat aircraft and missiles which maximizes beyond-visual-range engagements and helps to avoid fratricide. Identifying and classifying friendly air platforms from enemy assets while employing various weapon systems is a highly complex task. For example, visual identifications may be different from identifications provided earlier by alerting sensors and supporting systems. Additional considerations include time of detection, range, projected flight path, and possible battlefield effects. Tasks will differ based on the type of threat platform detected, whether manned or unmanned. While the ROE for aircraft may allow additional time before engagement, this will not always be the case with UAS.

Hostile Criteria

Hostile criteria are basic rules that assist in the identification process of friendly or hostile air platforms. These rules are promulgated through subordinate commanders from the joint force commander or AADC when so authorized. The commander who establishes hostile criteria parameters may consider the factors of speed, altitude, and heading or other requirements within specified volumes of airspace. The local commander may also consider specific enemy threat characteristics or hostile acts.

Echelons having engagement authority use hostile criteria to determine the identification of detected air targets. The highest echelon capable of managing engagement operations normally retains identification and engagement authority. Upon target detection, fire units and sensors with real time and near-real-time data transmission capability assist the controlling authority by forwarding targeting information. The controlling authority makes the final target identification and delegates engagement authority. Engagement authority may be given to the lowest echelon, such as a crew, capable of directing an engagement, depending upon such factors as the type, intensity, and area of operations of the air threat.

C. Fire Control Orders

Fire control orders are commands used to control engagements on a case-by-case basis, regardless of the prevailing weapon control status. Higher ADA echelons, when monitoring the decentralized operations of subordinate units, most often use these commands. Fire control orders can be transmitted electronically or verbally; however, not all the fire control orders shown below can or will be used by every type of ADA unit.

The orders most applicable to C-UAS operations are-

- **Engage**. In air and missile defense, a fire control order used to direct or authorize units and/or weapon systems to attack a designated target. (JP 3-01) Engage is an order issued by the engagement authority.
- **Hold Fire.** An emergency fire control order to stop firing on a designated target and destroy missiles in flight (ATP 1-02.1). • Cease Engagement or Cease Fire. A fire control order used to direct air defense units to stop tactical action against a specified target. Missiles in flight will continue to intercept (ATP 1-02.1).

D. Weapons Control Status

A weapon control status is an air and missile defense control measure declared for a particular area and time by an area air defense commander, or delegated subordinate commander, based on the ROE, that establish conditions under which fighters and surface air defense weapons are permitted to engage threats (JP 3-01). Weapon control status establishes procedures for forces using surface air defense weapons (including small arms weapons) to engage threats. Weapon control statuses can apply to weapon systems, volumes of airspace, or types of air platforms. The tactical situation normally determines the degree or extent of control necessary over particular weapon systems. ADA officers at headquarters planning cells or AMD coordination and liaison teams typically establish or recommend separate weapon control statuses for various air threats, including fixed and rotary-wing aircraft, or UAS, at their supported headquarters. Air defense is optimized when all forces have the ability to rapidly receive and disseminate weapon control statuses for all air platforms.

The three weapon control statuses are:

- **Weapons free** is the least restrictive status: used to indicate when any target not positively identified as friendly in accordance with current ROE may be engaged (JP 3-01).
- **Weapons tight** is the normal status. Units may only fire on targets identified as hostile in accordance with current ROE (JP 3-01).
- **Weapons hold** is the most restrictive status. Units may only fire in self-defense or when ordered by proper higher authority (JP 3-01).

Any unit can receive ADW or directed alert of possible or immediate threats to their forces or to a specific area of the battlefield. The AMD cell provides the air defense expertise for integration into the corps/division air defense plan. The ADAM personnel contribute to the brigade and below units' airspace plans. The airspace control element reviews and deconflicts the corps or division air defense plan with other division control measures. The control measures for the air defense plan are normally sent to higher headquarters through AMD channels.

E. Level of Control

Level of control describes the ADA echelon at which positive management of the air battle is conducted. This can be an AADC, regional air defense commander, sector air defense commander, ADA brigade, battalion, or the individual unit. Different levels of control are normally established for engagements of fixed-wing aircraft, rotary-wing aircraft, and UAS.

F. Modes of Control

Combined arms forces must be familiar with the overall area air defense plan that specifies the modes of control, trigger events and when they should be changed, and who has the authority to change them. The mode of control selected may depend upon the capabilities of the communications system, the weapons systems employed, and both the friendly and enemy air situations. There are three modes of control in AMD: centralized, decentralized, and autonomous.

- **Centralized Control Mode.** In this mode, a higher echelon must authorize target engagements by a firing unit.
- **Decentralized Control Mode.** In this mode, a higher echelon monitors unit actions and makes direct target assignments on a management by exception basis, to ensure proper fire distribution, prevent engagement of friendly air platforms, and prevent simultaneous engagements of hostile air targets.
- **Autonomous Operations**. A unit assumes autonomous operations after it has lost all communications with higher and adjacent echelons. The unit commander assumes full responsibility for control of weapons and engagement of hostile targets. Normally, the ROE and supplemental fire control measures in effect at the time of communications loss remain in effect until communications are regained. Changes to ROE and supplemental fire control measures scheduled to go into effect after communications are lost will be implemented as scheduled.

Chap 7 Small-Unmanned Aircraft Systems (sUAS)

I. DoD Unmanned Aircraft Systems (UAS): Roles, Missions, and Future Concepts

Ref: CRS Report R47188, Unmanned Aircraft Systems: Roles, Missions, and Future Concepts (Jul '22), Kelley M. Sayler and Michael E. DeVine.

Unmanned aircraft systems (UAS) have gained increased prominence in U.S. military operations. The Department of Defense (DOD) is currently developing advanced UAS, along with optionally crewed aircraft, as part of its modernization strategy. The roles and missions of UAS are relevant to Congress in authorizing, appropriating, and providing oversight to DOD and the military services for these systems.

Over the past decades, military forces have used UAS to perform various tasks, including

- intelligence, surveillance, and reconnaissance;
- close air support;
- cargo and resupply; and
- communications relay.

Analysts and DOD argue that UAS could replace crewed aircraft for a number of missions, including

- aerial refueling;
- air-to-air combat;
- strategic bombing;
- battle management and command and control (BMC2);
- suppression and destruction of enemy air defenses; and
- electronic warfare (EW).

In addition, DOD is developing several experimental concepts—such as aircraft system-of-systems, swarming, and lethal autonomous weapons—that explore new ways of employing future generations of UAS.

See following pages (pp. 7-2 to 7-3) for an overview and further discussion.

II. The U.S. Army's Small Uncrewed Aircraft Systems (sUAS)

Ref: CRS Report IF12668, The U.S. Army's Small Uncrewed Aircraft Systems (Nov '24), Daniel M. Gettinger.

The U.S. Army is acquiring a family of small uncrewed aircraft systems (sUAS) for ground maneuver elements at the battalion level and below to provide real-time reconnaissance, surveillance, and target acquisition (RSTA) capabilities. For the past two decades, this role was largely filled by the AeroVironment RQ-11 Raven. On February 8, 2024, Secretary of the Army Christine Wormuth and General Randy George, Army Chief of Staff, announced Army plans to phase the RQ-11 Raven out of service as part of a broader "rebalance" of the Army's aviation investments. This product covers the Group 1 and 2 UAS—those that weigh less than 55 pounds and fly at or below 3,500 feet above ground level—intended to serve as successors to the Army's legacy RSTA sUAS.

Current Roles of DoD UAS (Overview)

Ref: CRS Report R47188, Unmanned Aircraft Systems: Roles, Missions, and Future Concepts (Jul '22), John R. Hoehn, pp. 1 to 10.

Since the introduction of more sophisticated UAS in the 1990s, UAS have performed several types of missions that were previously performed solely by crewed platforms. These missions include cargo and resupply; intelligence, surveillance, and reconnaissance; close air support; communications relay; and aerial refueling.

Intelligence, Surveillance, and Reconnaissance (ISR)

UAS initially gained prominence in U.S. military operations by performing intelligence, surveillance, and reconnaissance (ISR) missions. Although the United States has operated ISR UAS since the 1960s, the United States' use of UAS for such missions became more widespread with General Atomics' development of the MQ-1 Predator—then designated as the RQ-1 Predator—in the 1990s. The military's deployment of UAS in conflicts such as Kosovo (1999), Iraq (2003-present), and Afghanistan (2001-present).

The United States military primarily used MQ-1 UAS for reconnaissance and acquisition of potential ground targets in these conflicts. To accomplish this mission, the aircraft operated with a 450-pound surveillance payload, which included two electro-optical (E-O) cameras and one infrared (IR) camera for use at night. Since the MQ-1, several iterations of UAS, including the MQ-9 Reaper, have included more sophisticated ISR sensors. Uncrewed aircraft, like the RQ-4 Global Hawk UAS, also fly at higher altitudes and use sophisticated signals intelligence payloads and synthetic aperture radars. Demand for ISR UAS has increased as the Department of Defense (DOD) has procured larger fleets of aircraft. DOD officials and defense analysts state that combatant commanders routinely request more ISR aircraft—in particular UAS, due to their ability to fly for long durations. As the United States has withdrawn from conflicts around the world, the military has increasingly relied on UAS to provide ISR and targeting.

Close Air Support (CAS)

DOD has expanded the air-to-ground strike mission of UAS to include close air support (CAS). DOD defines CAS as "[a]ir action by aircraft against hostile targets that are in close proximity to friendly forces and that require detailed integration of each air mission with the fire and movement of those forces." Because of the high risks that CAS poses to friendly forces on the ground due to such proximity, CAS was traditionally performed by crewed platforms, such as the A-10 Thunderbolt II. During operations in Iraq, Afghanistan, and Syria over the past decade, however, the mission set of MQ-9 Reapers has evolved to perform CAS.

As an example of UAS performing CAS, on February 21, 2001, the Air Force launched an AGM-114 Hellfire missile from what was then called an RQ-1 Predator UAS, marking the first use of strike UAS. A strike UAS has the capability to launch weapons, such as precision guided missiles against a target. The Air Force's use of the MQ-1 expanded considerably during the wars in Iraq (2003-2010) and Afghanistan (2001-2021).

Communications Relay

During Operation Enduring Freedom, ground forces identified challenges with operating their line-of-sight communications systems in mountainous terrain, which limited the distance that radio frequencies can travel. In response, in 2008, the Air Force fielded the Battlefield Airborne Communications Node (BACN) to provide a relay for both voice communications and tactical datalinks. The Air Force selected the crewed E-11A—a modified Bombardier Global XRS/6000 series aircraft—as well as older RQ-4 Global Hawk aircraft (subsequently designated the EQ-4) to perform this function due to their long endurance and ability to fly at relatively high altitudes.

Potential Roles and Missions for UAS

Aerial Refueling

Using UAS to refuel other aircraft while airborne, especially in locations far away from airbases, can potentially reduce the threat to crewed tankers. The Navy is procuring the MQ-25 Stingray carrier-based UAS for such operations. In addition, the Defense Advanced Research Projects Agency (DARPA) has also conducted research to test the suitability of RQ-4 Global Hawk UAS aircraft as long endurance refueling platforms, but the military services have yet to develop formal requirements for this role.

Air-to-Air Combat

Some relatively early concepts of military UAS sought to carry out fighter operations, engaging with and destroying enemy aircraft. For example, the Navy's Unmanned Carrier-Launched Surveillance and Strike (UCLASS) program intended to develop light-strike capabilities (i.e., using small aircraft to attack targets) in a fighter-sized UAS. More recently, demonstrator programs like the Boeing Loyal Wingman seek to add air combat into the mix of fighter aircraft.

Combat Search and Rescue/Casualty Evacuation

Locating and recovering downed personnel, known as search and rescue, as well as aerial casualty evacuation, may involve extended periods of time in high-threat environments. Due to their endurance and reduced size (i.e., no crew), UAS may be well suited for carrying out search missions without being detected or requiring crew relief. In addition, military leaders may potentially accept a greater risk of attrition for UAS in these situations.

Cargo and Resupply

DOD has also used UAS in cargo and resupply missions. Notably, the Marine Corps used a UAS helicopter called the K-MAX for operations in Afghanistan. The K-MAX, which was intended to replace ground convoys, flew autonomously to remote outposts carrying up to 6,000 pounds of cargo or supplies.

Strategic Bombing

The Air Force intends for the next generation of U.S. bombers—such as the B-21 Raider Long-Range Strike Bomber—to be optionally crewed, allowing remote operation for conventional strike missions into heavily defended areas.

Battle Management Command and Control (BMC2)

Current battle management command and control (BMC2) platforms, such as the E-3 Sentry Airborne Warning and Control System (AWACS) and E-8 Joint Surveillance Target Attack Radar System (JSTARS),carry an array of sensors and a crew to translate those sensor inputs and direct friendly aircraft and forces.

Suppression and Destruction of Enemy Air Defenses (SEAD)

Suppression and destruction of enemy air defenses is the first and most dangerous part of an air campaign, because attacking aircraft generally face enemy air defense networks at their full capability. In addition to expendability, UAS offer a number of advantages for such missions. Large numbers of small UAS engaging enemy air defenses could overwhelm these defenses and compel an adversary to use many of its weapons against small, comparatively low-value targets. Such swarming tactics could also direct many attacking aircraft to the target and confuse responders.

Electromagnetic Warfare (EW)

Uncrewed aircraft could also serve as electromagnetic warfare platforms, designed to be outfitted with radio frequency jamming equipment capable of interfering with adversary electronics.

Background

In 1988, the Department of Defense (DOD) Unmanned Aerial Vehicle Joint Program Office (UAV JPO) published its first "Master Plan" for uncrewed aircraft, which established the requirements and acquisition strategy for UAV systems. The Master Plan recommended a "close range" UAS for "lower-level tactical units," one that could be acquired in large numbers and at low cost. In the late 1990s and early 2000s, the Military Operations in Urban Terrain Advanced Concept Technology Demonstration (MOUT ACTD), an Army-led experimentation program, showed how a man-portable sUAS could provide ground units with enhanced situational awareness and force protection. The MOUT ACTD led the Army and Special Operations Command (SOCOM) to work with AeroVironment to develop in 2002 the fixed-wing, 4-pound RQ-11 Raven, a smaller, more advanced version of the Gulf War-era AeroVironment FQM-151 Pointer. The Army introduced the RQ-11 in 2003 and, by 2010, fielded close to 4,000 Raven aircraft. The Marine Corps, SOCOM, and the Air Force also adopted the Raven.

Beginning in the early 2010s, Army officials devised plans for expanding the service's small UAS capabilities to include a family of short-, medium-, and long-range sUAS platforms. Under the Rucksack Portable UAS Increment II Capability Production Document (RPUAS CPD), approved in 2013, the Army provided a modified Raven and limited quantities of the AeroVironment RQ-20 Puma to companies and battalions, respectively, for medium- and long-range RSTA capabilities on an interim basis pending the development of new platforms for these roles. Platoons, meanwhile, would be equipped with a short-range drone, the development of which the Army planned to begin later in the decade. Separately, in 2017, the Army approved the Soldier Borne Sensor (SBS) program to acquire miniature drones for infantry squads. The Army selected the FLIR Systems Black Hornet as the SBS and awarded FLIR a contract for the first batch of SBS systems in May 2018.

The DOD Defense Innovation Unit (DIU) issued a request for information (RFI) in November 2018 for a quadcopter drone to serve as a short-range sUAS; DIU partnered with the Army on the platoon-level drone program the following April. For medium- and long-range sUAS, the Army did not begin to seek successors to the Raven and Puma until the early 2020s.

In June 2023, the Army transitioned the RPUAS CPD requirements to the Joint Small Uncrewed Aircraft Systems Capability Development Document (J-sUAS CDD), guidance that specifies key system and performance attributes and an acquisition timeline for the Army's planned family of RSTA sUAS. Similar to the Army's 2013 RPUAS CPD, the J-sUAS CDD describes Army plans to field what the service now refers to as short-range reconnaissance (SRR), medium-range reconnaissance (MRR), and long-range reconnaissance (LRR) UAS at the levels of platoon, company, and battalion, respectively. Additionally, the J-sUAS CDD contains three new initiatives—first-person view (FPV) drones for squads and platoons, tethered UAS for platoons, and a drone swarm capability for companies—the requirements for all of which remain in various stages of development. Including the SBS, the J-sUAS architecture encompasses seven programs.

The requirements for the aircraft in the J-sUAS CDD differentiate them from the Raven and Puma in several ways. The Raven and Puma sUAS feature a conventional fixed-wing configuration, which potentially hampers their usefulness in restrictive terrain, such as urban or forested areas. With SRR, MRR, and LRR UAS, the Army appears to be prioritizing a vertical takeoff and landing (VTOL) capability in the form of either multirotor or hybrid-VTOL configurations. The Army's RFIs for future sUAS indicate that the service plans to acquire aircraft that can deliver lethal payloads, such as air-dropped grenades or glide munitions. In contrast to the Raven and Puma, which were designed largely to conduct surveillance and reconnaissance, the Army may require the next generation of small UAS to conduct a greater variety of missions, including launching lethal strikes and relaying communications for other drones and ground units.

III. Pentagon to Increase Low-Cost Drone Production in U.S.

Ref: DoD News article by David Vergun (Jul. 16, 2025). https://www.war.gov/News/News-Stories/Article/Article/4246987/pentagon-to-increase-low-cost-drone-production-in-us/ (accessed 19 Oct 25.)

The Defense Department, with help from industry, will ramp up production and fielding of drones to maintain battlefield superiority.

Defense Secretary Pete Hegseth, who toured the displays, said the drones that are manufactured using off-the-shelf components for rapid production are examples of disruptive thinking.

Emil Michael, undersecretary of defense for research and engineering, said the prototypes on display went from concept to development in just an average of 18 months, a process that normally takes up to six years.

Defense Secretary Pete Hegseth and another civilian stand and talk in front of an unmanned aerial vehicle on display outdoors.

The department will continue to rapidly innovate and scale up production of drones and other systems using cost, resilience, firepower and range as driving factors, which are areas DOD wants to improve upon, Michael said.

Hegseth said in a July 10, 2025, memorandum that he's rescinding restrictive policies that hindered drone production.

The memorandum can be downloaded at: https://media.defense.gov/2025/Jul/10/2003752117/-1/-1/1/UNLEASHING-U.S.-MILITARY-DRONE-DOMINANCE.PDF

"Drones are the biggest battlefield innovation in a generation, accounting for most of this year's casualties in Ukraine. Our adversaries collectively produce millions of cheap drones each year," he said, noting the U.S. military is lacking needed quantities of lethal small drones.

The secretary said there are three goals:

- Prioritizing the purchase of American-made drones and parts with help from industry's private capital;
- Arming combat units with low-cost drones made by America's world-leading engineers and artificial intelligence experts; and
- Training with drones in realistic battlefield scenarios, led by leaders who are not risk averse.

President Donald J. Trump signed a June 6, 2025, executive order to speed up U.S. drone production using the latest innovative industry technologies.

The executive order can be downloaded at: https://www.whitehouse.gov/presidential-actions/2025/06/unleashing-american-drone-dominance/

The president said he supports reducing regulatory uncertainty and streamlining approval and certification processes for safe and secure drone production.

A man points out details of a drone on display to two other men. There is a crowd of onlookers behind them.

Also, the Federal Aviation Administration and DOD will coordinate to streamline the approval processes to expand access to airspace for conducting drone training, Trump said.

Small UAS in the Army's FY2025 Budget Request

In its proposed FY2025 budget, the Army requested approximately $46.1 million in procurement and $26.8 million in research, development, test, and evaluation (RDT&E) funding for the SRR, MRR, and LRR UAS. Additionally, the Army's list of unfunded budget priorities for FY2025 includes $70.5 million in procurement funding for the SRR and MRR UAS. For the Soldier Borne Sensor program, the Army requested $22 million and $1.6 million in procurement and RDT&E funds, respectively. Excluding the unfunded priorities list, the Army's FY2025 budget request for the aforementioned small UAS is roughly 21% greater than that of FY2024.

The SRR UAS is the Army's first program of record for a small quadcopter drone. DIU and the Army argued that the SRR UAS should capitalize on the increasing sophistication of the drones available on the commercial market. By executing the program in phased tranches, the Army has indicated that it aims to maintain the flexibility to respond to advances in technology and user feedback. In April 2019, DIU and Army officials selected six companies to compete for Tranche 1 of the SRR UAS program, before awarding Skydio a contract in February 2022 for the RQ-28A, a militarized version of the Skydio X2D. In October 2024, the Army confirmed that it had selected Teal Drones, a subsidiary of the firm Red Cat, for Tranche 2 of the SRR program.

In its proposed FY2025 budget, the Army requested $21.1 million to acquire 270 Tranche 2 systems, or 540 aircraft (two air vehicles constitute each SRR system), as well as $1.15 million in RDT&E funding. The cost of one SRR system is expected to rise from $39,800 in Tranche 1 to $65,000 in Tranche 2. Army budget justification documents attribute this potential increase to the Tranche 2 version's improved obstacle avoidance, communications, and electro-optical and infrared sensors, among other enhancements. It is unclear as of this writing whether the Army intends to engage in additional tranches of the SRR program in the future.

The MRR UAS would provide an organic RSTA capability to Army companies. In 2023, the Army Futures Command approved a Directed Requirement for a "Company-Level sUAS" that is intended to provide an initial Tranche 1 capability for the MRR UAS, as well as inform the requirements of that future system. In its proposed FY2025 budget, the Army requested $25 million in procurement funding for the Company-Level sUAS. Following a March 2024 RFI, the Army selected the Anduril Industries Ghost X, a helicopter-type drone, and the Performance Drone Works C-100 UAS quadcopter drone for Tranche 1 of the Company-Level sUAS Directed Requirement in September 2024. Then, in November 2024, Deputy Secretary of Defense Kathleen Hicks announced that the Army's selections for Tranche 1 of its Company-Level sUAS would participate in DOD's Replicator program.

In January 2023, the Army issued an RFI for the LRR UAS that expressed an interest in a platform that is capable of taking off and landing vertically and of flying for a minimum of five hours and at a line-of-sight range of 30 kilometers (18.6 miles). In its proposed FY2025 budget, the Army requested approximately $25.6 million in RDT&E funds for work on the LRR UAS. According to Army projections in its budget justification documents, the Army expects to evaluate prototypes for the LRR through FY2026 and to begin procuring the aircraft the following fiscal year.

IV. Army Producing Second Tranche of Short-Range Reconnaissance Drones

Ref: DoD News article by Matthew Ryan (Aug. 27, 2025). https://www.war.gov/News/News-Stories/Article/Article/4287649/army-producing-second-tranche-of-short-range-reconnaissance-drones/ (Accessed 19 Oct 25.)

The Army has initiated production of the second tranche of its short-range reconnaissance unmanned aircraft systems, marking a significant step forward in modernizing battlefield capabilities. The Army selected two vendors to manufacture the SRR system, which will equip the Army's Transformation in Contact units with advanced, networked communication systems designed to address emerging threats.

Initial fielding of SRR tranche one began in September 2022, and, to date, the Army has fielded over 16 brigades with this capability. Critical lessons learned and soldier feedback from tranche one were incorporated into tranche two. This strategy of integrating new technologies into future tranches will continue to provide the best UAS capabilities on an accelerated schedule.

A short-range reconnaissance unmanned aircraft system flies through the air May 9, 2024, at Joint Base McGuire-Dix-Lakehurst, N.J. The system is used to conduct surveillance during combat operations, providing warfighters with enhanced situational awareness.

The TiC initiative is part of a broader Army effort to evaluate how personnel, tactics and technologies align to meet the demands of modern warfare. The program manager for uncrewed aircraft systems remains committed to delivering essential equipment to soldiers in time to support the Army's strategic objectives.

These systems act as force multipliers, enabling precise fires, supporting ground operations with timely intelligence and enhancing operational flexibility and agility. They use cutting-edge sensors to integrate with uncrewed vehicle control, providing real-time intelligence, reconnaissance and target acquisition, empowering commanders with the tools to make rapid and informed decisions.

Additionally, SRR systems continue to improve soldier survivability by detecting threats, identifying safe routes and minimizing exposure to hazards, making them indispensable assets on the modern battlefield.

The Army's investment in SRR systems underscores its commitment to advanced technologies that improve combat effectiveness and protect soldiers in increasingly complex operational environments.

V. Blue UAS Cleared List (DoW Defense Innovation Unit)

Ref: (Updated Continuously) https://www.diu.mil/blue-uas-cleared-list (Accessed 19 Oct 25.)

The Defense Innovation Unit (DIU) strengthens national security by accelerating the adoption of commercial technology throughout the military and bolstering our allied and national security innovation bases. DIU partners with organizations across the Department of War (DoW) to rapidly prototype and field dual-use capabilities that solve operational challenges at speed and scale.

The following is the Blue UAS Cleared and Select lists. The Blue UAS Cleared List are those approved drones providing options for the evolving mission needs of DoD users. These drones are compliant with current law and policy, validated as cyber-secure (by either being added from AUVSI's Green List or by the third-party recognized assessors), and are available for DoD purchase and operation.

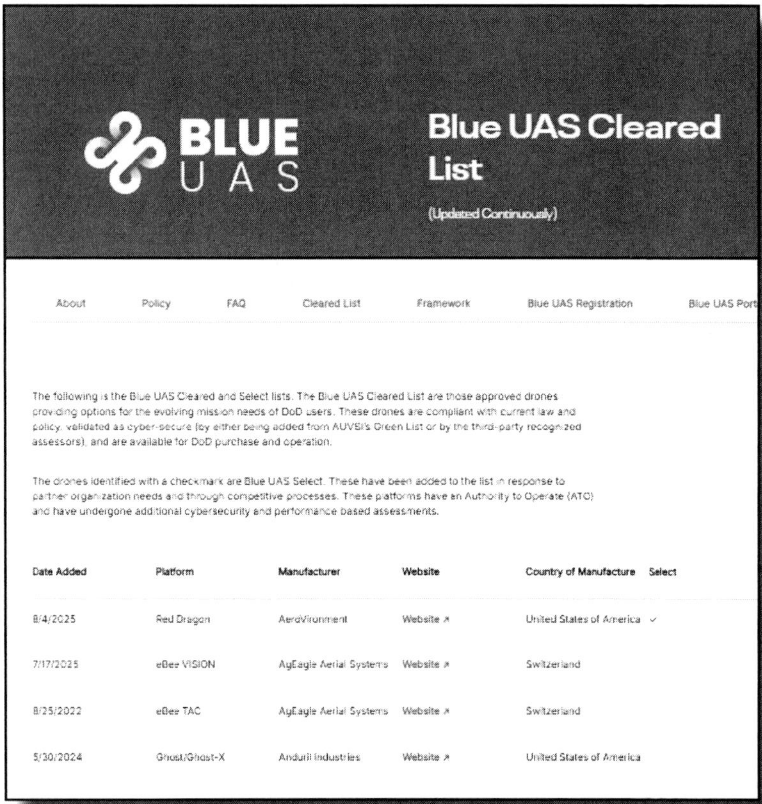

BLUE UAS

Blue UAS Cleared List

(Updated Continuously)

About | Policy | FAQ | Cleared List | Framework | Blue UAS Registration | Blue UAS Port

The following is the Blue UAS Cleared and Select lists. The Blue UAS Cleared List are those approved drones providing options for the evolving mission needs of DoD users. These drones are compliant with current law and policy, validated as cyber-secure (by either being added from AUVSI's Green List or by the third-party recognized assessors), and are available for DoD purchase and operation.

The drones identified with a checkmark are Blue UAS Select. These have been added to the list in response to partner organization needs and through competitive processes. These platforms have an Authority to Operate (ATO) and have undergone additional cybersecurity and performance based assessments.

Date Added	Platform	Manufacturer	Website	Country of Manufacture	Select
8/4/2025	Red Dragon	AeroVironment	Website ↗	United States of America	✓
7/17/2025	eBee VISION	AgEagle Aerial Systems	Website ↗	Switzerland	
8/25/2022	eBee TAC	AgEagle Aerial Systems	Website ↗	Switzerland	
5/30/2024	Ghost/Ghost-X	Anduril Industries	Website ↗	United States of America	

The drones identified with a checkmark are Blue UAS Select. These have been added to the list in response to partner organization needs and through competitive processes. These platforms have an Authority to Operate (ATO) and have undergone additional cybersecurity and performance based assessments.

I. Army Airspace

Ref: Center for Army Lessons Learned (CALL) Small Unmanned Aircraft System (sUAS) Airspace Management and Control handbook #25-14, chap. 1.

The Center for Army Lessons Learned (CALL) Small Unmanned Aircraft System (SUAS) Airspace Management and Control handbook was written to give Army leaders, planners, and sUAS team members a relevant, up-to-date guide on managing and controlling the Army's increasingly complex and congested airspace.

See pp. 1-16 to 1-17 for an overview of unmanned aircraft categories.

For the purposes of this handbook, sUAS are defined as systems in Department of Defense (DOD) UAS groups 1 and 2—those that have a maximum takeoff weight of less than 55 pounds and operate at altitudes up to 3,500 feet above ground level (AGL). These systems are concentrated at brigade-level-and-below echelons and are increasingly heavily represented in maneuver units of all types. Though the term sUAS is not yet doctrinal, it is widely recognized across the Army and joint communities.

I. Airspace Operational Environment

Understanding the airspace operational environment (OE) is a critical first step to successful sUAS planning and operations at brigade-level-and-below echelons. The Army's multi-domain concept demands an unprecedented level of cooperation and synchronization between Army, joint, and multinational organizations—a complex and challenging task that will only increase in importance.

Over the last decade, OEs worldwide experienced a remarkable surge in the number, capability, and significance of sUAS supporting ground combat formations. Although an increase in general unmanned aircraft system use was widely anticipated, actual growth—especially of sUAS—far exceeded even the wildest predictions. sUAS are now employed in huge numbers and a wide range of roles, fundamentally altering the conduct of military ground operations. The enormity of this shift emphasizes the importance of managing and controlling the airspace in which sUAS operate.

Massed proliferation of sUAS within Army formations will continue. Successful integration of these systems demands Army leaders and planners who have a proficient working knowledge of basic airspace processes and who properly prioritize developing and implementing unit airspace plans (UAPs) within their formations.

II. Army Airspace Control and Management

Airspace control is the exercise of delegated authority over designated airspace and users through control procedures and coordination measures to maximize operational effectiveness (JP 3-52).

Army operations of all types are inherently linked to the air domain. The authority to control and manage airspace and airspace users within that volume, however, is not automatically granted to the ground commander. Although Army commanders "own" the ground within their area of operations (AO), they do not own the airspace above it. The airspace control authority (ACA), normally the joint force air component commander (JFACC), may allocate a volume of airspace to Army commanders that request it, and who meet certain criteria. This allocation delegates certain authorities and responsibilities to Army and joint airspace elements for local control and management as a part of the theater airspace control system (ACS).

Joint doctrine describes airspace as a shared resource used by joint air operations, air and missile defense, and joint fires. sUAS operations must integrate with these other airspace users. This integration begins during planning. Army operations of all types, including sUAS operations, require awareness of the broader joint operations area (JOA) and potential conflicts with concurrent joint activities. These conflicts may involve both physical airspace and electromagnetic spectrum (EMS).

> **Airspace management** is the planning, coordination, integration, and regulation of airspace and users by airspace control elements in support of airspace control (JP 3-52).

Although Army air assets are generally kept organic and not taskable by the JFACC, compliance with JFACC and ACA orders is a requirement for Army operations. The JFACC develops and publishes the theater airspace control plan (ACP), daily airspace control orders (ACOs), air tasking orders (ATOs), and special instructions (SPINS). In addition, as the area air defense commander (AADC), the JFACC publishes the area air defense plan (AADP). These documents are the "playbook" for joint air operations across the AOR. Army planners must use their understanding of the ACP, ACO, ATO, airspace coordinating measures (ACMs) and other coordination measures to synchronize and integrate sUAS operations with Army and joint operations.

III. Army Airspace Elements and Entities

This section provides a description of Army airspace elements and entities at echelon.

A. Theater Army

The theater army implements theater-level airspace policy as issued by the ACA, coordinates with the joint force and other services, distributes ACPs and SPINS, and manages airspace, as required. The theater army also operates or contributes to these airspace elements:

- **Army Air and Missile Defense Command (AAMDC)**: Plans, coordinates, and executes Army air defense operations.
- **Theater Fires Element**: Plans and synchronizes multi-domain fires, including field artillery, air defense artillery, and other fire support assets.
- **Air Traffic Service (ATS)**: Provides air traffic control services for Army airfields and landing zones.
- **Battlefield Coordination Detachment (BCD)**: Serves as the primary Army liaison to the air operations center (AOC), integrating Army airspace requirements into joint air operations. Note: The BCD's airspace management section processes the Army's UAP and airspace coordinating measures requests (ACMREQs) for adjudication and integration into the ACO.

B. Corps and Division

The corps airspace element contributes to airspace control policy, standardization of procedures, and integration with joint airspace plans. It provides airspace expertise to the corps commander and coordinates with subordinate divisions but does not actively control airspace.

The division actively manages and controls airspace in its AO through a mix of Army and joint airspace elements. These include—

- **Fire Support Element (FSE)**: Coordinates fires and fire support coordination measures (FSCMs) across the division AO.
- **Joint Air-Ground Integration Center (JAGIC)**: Coordinates fires and fire support coordination measures (FSCMs) across the division AO.

sUAS Airspace Example

Ref: Center for Army Lessons Learned (CALL) Small Unmanned Aircraft System (sUAS) Airspace Management and Control handbook #25-14, p. 7.

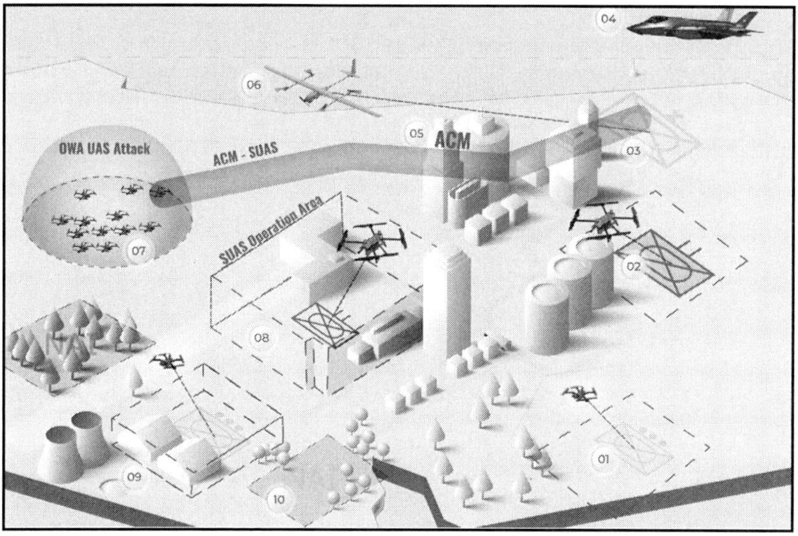

Figure 1-3. Reconnaissance platoons (01 and 09) operate short-range reconnaissance (SRR) intelligence, surveillance, and reconnaissance (ISR) unmanned aircraft systems (UASs) at the treetop level to assess enemy activity at named areas of interest (NAIs) (10). These systems do not typically require additional coordination past the brigade's sUAS standard operating procedures (SOPs).

Infantry battalions (02 and 08) operate medium-range reconnaissance (MRR) UASs to support their subordinate companies. These operate up to 500 feet above ground level (AGL) and require ACMREQs processed through the ADAM/BAE.

The infantry brigade (03), using intelligence provided by its long-range reconnaissance (LRR) UASs (06), executes a mass one-way attack (OWA) UAS attack on a high-value enemy target (07). This requires a special ACM (05) to ensure deconfliction from friendly fixed-wing aircraft (04) operating in the same vicinity, but above the coordination line.

sUAS Integration

- **Air and Missile Defense (AMD) Element:** Integrates air track data from various sources and contributes to the joint air picture, providing situational awareness of aerial threats and friendly aircraft.
- **Air Defense Coordinator (ADCOORD):** Serves as the corps/division commander's senior air defender and coordinates all air and missile defense activities in the unit.
- **Air Support Operations Center (ASOC):** Responsible for the direction and control of air operations directly supporting the ground combat element. Processes and coordinates requests for immediate air support and coordinates air missions requiring integration with ground forces.
- **Tactical Air Control Party (TACP):** A team of specially trained personnel aligned with ground combat units to advise ground commanders on the capabilities and limitations of air power and assist in planning, requesting, coordinating, and controlling air effects.

C. Brigade/Brigade Combat Team

Brigades are not typically delegated full control of airspace but are responsible for managing airspace users in their AO. The brigade is the lowest echelon required to submit a UAP and publish an appendix 10 (airspace). Increasingly, this echelon will be the primary touchpoint for sUAS operations across the Army. Airspace elements in the brigade include—

- **Air Defense Airspace Management (ADAM)/Brigade Aviation Element (BAE):** Manages airspace within the brigade AO, develops the UAP, coordinates with the division JAGIC, and deconflicts airspace for fires and aviation. Note: ADAM/BAE will be renamed the air defense support element (ADSE)/air ground integration element (AGIE) in an upcoming doctrinal update.
- **FSE:** Coordinates fires and ACMs within the brigade AO, ensuring synchronization and deconfliction of indirect fires.
- **TACP:** See previous description.
- **ATS:** See previous description.

D. Battalion and Company

Battalion operations sections are a key link between the ADAM/BAE and sUAS operations. They provide input to their brigades' airspace working group to help shape brigades' UAPs while supporting subordinate units with planning and operating their organic sUAS. Battalions collects airspace requests from companies, consolidates them, and submits them to the ADAM/BAE, then monitors certain high-profile sUAS missions. See figure 1-3. Company commanders are responsible for ensuring all company-level airspace requirements are communicated to the battalion, including those supporting sUAS operations. Key airspace contributors at the battalion and company include the following:

- **Battalion S-3:** Serves as a liaison between subordinate companies and the ADAM/BAE. Consolidates airspace requests from companies and submits them for deconfliction while also assisting as needed with planning and operation of organic sUAS and monitoring higher-echelon missions.
- **Company Commander:** Integrates sUAS into company operations and supervises communication and implementation of airspace requirements.
- **sUAS Operators:** Assists the company commander by providing advice on sUAS operations, preparing requirements for approval, and implementing plans and control measures.

IV. Methods of Airspace Control and Types Of Separation

Ref: Center for Army Lessons Learned (CALL) Small Unmanned Aircraft System (sUAS) Airspace Management and Control handbook #25-14, pp. 8 to 9.

The Army uses a combination of positive control and procedural control methods to manage aircraft in its assigned airspace. While emphasizing procedural control for broader airspace management, the Army maintains the capability to implement positive control, as required, in specific areas.

Positive Control

This method relies on real-time identification, tracking, and continuous communication with aircraft. Positive control is typically employed by ATS around airfields and other landing zones or in other areas with high concentrations of manned aircraft. Positive control requires the ability to locate/identify aircraft electronically and maintain constant communication.

Procedural Control

This method uses pre-arranged orders, procedures, and coordination to control aircraft. Procedural control is the primary method used by ASOCs and JAGICs to synchronize maneuver and fires. It relies on clear understanding of established plans like the ACP, appendix 10 (airspace), UAP, ACO, and higher echelon operations. Procedural control is particularly important when communications or situational awareness are limited.

The method of airspace control used for sUAS depends on the complexity of the operation and capabilities of the aircraft. Positive control might be used for larger sUAS operating near established airfields or on flights deliberately coordinated with manned aircraft, where precise tracking and communication are critical. Procedural control is more likely for routine sUAS missions, relying on pre-defined flight paths, altitude restrictions, and coordination with other airspace users. sUAS missions originating outside of a ground commander's AO require specific coordination.

Brigades can achieve a more permissive airspace environment by participating in airspace planning and conducting effective airspace management.

At the brigade level, the freedom to manage airspace relies on well-rehearsed SOPs, a trained ADAM/BAE, an approved brigade UAP and appendix 10 (airspace), current ATOs and ACOs integrated into the brigade's command information systems, and sustained communications with the brigade's higher-echelon headquarters.

V. sUAS Responsibilities at the Brigade Level and Below

At the brigade level and below, an ADAM/BAE plays the most significant role supporting sUAS operations. The main tool for this effort is the UAP, which describes how airspace is managed and deconflicted within a brigade's AO. 10 An ADAM/BAE approves and integrates sUAS missions, coordinates ACMs to ensure safe operations, and disseminates critical airspace information to subordinate units. Its ability to effectively manage airspace directly enables freedom of maneuver for sUAS teams.

Proactive planning and effective airspace management enable brigade combat teams to more freely operate their sUAS.

At the battalion and company levels, cooperation between the airspace element of the battalion's operations section and company commanders is crucial for successful sUAS operations. The operations section links ADAM/BAE and line companies, informing commanders and sUAS operators of airspace conditions and processing airspace requests to the brigade.

Company commanders, supported by their sUAS operators, integrate sUAS operations into the company's mission and higher-level UAPs. They must understand the capabilities and limitations of their organic sUAS assets and ensure company-level sUAS operations properly integrate with the battalion and ADAM/BAE. Effective communication and coordination between these elements are essential for safe and effective sUAS operations.

TACPs also has a role supporting sUAS operations, particularly if close air support (CAS) aircraft are operating in the same airspace as sUAS. They help deconflict airspace to ensure sUAS and CAS assets operate safely in the same vicinity.

VI. Unmanned Aircraft System Airspace and the Operations Process

Airspace management is integral to the Army's successful employment of sUAS. Airspace elements work with the commander and operations section to plan, coordinate, and regulate airspace use for all airspace users into the unit's overall operational plan.

Airspace planning is crucial to establishing conditions that enable procedural control. This involves incorporating sUAS ACMs into the overall airspace plan while considering factors like flight paths, altitude restrictions, and potential conflicts with manned aircraft. Airspace elements contribute subject matter expertise to staffs during planning, ensuring sUAS operations are integrated and aligned with commanders' intent. sUAS operations and airspace plans must change and adapt as the situation on the ground evolves.

During the preparation phase, airspace elements help refine the plan and conduct rehearsals, validating airspace procedures and coordination measures specific to sUAS missions. Execution relies on pre-coordinated measures outlined in orders, graphics, and unit SOPs, creating a predictable airspace environment that facilitates procedural control. Well-planned coordination measures and thoroughly trained SOPs enable rapid adjustments to sUAS routes and altitudes, as needed.

Continuous assessment is vital. Airspace elements monitor current and projected airspace use, track changes in higher-level orders, and maintain situational awareness of the theater ACP, ACO, and SPINS. Airspace elements should strive to ensure leaders and operators at echelons below brigade have the most up-to-date information governing sUAS employment and airspace.

II. (sUAS) Airspace & Mission Planning

Ref: Center for Army Lessons Learned (CALL) Small Unmanned Aircraft System (sUAS) Airspace Management and Control handbook #25-14, chap. 2.

This section introduces essential small unmanned aircraft system (sUAS) airspace planning and coordination concepts focused on lower tactical echelons—battalion, company, platoon, and squad. Although these echelons do not formally control airspace, leaders and sUAS teams must understand the established airspace control framework and airspace management procedures to operate their systems effectively and safely.

Understanding Small Unmanned Aircraft System Operations

Successfully employing sUAS at all echelons means integrating these assets into a battlespace shared with other manned and unmanned aircraft, air defense systems, sensors, and indirect fire systems. This integration relies on understanding and adhering to airspace instructions passed down from higher headquarters, typically through the corps/division appendix 10 (airspace) and air defense airspace management (ADAM)/brigade aviation element (BAE).

I. The Operational Framework for Small Unmanned Aircraft Systems (sUAS)

All sUAS operations begin with orders from air and ground commanders. These directives are officially communicated through documents like the daily/weekly airspace control order (ACO) and air tasking order (ATO), supporting unit airspace plans (UAPs) and Army operation orders (OPORDs). Personnel at the battalion level and below typically receive only relevant excerpts or specific instructions derived from the ACO and ATO, as these are often classified. The ADAM/BAE is responsible for transforming these higher-level orders into actionable guidance for their subordinate units.

ACMREQ Process

Operators initiate these requests, which are processed through the battalion S-3 and the ADAM/BAE. This ensures the sUAS mission is integrated, approved by the appropriate airspace control element, and entered into the Tactical Airspace Integration System (TAIS) for visibility across the battlefield. TAIS is typically managed at the BCT; battalions should coordinate with their ADAM/BAE to ensure their ACMREQs are entered and tracked.

Understanding the Joint Air Tasking Cycle

The ATO and ACO can be thought of as the joint "playbook" for the air domain. They are published by the joint force air component commander (JFACC) and airspace control authority (ACA) respectively on a 72-or 96-hour cycle and direct all air operations in the theater.

- The ATO tasks aircraft and other assets with specific missions and provides additional required details.
- The ACO is like the "rules of the road" for the sky. It defines safe corridors, restricted areas, and coordination altitudes to integrate joint air operations, prevent collisions and fratricide, and optimize weapons employment.

II. Battalion and Company Airspace Operations in Practice

sUAS operations in support of battalion and company missions are detailed in OPORDs and must align with the brigade UAPs, along with the current airspace control plan (ACP) and ACO. The ADAM/BAE is the key entity for coordinating formal and informal ACMs. It processes ACM requests for MRR/LRR sUAS and ensures approved measures are reflected in TAIS, making them visible to other airspace users. For SRR sUAS, unit-level adherence to unit SOPs and informal ACMs is the more typical method of control. The following describes sUAS operator coordination and deconfliction responsibilities:

Internal

Company/platoon leadership and sUAS operators are responsible for deconflicting sUAS operations with other unit activities (e.g., other sUAS and indirect fires observation).

External

For MRR and LRR sUAS, or when operating near boundaries or controlled airspaces, coordination is passed from the sUAS operator/team leader, through company leadership, then to the battalion S-3, who liaises with the ADAM/BAE.

(sUAS) Planning Considerations for Battalion/Company/Platoon Leadership

Commander's Intent

Consider how sUAS operations support the commander's overall intent for the operation.

sUAS Capabilities and Limitations

Leaders must understand the specific capabilities (range, endurance, and sensor package) and limitations (weather, terrain masking, and battery life) of their sUAS to provide reasonable and actionable objectives for sUAS operators.

Risk Management

Leaders must assess and mitigate risks. Key risks include—

- **Weather**: Wind, precipitation, and visibility.
- **Terrain**: Line-of-sight for control, terrain masking for targets or threats, and suitable launch/recovery areas.
- **Enemy Activity**: Air defense capabilities and electronic warfare (EW) threats.
- **Electromagnetic spectrum Congestion**: Interference with control links or data feeds.
- **Friendly Airspace Users**: Potential conflicts with other airspace users.

Contingency Planning

Units must have simple, actionable contingency plans for common issues like lost or degraded communications, system malfunctions, emergency recovery procedures, and unexpected changes in the tactical or airspace situation.

III. (sUAS) Missions

Ref: Center for Army Lessons Learned (CALL) Small Unmanned Aircraft System (sUAS) Airspace Management and Control handbook #25-14, pp. 10 to 11.

sUAS operations at lower-echelon levels typically involve two distinct mission types, each with different airspace requirements. These terms are not currently formalized. Commanders may define them locally based on their unit's standard operating procedures (SOPs):

A. Reconnaissance Missions

Short-Range Reconnaissance (SRR) *(See p. 7-18.)*

These missions are flown at low altitudes and near the supported unit. Operations are often governed by unit SOPs or airspace coordinating measures (ACMs) established at the unit level (e.g., "sUAS operations will remain below 200 feet above ground level (AGL) and within 500 meters of the platoon's location.").

Because of their limited range and altitude, SRR missions typically operate well below manned aircraft or established coordination levels (CLs) and rarely require formal, higher-level airspace deconfliction unless specific local procedures dictate otherwise. Although SRR missions are typically governed by unit SOPs, this does not imply their operations are any less stringently controlled; safe operation depends on trained and certified operators maintaining constant airspace awareness.

Medium-Range Reconnaissance (MRR) *(See p. 7-19.)*

These missions are flown at greater altitudes and distances, potentially extending near or above the CL or into airspace used by manned aircraft. Operations with MRR sUAS, especially when flying near the CL, across unit boundaries, or in areas with other air traffic, often necessitate formal airspace coordinating measures requests (ACMREQs). MRR missions typically fly at or around 1,000 feet AGL, with some variance based on terrain and noise abatement factors. MRR platforms range from 10 to 25 kilometers and beyond; therefore, MRR missions may extend appreciably beyond the forward line of own troops (FLOT). Once past the FLOT, MRR management and control may become more permissive because of the lower density of friendly aircraft in this airspace.

Long-Range Reconnaissance (LRR) *(See p. 7-19.)*

These missions may be flown below or above the CL out to maximum ranges of sUAS, 15 kilometers or more. Group 2 may fly LRR, but the majority of LRR is flown by group 3. LRR typically requires ACMREQs to deconflict with manned aircraft along with air defense units and other airspace users.

B. Attack/Strike Missions

Launched Effects (LE)

These missions encompass all UAS attack missions, including:

- short-and long-range one-way attacks (OWAs),
- loitering munitions, and
- payload delivery systems.

These systems may be recoverable or attritable, and feature autonomous and human-in-the-loop systems. Control of short-range, low-altitude LE platforms likely falls under unit SOPs; larger systems require specific coordination to deconflict.

IV. (sUAS) Reconnaissance Roles

Ref: Center for Army Lessons Learned (CALL) Small Unmanned Aircraft System (sUAS) Airspace Management and Control handbook #25-14, app. C.

This section provides a general overview of small unmanned aircraft system (sUAS) platforms employed by the Army, categorized by their operational roles: short-range reconnaissance (SRR), medium-range reconnaissance (MRR), and long-range reconnaissance (LRR). These role-based categories generally align with traditional Department of Defense (DOD) unmanned aircraft system (UAS) groupings (groups 1 through 5) that are based on factors such as maximum takeoff weight (MTOW), operating altitude, and speed. Typically, SRR systems correspond to group 1, while MRR and LRR systems often align with group 2, with some advanced LRR capabilities nearing group 3.

Note: Specific system capabilities, payloads, and operating procedures vary based on model, configuration, and software versions. Always consult official technical manuals (TMs), training circulars (TCs), and program of record (POR) documentation for the most current and authoritative information.

Note: Continued use and development of commercial off-the-shelf (COTS) sUAS is likely to expand significantly in the short term. Rapid prototyping and fielding of these systems may evolve these descriptions and definitions.

Short-Range Reconnaissance Small Unmanned Aircraft Systems (SRR)

Typical Platforms: RQ-28A (Skydio), Black Widow, and similar micro or small hand-launched systems (some previously categorized as handheld sUAS or lower-end group 1).

Primary Roles: Very short-range to short-range reconnaissance, direct situational awareness, and security for dismounted troops, squads, platoons, and companies.

Key Features

- Extremely portable, often pocket-sized or backpackable, capable of being carried and launched by a single Soldier or small team.
- Characterized by quiet operation and small size, offering a stealthy intelligence-gathering capability, especially for micro systems.
- Range typically varies from a few hundred meters for micro systems up to approximately 10 miles for larger SRR platforms.
- Primarily equipped with electro-optical (EO) and/or infrared (IR) cameras for real-time video.
- Relatively simple to operate and maintain.

Employment Considerations

- Ideal for squad-, platoon-, and company-level operations, including urban environments, complex terrain, or other confined spaces.
- Provides immediate "over-the-hill" or "around-the-corner" reconnaissance directly to the point of need.
- Micro systems often operate below treetop level. Although all sUAS require airspace awareness, the smallest SRR systems may operate under less restrictive airspace coordination procedures, though adherence to local standard operating procedures (SOPs) is mandatory.

Medium-Range Reconnaissance Small Unmanned Aircraft Systems (MRR)

Typical Platforms: Anduril Ghost systems and similar platforms.

Primary Roles: Reconnaissance, surveillance, and target acquisition support for company-and battalion-level operations.

Key Features

- Greater range, endurance, and altitude capabilities than SRR systems.
- May carry more advanced or modular payloads, potentially including multi-sensor options.
- Can be hand, mechanical, or vertical takeoff and landing (VTOL) launched, depending on the specific system.
- Improved datalinks and navigational capabilities.

Employment Considerations

- Suitable for providing tactical intelligence over broader areas for company and battalion operations.
- Bridges the gap between the immediate reconnaissance of SRR and the wider area coverage of LRR.
- Requires more deliberate airspace planning and coordination than most SRR systems because of their increased operational envelope (altitude and range).

Long-Range Reconnaissance Small Unmanned Aircraft Systems (LRR)

Typical Platforms: Stalker, P550, and similar class systems.

Primary Roles: Extended-range reconnaissance, surveillance, target acquisition, and force protection over wider operational areas.

Key Features

- Significantly longer range and endurance than MRR systems (often 20 miles or more).
- Capable of carrying heavier and more sophisticated payloads, which may include multi-sensor EO/IR systems, laser pointers/designators (platform dependent), and advanced communications relays.
- May require catapult launch, paved, or runway-independent VTOL systems.

Employment Considerations

- Typically employed at battalion and brigade levels to support broader operational objectives.
- Offers a balance between tactical responsiveness and increased intelligence, surveillance, and reconnaissance (ISR) capabilities.
- Airspace coordination is critical and more complex because of their extended range, higher operating altitudes, physical size, and potential to interact with other manned and unmanned aircraft.

Note on Other UAS Categories (Groups 3 - 5)

The Army historically employed groups 3 and 4 UASs that operated at higher echelons and altitudes, possessed greater capabilities and endurance, and demanded more rigorous and detailed airspace management. Following the divestment of the RQ-7 Shadow and MQ-1C Grey Eagle, the Army currently does not have groups 3 or 4 UAS PORs, but future systems in one or both categories are likely.

V. (sUAS) Mission Planning and Troop Leading Procedures

Ref: Center for Army Lessons Learned (CALL) Small Unmanned Aircraft System (sUAS) Airspace Management and Control handbook #25-14, pp. 13 to 16.

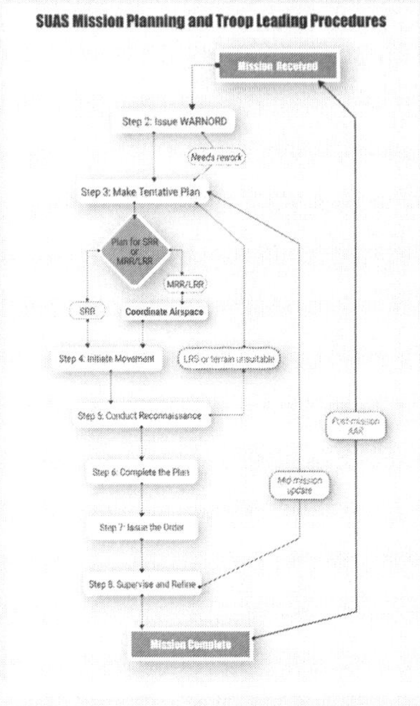

This section describes how sUAS missions are nested inside troop leading procedures (TLPs) at battalion-and-below echelons. Actions and outputs are provided for each step of the TLP to better inform leaders and sUAS operators how to best integrate sUAS operations with other unit activities. (Ref: Figure 2-2. sUAS mission planning and TLPs.)

1. Receive the Mission

- **Actions:** Upon receiving a mission from higher headquarters, the unit command and sUAS team leader(s) identify potential sUAS support requirements. Review the current sUAS status, known friendly locations, and any standing airspace instructions or restrictions the ADAM/BAE relays.
- **End State:** Understand the task and commander's intent, initial sUAS concept of support, and awareness of current airspace limitations.

2. Issue a Warning Order (WARNORD)

- **Actions:** Alert sUAS operator(s) of the pending mission. Include basic information about the area of operations (AO) and potential tasks.
- **End State:** sUAS team begins initial preparations (system checks, route, and loiter area planning).

3. Make a Tentative Plan

- **Actions:**
  - Determine specific sUAS tasks (e.g., route reconnaissance, overwatch for an assault, and specific named area of interest [NAI] surveillance).
  - Analyze terrain for optimal launch/recovery sites (LRSs), flight routes, and potential dead zones for communication.

sUAS Integration

- For SRR: Plan operations consistent with unit SOPs and within the unit's immediate operational area, ensuring deconfliction with other local airspace users.
- For MRR/LRR: Identify preliminary airspace requirements (desired area, altitude, and time on station). Communicate these needs up to the ADAM/BAE for integration and coordination.
- Assess risks (weather, enemy, and terrain) and develop mitigation measures.
- Develop contingency plans (lost link and alternate LRS).

• **End State:** Tentative sUAS flight plan(s), primary and alternate LRS, identified risks and mitigation measures, coordination requests for MRR/LRR operations are submitted to higher.

4. Initiate Movement

• **Actions:** sUAS team commences movement to planned LRS and begins to prepare systems for flight.

• **End State:** sUAS teams are moving and sUAS are ready for action.

5. Conduct Reconnaissance

• **Actions:** sUAS team leader confirms final suitability of LRS, verifies line-of-sight, and checks for local hazards. Final check for any updated airspace instructions from the battalion or ADAM/BAE. Conduct a map reconnaissance of the flight plan/routes.

• **End State:** Confirmed LRS and updated situational awareness of local airspace and operational environment (OE).

6. Complete the Plan

• **Actions:** Finalize sUAS mission details based on other unit reconnaissance activities and any additional updated information. For MRR/LRR, confirm airspace clearance and parameters received from ADAM/BAE.

• **End State:** Finalized sUAS mission plan, including objectives, routes, altitudes, timings, and contingencies.

7. Issue the Order

• **Actions:** Unit leadership and/or the sUAS team leader briefs the sUAS operator(s) on final mission specifics: objectives, flight path, altitudes, communication plan (callsigns and frequencies), emergency procedures, no-fly areas, and any additional specific instructions.

• **End State:** sUAS operator(s) understand the mission and are ready for execution.

8. Supervise and Refine (Execution Phase)

• **Actions:**
- Monitor sUAS operations through continuous contact with sUAS operators.
- For MRR/LRR: Maintain communication with the ADAM/BAE to ensure operations remain integrated with other airspace users. Be prepared to adjust flight profiles, timings, or mission objectives based on real-time conditions or instructions.
- For SRR: Maintain communication with the battalion for dynamic changes to the local tactical situation. Be prepared to adjust flight profiles, timings, or mission objectives based on real-time conditions or instructions.
- Conduct after action reviews (AARs) post-mission or post-operation to identify lessons learned for future sUAS employment.

• **End State:** Successful mission execution, continuous situational awareness, dynamic adjustments as needed, and lessons learned.

VI. (sUAS) Preplanned Operations

Ref: Center for Army Lessons Learned (CALL) Small Unmanned Aircraft System (sUAS) Airspace Management and Control handbook #25-14, pp. 16 to 17.

Preplanned sUAS missions are those that allow leaders and operators ample lead time to plan and execute sUAS operations. These operations are common when supporting larger battalion or brigade missions. Key factors supporting successful preplanned sUAS operations include—

Early Integration with Troop Leading Procedures (TLP)

Include sUAS considerations from the "Receive the Mission" step of TLP. This ensures sUAS capabilities are effectively factored into the overall operational plan. For MRR/LRR, this allows timely and detailed submission of airspace requests to the ADAM/BAE.

Detailed Preparation

- **Intelligence Focus**: Obtain information from the S-2 on known/suspected enemy air defenses, potential electronic attack, and terrain that impacts sUAS line-of-sight.
- **Airspace Coordination (MRR/LRR)**: Submit detailed airspace requests to the ADAM/BAE well in advance. Clearly define operating areas, altitudes, times, and how sUAS support the mission. sUAS leadership/operators should be prepared to provide details to battalion or brigade planners.

Synchronize with Other Elements

- **Within the Platoon/Company**: Ensure sUAS plans are deconflicted with other local airspace users.
- **Army/Joint Fires Integration (Primarily MRR/LRR)**: If sUAS are supporting fires, coordinate closely with the forward observer (FO) or company/battalion fire support elements for targeting and battle damage assessment.

Execute and Adapt

- **Thorough Pre-Mission Checks and Briefs**: Ensure sUAS operators understand objectives, adhere to SOPs, and are aware of all coordination measures (especially altitude restrictions and boundaries), contingency plans, and communication protocols. For SRR, execution is often in accordance with SOP and training; deviations are briefed by exception. For MRR/LRR, any deviations from standard flight profiles are briefed on contact to the ADAM/BAE.
- **Communication Procedures**: Use established communication frequencies, call signs, and reporting formats per unit SOPs and mission orders for clear communication with the controlling element and for passing information to the battalion/ADAM/BAE, as required.
- **Real-time Coordination (Primarily MRR/LRR)**: Maintain situational awareness of any changes in the airspace or mission requirements. Proactively coordinate adjustments when operating in airspace shared with rotary-wing aircraft.

Enforcing a robust lessons learned cycle ensures unit SOPs continuously evolve to meet new operational challenges.

VII. (sUAS) Dynamic Operations

Ref: Center for Army Lessons Learned (CALL) Small Unmanned Aircraft System (sUAS) Airspace Management and Control handbook #25-14, pp. 17 to 18.

Dynamic sUAS operations emphasize speed and adaptability, often with minimal planning time. SRR operations, in particular, are frequently dynamic and governed largely by unit SOPs and tactics, techniques, and procedures (TTPs).

Launch and Coordination Procedures

- **On-Call Posture**: Maintain a high state of readiness: preflighted aircraft (as much as SOPs allow), charged batteries, and available crews. Pre-identify potential LRS.
- **Immediate Communication (Upon Receiving Tasking)**:
  - **SRR**: Unit leadership informs sUAS operator(s) and begins deconflicting with other organic elements in the unit.
  - **MRR/LRR**: sUAS team or unit leadership immediately contacts the ADAM/BAE or S-3 with essential information: requesting unit, desired support, target area/NAI, and desired time/duration.
- **Real-time Airspace Clearance (MRR/LRR)**: The ADAM/BAE rapidly deconflicts the requested airspace and issues a concise clearance, including approved altitude, flight restrictions, and coordination points.
- **Dynamic Retasking**: Be prepared to receive new or adjusted missions while airborne, especially for MRR/LRR systems. Updates to mission will typically come from unit leadership based on the situation and evolving tactical needs. Updates to airspace coordination typically come from the ADAM/BAE.

Data Exploitation and Dissemination

- **Streamlined Analysis**: Focus on rapid identification, support to targeting, and reporting of priority intelligence requirements (PIRs).
- **Rapid Dissemination**: Use voice calls, digital messages, or pre-formatted reports to transmit critical information quickly to unit staff members and leadership.
- **Marked Imagery**: If capabilities allow, quickly mark and transmit key imagery (enemy positions or obstacles) to provide actionable intelligence.

A well-rehearsed unit SOP allows for more rapid and effective sUAS employment during unexpected situations.

VIII. Challenges for sUAS Operations

Communications Reliability

Operating in a congested or contested electromagnetic spectrum (EMS) can disrupt control links or data feeds. Plan for primary, alternate, contingency, and emergency (PACE) communications; use terrain for masking, when possible.

Operator Workload and Fatigue

Dynamic and high-stress environments can overwhelm operators. Leaders must be cognizant of work/rest cycles for sUAS operators and mitigate risks associated with fatigue, especially during sustained operations. Use crew rotations if multiple qualified operators are available. Training for high-stress scenarios and adherence to SOPs can reduce cognitive load.

Survivability

Because of their EMS and visual signature, relative lack of mobility, and impact on the battlefield, sUAS operators and ground stations are vulnerable and high-value targets. Leaders and operators must prioritize survivability measures, such as EMS and visual signature discipline and regular survivability movement.

Chap 7

III. (sUAS) Tactics, Techniques, & Procedures

Ref: Center for Army Lessons Learned (CALL) Small Unmanned Aircraft System (sUAS) Airspace Management and Control handbook #25-14, chap. 3.

This section outlines select tactics, techniques, and procedures (TTPs) and best practices to improve airspace coordination at the unit level and improve the efficiency of small unmanned aircraft system (SUAS) operations.

I. Planning and Coordination

A well-rehearsed and extensively validated brigade standard operating procedure (SOP) is the center of gravity for successful sUAS operations. This SOP must be relevant, easily understood, and practical, focusing on information that directly enables lower echelon leaders and sUAS crews.

> Realistic training using live aircraft, tactical software, and multiple simultaneous airspace users is a best practice to validate and evolve unit sUAS SOPs.

Rigorous training is paramount for rehearsing and validating the SOP. This training should establish and test procedural controls, such as standardized airspace measures and critical notification timelines in realistic, high-intensity training environments. Training must also include elements, such as electronic warfare (EW) and degraded communications to ensure the SOP functions as needed, even in sub-optimal conditions. Other key planning and coordination TTPs include—

A. Brigade-Level Airspace Working Groups

Regular participation of sUAS representatives in forums like brigade airspace, information collection, and targeting working groups is crucial. These meetings, involving the air defense airspace management (ADAM)/brigade aviation element (BAE), aviation elements, fire support elements, and maneuver units, ensure all stakeholders understand the commander's intent, deconflict airspace, and establish clear communication for routine and contingency operations, especially for medium-range reconnaissance (MRR) and long-range reconnaissance (LRR) integration.

B. Standardized Airspace Coordinating Measures Requests (ACMREQs) Procedures

Using standardized templates and a well-rehearsed unit SOP for ACMREQs streamlines the airspace approval process. Templates ensure consistent and clear communication of necessary information (operating areas, altitudes, times, and justifications) to the ADAM/BAE. A practiced SOP ensures all stakeholders understand their roles and receive critical information.

C. Routine Airspace Coordination Areas

Establishing routine airspace coordination areas within the unit's area of operations (AO) reduces risk, accelerates planning for sUAS missions, and allows ADAM/BAE resources to focus on more complex airspace management. An example UAP establishes a regular 24-hour sUAS restricted operations zone (ROZ) that is tied to each maneuver battalion's boundaries. The sUAS ROZ should be approximately 3 kilometers behind the forward line of own troops (FLOT) and out 3 kilometers beyond the battalion's objective or engagement area, with a ceiling up to 300 feet AGL, enabling the majority of SRR missions to fly without separate ACMREQs.

D. Shared Airspace Visualization

Although direct access to advanced digital mapping tools by all sUAS operators may be limited, the ADAM/BAE should leverage systems like the Tactical Airspace Integration System (TAIS) and Advanced Field Artillery Tactical Data System (AFATDS) whenever possible to improve situational awareness for sUAS users and commanders. These tools facilitate a shared understanding of the airspace, enabling the visualization of planned sUAS flight paths alongside other air activities like aviation elements and fire support trajectories. This supports collaborative planning efforts and builds shared situational awareness among all airspace users.

II. Execution and Deconfliction

A. Altitude Separation

Assigning specific altitude blocks for sUAS operations is a fundamental deconfliction technique. This practice, detailed within the unit SOP and relevant annexes (e.g., division/brigade annex C, appendix 10), accounts for mission requirements, risk, and system capabilities to assign optimal and standardized altitude layers to each sUAS type throughout the unit. System capability is the most important planning factor; smaller and/or less capable systems likely require lower operating altitudes because of shorter operational ranges and less capable sensor systems. Well-planned altitude separation simplifies deconfliction and minimizes collision risk with other aircraft, especially in congested airspace.

See facing page for an overview and further discussion.

B. Communications and Targeting

Dedicated sUAS Frequencies

Using designated, preplanned frequencies specifically for sUAS operations minimizes interference with other battlefield emitters and can work to minimize enemy EW impact. Frequencies for sUAS should be coordinated through the unit's spectrum manager and validated against the joint restricted frequency list (JRFL).

Standardized Reporting Formats

Employing standardized reporting formats for sUAS observations, target information, and airspace situation updates reduces ambiguity and communication errors, while speeding action.

sUAS, EW, and Targeting Integration

Ensuring active and practiced links between sUAS operators, EW personnel, and targeting personnel works to offset enemy EW effects on friendly sUAS operations and facilitates rapid targeting of enemy emitters in the AO.

C. Operational Constraints and Contingencies

Weather Impact and Standardized Responses

sUAS are particularly sensitive to weather. Continuous monitoring and coordination with meteorological staff members are essential to assessing conditions such as winds; cloud floors, layers, and ceilings; prevailing visibility (including the effects of precipitation or other obscurants); precipitation (type and intensity); and temperature extremes. Unit SOPs must include a branch plan for varying conditions. These procedures should define operational thresholds and actions based on how specific weather elements impact sUAS capabilities. Commanders of all types must be aware of weather-specific impacts on the sUAS supporting them.

Altitude Separation Guidelines

Ref: Center for Army Lessons Learned (CALL) Small Unmanned Aircraft System (sUAS) Airspace Management and Control handbook #25-14, pp. 20 to 21.

The following are examples of separation guidelines:

Very Short-Range Systems

Operations are generally confined to the treetop/obstruction level or below. No prior coordination is typically required, though operators may immediately be required to "go to ground" if any rotary-wing traffic is observed.

Short-Range Reconnaissance (SRR) Systems

Operated at 300 feet above ground level (AGL) and below. Although direct ADAM/BAE coordination may not be mandatory, a strong SOP involves coordinating with the battalion/brigade S-3 or ADAM/BAE for optimal flight times and leveraging its awareness of planned rotary-wing routes. SRR operators must also immediately "go to ground" if rotary-wing aircraft are observed.

MRR Systems

Operated at 1,000 feet AGL and below. Coordination with the ADAM/BAE is required. Unit SOPs should establish standard route parameters and mission area ROZ dimensions for MRR flights.

LRR Systems

Typically flown 2,000 to 4,000 feet AGL, depending on the mission and sensor payload. ADAM/BAE coordination is mandatory. A minimum of 30 minutes prior notification is recommended, especially if operating above the coordination level (CL), to allow the ADAM/BAE sufficient time to submit an ACMREQ to the division for approval. Standard route and mission area ROZ parameters for LRR systems must also be defined in the unit SOP.

Lateral Separation

Making use of natural (terrain features) and artificial (unit boundaries and grid lines) offers additional opportunity for physical separation of UAS. Planners should develop SOPs that make the best use of tools readily available and useful to operators; terrain features, such as rivers and roads make ideal lateral boundaries, easily visible to most UAS platforms while in flight. Artificial lateral boundaries are most useful for systems operated in conjunction with digital airspace management tools.

Route Planning

sUAS operators flying in rear areas should plan flight paths that avoid known high-traffic areas for Army aviation, artillery firing points, and potential close air support (CAS) ingress/egress routes. This technique reduces the likelihood of encountering other aircraft and simplifies airspace coordination.

Time-Phased Operations

Phase sUAS launch, transit, and recovery times to avoid periods of peak activity in busy areas, such as air assaults, resupply, or joint air pulses. This procedural approach ensures sUAS operations do not impact other critical efforts, while simultaneously reducing risk.

III. (sUAS) Mission Flow Guide

Ref: Center for Army Lessons Learned (CALL) Small Unmanned Aircraft System (sUAS) Airspace Management and Control handbook #25-14, app. A.

This guide provides a basic overview of small unmanned aircraft system (sUAS) mission execution, intended to inform commands and staff members about the fundamentals of sUAS operations supporting their unit. See A-1.

Phase 1: Mission Planning and Airspace Coordination

1. Mission Receipt and Assessment

Receive mission tasking (objective, area of interest, duration, and intelligence requirements). Assess feasibility, asset availability, and preliminary airspace considerations. Advise commands and staff members of assessment results and any changes to risk level.

2. Coordinate with Air Defense Airspace Management (ADAM)/Brigade Aviation Element (BAE)

If required for the mission, contact the ADAM/BAE to request airspace guidance. Provide preliminary mission details and determine if an airspace coordinating measures request (ACMREQ) or coordination meeting is needed.

3. Airspace Approval

Submit an ACMREQ (if required) and participate in coordination meetings to deconflict with other airspace users. Ensure all crews adhere to the final unit airspace plan (UAP).

Phase 2: Preflight and Launch

1. Preflight Preparation

Conduct thorough inspections of the sUAS, ground control station, and equipment (battery, communications, and sensors). Program return-to-home (RTH) and emergency procedures, briefing the crew accordingly.

2. Communication Check

Establish and verify communication links with the ADAM/BAE and supported ground units.

3. Launch

Launch the sUAS, adhering to approved local airspace controls while paying special attention to launch and recovery areas in the vicinity.

Phase 3: Mission Execution and Recovery

1. Execute Mission

Conduct the mission, providing updates and reports, as required by standard operating procedures (SOPs). Be prepared for dynamic retasking.

2. Recovery and Post-Flight

Return the sUAS to the designated recovery point. Conduct post-flight inspections, secure mission data, and debrief the crew, ADAM/BAE, and supported units, as required by SOPs.

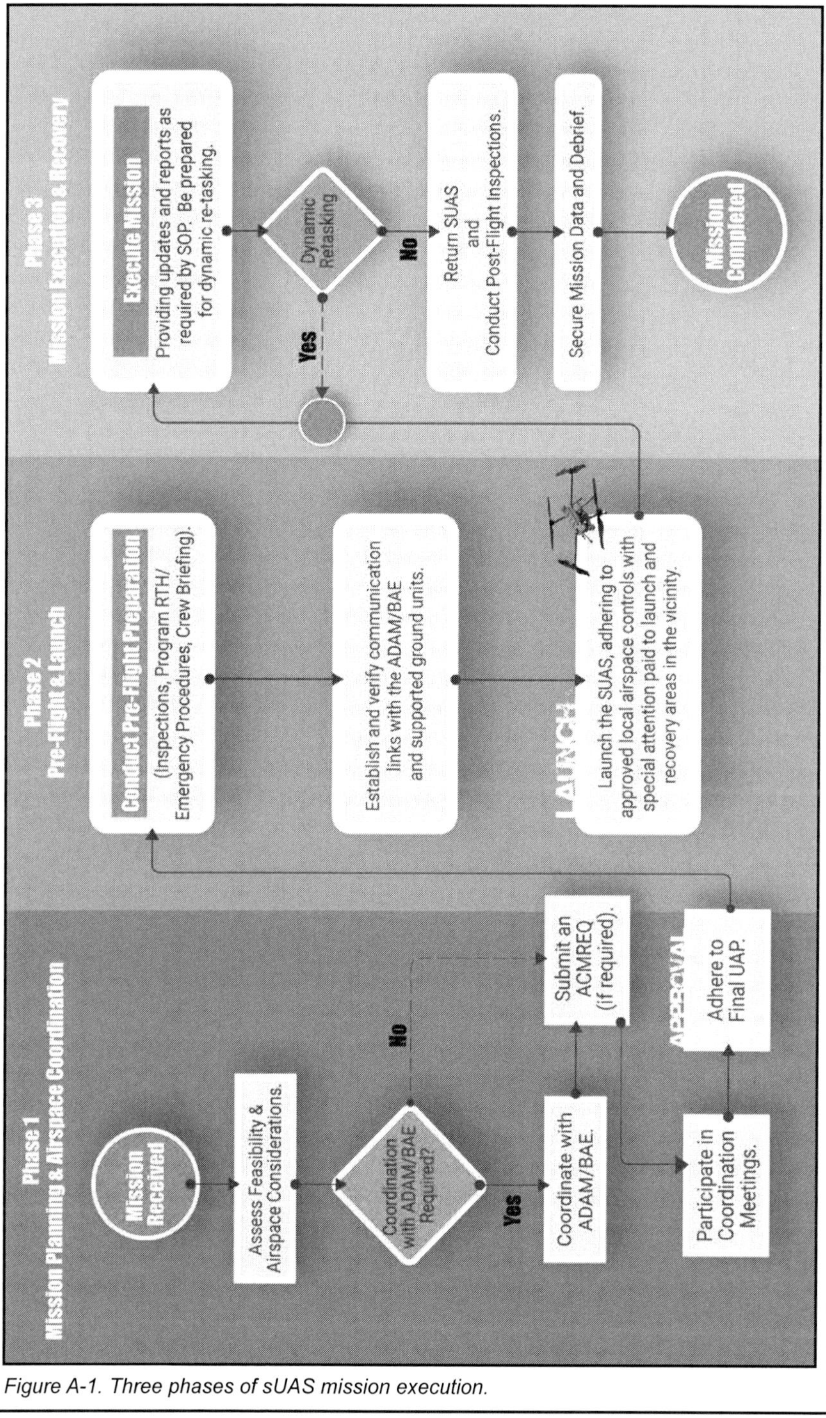

Figure A-1. Three phases of sUAS mission execution.

sUAS Integration

Emergency Recall Procedures

An actionable and reliable system must be in place for the rapid recall or flight termination of all sUAS operating in a given airspace volume. Such procedures are critical for responding to unexpected airspace closures, dynamic changes in the tactical situation, or any safety concerns that necessitate the immediate grounding or recovery of sUAS assets.

Deconfliction With Mortars and Artillery

Commanders, enabled by staff input, decide which sUAS or mission types do not need to deconflict with indirect fires munition flight paths and firing points. Particularly during high-intensity large-scale combat operations (LSCO), commanders must balance the risk of excessively constricting operations versus the risk of sUAS colliding with munitions.

Return Procedures and Survivability

Crews should consider measures, such as multiple or offset landing points to improve survivability. Crews should actively monitor enemy surveillance of landing points.

D. Procedural Airspace Control TTPs

Visual Flight Rules (VFRs) Operations

For low-risk missions in uncongested airspace, leverage VFR procedures where sUAS operators maintain visual contact with their aircraft on launch and recovery to ensure separation from other air traffic. Clearly define altitudes and lateral boundaries for VFR sUAS flights to minimize the risk of conflict with manned aircraft.

Lateral Boundary Designators

When possible, leverage large, readily identifiable terrain features—such as roads, rivers, bridges, or buildings—to define lateral boundaries for sUAS operating areas. This simplifies airspace visualization for sUAS operators and other airspace users.

Altitude Reservation System

Implement a system that routinely assigns specific altitude blocks to specific units or systems within the unit's airspace. A well-rehearsed SOP ensures shared understanding regarding which systems habitually operate at which altitudes.

Designated sUAS Training Areas

Establish habitual airspace areas within unit AOs designated for sUAS training and proficiency flights that do not require the same level of airspace coordination as operational missions.

E. Loss Of Link TTPs

Preprogrammed Return-to-Home (RTH) Procedures

If the system allows, program a designated RTH protocol. The RTH procedure should transit the aircraft to a safe location within friendly lines while, if possible, avoiding friendly units on the ground and restricted volumes of airspace. RTH flights should use a cleared altitude that avoids terrain and obstacles. If the sUAS cannot immediately land upon reaching the RTH point (e.g., because of a need to clear the area), program a contingent holding pattern or loiter procedure. This allows the ground crew to regain communication and prepare for a safe landing.

Visual Search Procedures

Train sUAS crews and Soldiers across the unit on standardized search procedures to locate grounded sUAS that deviate from a RTH path or are otherwise not immediately visible upon return.

IV. Sample Unmanned Airspace Coordinating Measures Request (ACMREQ) Format

Ref: Center for Army Lessons Learned (CALL) Small Unmanned Aircraft System (sUAS) Airspace Management and Control handbook #25-14, app. B.

1. Requested Airspace Coordinating Measure(s)

- Restricted Area (RA)
- Altitude Reservation
- Pre-Coordinated Flight Path
- Other: (specify)

2. Airspace Requirements

- Location: (grid coordinates, landmarks, or area description)
- Dimensions: (radius, length, and width [as applicable])
- Altitude(s): (feet above ground level [AGL] or mean sea level [MSL])
- Time Period(s): (start DTG to end DTG)

3. Supporting Justification

(Clearly articulate the operational justification for the requested airspace. Explain how the requested measure supports the commander's intent, the overall mission, and any specific intelligence or targeting requirements. If deconfliction with other airspace users is a concern, explain how the request mitigates those concerns.)

4. Supporting Information (as Applicable)

- sUAS Platform(s): (specify type and capabilities)
- Sensor Payload(s): (specify type and capabilities)
- Number of sUAS: (indicate quantity)
- Contingency Plan (lost link, etc.): (summarize)
- Primary, alternate, contingency, and emergency (PACE) Plan for Communications: (include encryption and link security measures)

Simplified ACMREQ Format (For Routine Operations)

Unit: (submitting unit name and call sign)

DTG: (YYYYMMDDHHMMZ)

sUAS Type: (type)

Location: (general area description or landmark)

Time: (start to stop)

Remarks: (any additional information or special instructions)

Submission and Approval (MRR and SRR)

- Submit ACMREQs to the air defense airspace management (ADAM)/brigade aviation element (BAE) through designated communication channels
- Ensure the request is submitted with sufficient lead time for processing and deconfliction.
- The ADAM/BAE will review the request and provide approval or guidance on any necessary modifications.
- Maintain communication with the ADAM/BAE throughout the mission for any changes or updates to airspace coordinating measures.

Emergency Landing Zones

Designate predetermined emergency landing zones within the brigade AO. These zones should be secure open areas with minimal obstacles and easy accessibility to crews or other personnel assisting with recovery.

Lost-Link Reporting

Establish clear reporting procedures for lost-link events. The sUAS operator should immediately notify the ADAM/BAE, providing the last known location, altitude, and time of the loss of link (LOL).

Chap 8

I. Fires Integration (C-UAS/sUAS)

Ref: ADP 3-19, Fires (Jul '19), chap. 1 and ADP 3-0, Operations (Mar '25), p. 19.

Air & missile defense *(see chap. 3)*, counter-unmanned aircraft system (C-UAS), and small unmanned aircraft (sUAS) operations -- including **strike coordination and reconnaissance (SCAR)** -- are all integrated within the fires warfighting function.

I. The Fires Warfighting Function

The fires warfighting function is the related tasks and systems that create and converge effects in all domains against the threat to enable actions across the range of military operations (ADP 3-0). These tasks and systems create lethal and nonlethal effects delivered from both Army and Joint forces, as well as other unified action partners. The fires warfighting function does not wholly encompass, nor is it wholly encompassed by, any particular branch or function.

Many of the capabilities that contribute to fires also contribute to other warfighting functions, often simultaneously. For example, an aviation unit may simultaneously execute missions that contribute to the movement and maneuver, fires, intelligence, sustainment, protection, and command and control warfighting functions. Additionally, air defense artillery (ADA) units conduct air and missile defense (AMD) operations in support of both fires and protection warfighting functions.

Space and cyberspace capabilities can provide commanders with options to defeat, destroy, disrupt, deny, or manipulate enemy networks, information, and decision making.

Commanders must execute and integrate fires, in combination with the other elements of combat power, to create and converge effects and achieve the desired end state. Fires tasks are those necessary actions that must be conducted to create and converge effects in all domains to meet the commander's objectives. The tasks of the fires warfighting function are:

Integrate Army, multinational, and joint fires through:

- Targeting *(See pp. 8-5 to 8-24.)*
- Operations process.
- Fire support.
- Airspace planning and management.
- Electromagnetic spectrum management.
- Multinational integration.
- Rehearsals.
- Air and missile defense planning and integration.

Execute fires across all domains and in the information environment, employing:

- Surface-to-surface fires.
- Air-to-surface fires.
- Surface-to-air fires.
- Cyberspace operations and electromagentic warfare (EW).
- Space operations.
- Multinational fires.
- Special operations.
- Information operations.

II. Useful (Restricted) Resources for Planning & Conducting C-UAS/sUAS Fires

Army Techniques Publications (ATPs) (Distribution Restricted/CAC Login Required)

The following is a listing of Army techniques publications (ATPs) related to planning and conducting C-UAS and sUAS, but with distribution restrictions. ATPs provide step-by-step methods and best practices for specialized tasks. They are designed to explain the practical execution of missions and activities. They are used to develop standard operating procedures (SOPs) and are a great starting point for units looking for best practices. CAC login required. (*https://armypubs.army.mil/*)

ATP 3-60.2 - SCAR Multi-Service Tactics, Techniques, and Procedures For Strike Coordination and Reconnaissance

This publication provides strike coordination and reconnaissance (SCAR) multi-Service tactics, techniques, and procedures (MTTP) to the military Services for coordinating interdiction.

ATP 3-01.15 - AMD, Multi-Service Tactics, Techniques, and Procedures for Air and Missile Defense

This publication provides multi-Service tactics, techniques, and procedures (MTTP) for planning and executing air and missile defense operations. Additionally, it provides warfighters and planners with a single-source reference for Service-specific air and missile defense capabilities and limitations.

ATP 3-60.1 - DT, Multi-Service Tactics, Techniques, and Procedures for Dynamic Targeting

This multi-Service tactics, techniques, and procedures (MTTP) publication for dynamic targeting (DT) is a single source, descriptive reference guide to assist component and Service commanders and their staffs in using DT to plan, target, and process respective targets in all domains and environments. This MTTP is used by Services and components in joint and multinational combined operations, including operating with other United States (US) departments and agencies.

ATP 3-09.34 - KILL BOX, Multi-Service Tactics, Techniques, and Procedures For Kill Box Planning and Employment

Multi-Service Tactics, Techniques, and Procedures (MTTP) for Kill Box Planning and Employment establishes tactics, techniques, and procedures for the application of kill boxes. Kill boxes are used to facilitate integrating joint fires as a permissive fire support coordination measure, in accordance with Joint Publication 3-09, Joint Fire Support. This publication standardizes kill box terminology and offers a detailed explanation of kill box planning, coordination, and employment.

ATP 3-52.1 - AC, Multi-Service Tactics, Techniques, and Procedures for Airspace Control

This multi-Service tactics, techniques, and procedures (MTTP) publication for airspace control is a single source, descriptive reference guide to facilitate multi-Service coordination, integration, and control of airspace during exercises, contingencies, and other operations where Service components must share airspace for operational use.

Center for Army Lessons Learned (CALL) Publications (CUI/CAC Login Required)

The following is a listing of CUI publications available from CALL related to C-UAS and sUAS. CALL gathers information from tactical, operational, and strategic levels through Army activities, training, and global operations and shares insights and best practices with units, training centers, and proponents through products like handbooks, forums, and multimedia content. Controlled unclassified information (CUI) is unclassified information the United States Government creates or possesses that requires safeguarding or dissemination controls limiting its distribution to those with a lawful government purpose. CUI may not be released to the public absent further review. CAC login required. (*https://www.army.mil/CALL)*

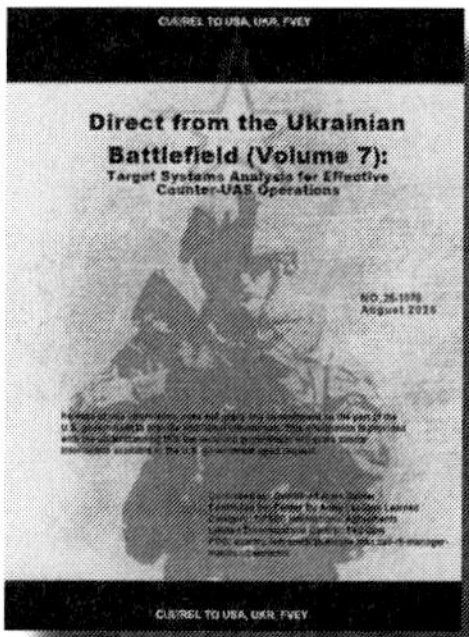

III. Fires Logic Diagram

Success in large-scale combat operations is dependent on the Army's ability to employ fires. Fires enable maneuver. Over the past two decades, potential peer threats have invested heavily in long-range fires and integrated air defense systems, making it even more critical that the U.S. Army possess the ability to maneuver and deliver fires in depth and across domains.

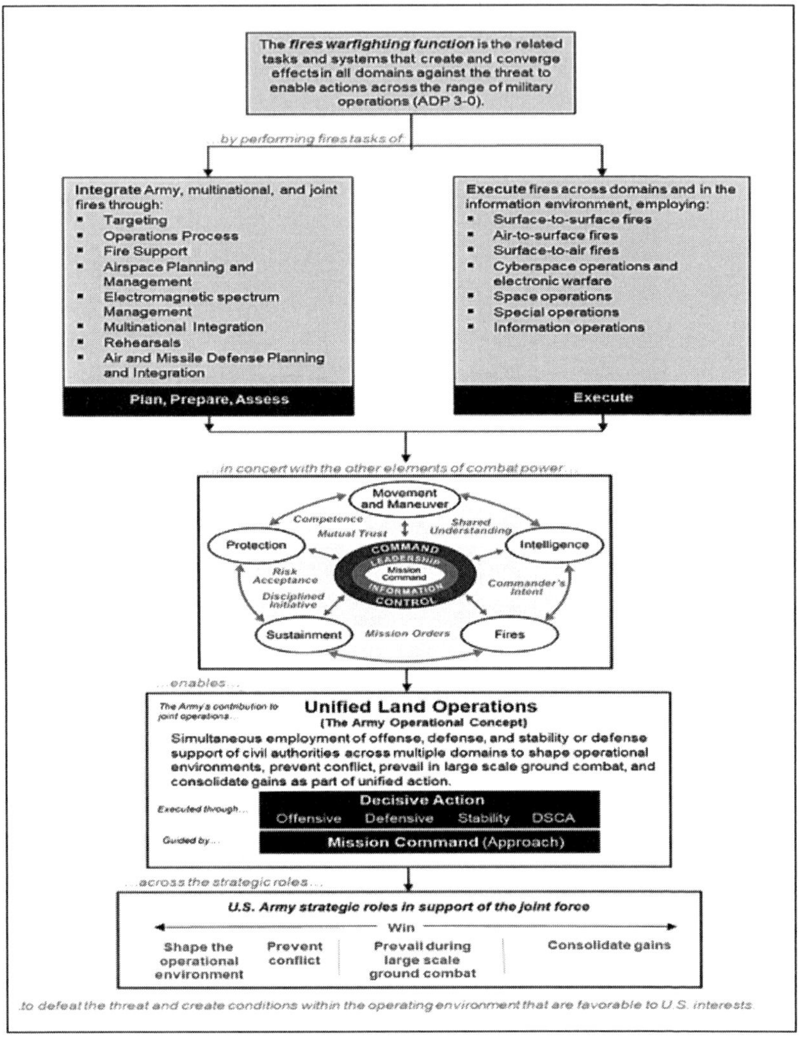

Ref: ADP 3-19, Fires (Jul '19), introductory figure, ADP 3-19 Logic chart.

Refer to AODS7-1: The Army Operations & Doctrine SMARTbook, 7th Ed. (w/SMARTupdate 1) for complete discussion of the fires warfighting function from ADP 3-19 (Jul '19). Sections include fires as a warfighting function; fires in unified land operations, execution of fires across the domains; and integrating Army, joint & multinational fires (to include airspace planning/integration and air and missile defense planning/integration).

Fires Integration (C-UAS/sUAS)

II. Targeting (D3A)

Ref: ATP 3-60, Army Targeting (Aug '23) and ADP 3-19, Fires (Jul '19).

Targeting is the process of selecting and prioritizing targets and matching the appropriate response to them, considering operational requirements and capabilities (JP 3-0). It is an integral part of the operations process that organizes the efforts of the commander and staff to integrate and synchronize fires in operations. Targeting seeks to create specific desired effects through lethal and nonlethal actions.

Targeting encompasses many processes, all linked and logically guided by the joint targeting cycle. Units may use the Army targeting process or the joint targeting cycle as appropriate according to organizational echelon to integrate and synchronize capabilities, across warfighting functions and information advantage, into operations, creating the desired effects in time and space. The targeting team recommends targeting guidance to the commander, develops targets, selects targets for attack, and coordinates, integrates, and assigns organic or allocated joint, interagency, and multinational fires to specific targets and target systems.

Targeting proceeds from the commander's objectives to an assessment of the results achieved by the executed course of action (COA). Participants in the targeting process should adhere to these targeting principles for creating the desired effects while diminishing undesired or adverse collateral effects.

Targeting Guidance

The targeting team ensures that the proposed commander's targeting guidance includes targeting priorities, operational objectives, HPTs, desired effects, target selection standards (TSS), authorized actions against targets, any delegated responsibilities for target validation and approval, and the desired endstate. The commander should leverage all available capabilities to create desired effects.

The commander can also provide restrictions as part of their targeting guidance. Restrictions at the tactical level are different than restrictions at the Joint level. Army forces use FS coordination measures to include no fire areas, target list worksheets, and ROE. Joint forces promulgate the joint force commander's (JFCs) no-strike list (NSL) and the restricted target list (RTL). RTLs apply to all subordinate echelons.

No-Strike List (NSL)

The NSL consists of objects or entities protected by—

- Law of war.
- International laws.
- ROE.
- Other considerations.

Restricted Target List (RTL)

An RTL is a valid target list with specific restrictions such as—

- Limit collateral damage.
- Do not strike during daytime.
- Strike only with a certain weapon.
- Proximity to protected facilities and locations.

Refer to FM 1-04, FM 3-57, JP 3-57, JP 2-0, and JP 3-60 for additional information on legal considerations and targeting restrictions.

Targeting Categories

Ref: ATP 3-60, Army Targeting (Aug '23),pp. 1-4 to 1-7.

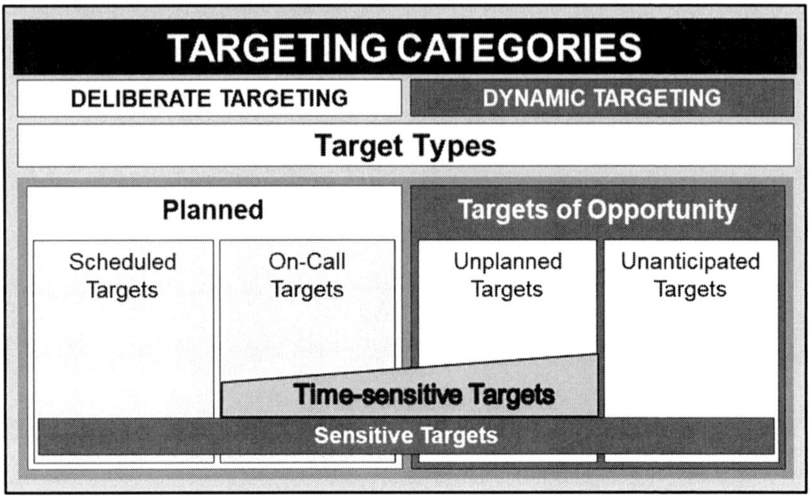

Ref: ATP 3-60 (Aug '23), fig. 1-1. Targeting categories.

There are two targeting categories: deliberate targeting and dynamic targeting.

A. DELIBERATE Targeting

Deliberate targeting prosecutes planned targets. These targets are known to exist in the area of operations (AO) and have actions scheduled against them. Examples range from targets on target lists in the applicable plan or order, targets detected in sufficient time to be placed in the joint air tasking cycle, mission type orders, or FS plans. There are two types:

- **Scheduled targets** in the AO and are located in sufficient time so that fires or other actions upon them are identified for engagement at a specific, planned time.
- **On-call targets** have actions planned, but not for a specific delivery time. The commander expects to locate these targets in sufficient time to execute planned actions.

Target Development. All Army forces conduct target development. Target development is the systematic examination of potential target systems— and their components, individual targets, and even elements of targets— to determine the necessary type and duration of the action that must be exerted on each target to create an effect that is consistent with the commander's specific objectives (JP 3-60).

Target Types (FIVE-O). Before beginning target development, the type of entity target must be established. There are five types of entity targets. Often referred to as the FIVE-O target types, they are **Facilities, Individuals, Virtual, Equipment, and Organizations.**

Target Vetting. Vetting occurs at all Army echelons. Staffs at echelon assess and analyze targets to determine and confirm location, function, and target type for execution. In principle, vetting in a joint environment is typically external to the organization and a formal process. Within Army processes, vetting is typically internal to the organization.

Target Validation. Target validation must occur to receive command approval to perform engagements against targets. Within Army processes, validation is a commander's approval of targets to a target list worksheet or a high-payoff target list (HPTL).

Target validation is presented by targeting personnel, in consultation with the planners, servicing judge advocate general, and other experts and agencies, as required. Validation in a joint environment is the approval of candidate targets to a joint target list (JTL) or an restricted target list (RTL).

Target Nomination. Target nominations in a joint environment are targets selected by the combatant command staff, components, subordinates, and task forces from the JTL and RTL to compile target nomination lists to nominate targets for engagement (see the current version of CJCSI 3370.01D). Within Army processes, target nominations from corps and below functioning as a tactical headquarters (HQ) is a lower echelon asking a higher echelon to execute a target on behalf of the requesting echelon. The higher HQ for all aspects of targeting to include development and execution owns approved target nominations.

Target Recommendation. Within Army processes, target recommendations from corps and below functioning as a tactical HQ are targets that meet execution criteria passed from the intelligence and other staff elements to the fire support element (FSE) for execution. The distinction between the joint and Army environment target recommendations is that candidate targets are not yet validated, whereas target recommendations are valid targets. Target recommendations in a joint environment are called candidate targets submitted by component commanders, appropriate agencies, or the JFC staff for further development and inclusion on the JTL or RTL or moved to the NSL (see the current version of CJCSI 3370.01D).

B. DYNAMIC Targeting

Dynamic targeting is targeting that prosecutes targets identified too late or not selected for action in time to be included in deliberate targeting (JP 3-60). Dynamic targeting accounts for the unpredictability of the threat and prosecutes targets of opportunity during current operations. Targets of opportunity are targets identified during execution that require the targeting team to account for changes to the previously planned actions, current assets available and risks to immediate execution. There are two types:

- **Unplanned targets** are known to exist in the AO, but no action has been planned against them. The target may not have been detected or located in sufficient time to meet planning deadlines. Alternatively, the target may have been located, but not previously considered of sufficient importance to engage.
- **Unanticipated targets are unknown or not expected to exist in the AO.**

Time-Sensitive Targets (TST). A time-sensitive target is a joint force commander-validated target or set of targets requiring immediate response because it is a highly lucrative, fleeting target of opportunity or it poses (or will soon pose) a danger to friendly forces (JP 3-60).

Component Critical Targets. Echelons designated as the land component commander (LCC) will identify and nominate component critical targets. Service component commanders may designate high-priority targets that present significant risks to, or opportunities for component forces and missions.

Sensitive Targets. Certain targets require special care or caution in treatment because failure to engage them or to engage them improperly can lead to major adverse consequences. Examples include leader targets (high-value individuals [HVI]) that must be handled sensitively due to potential political repercussions; targets located in areas with a high risk of collateral damage; and weapons of mass destruction facilities, where an attack can lead to major long term environmental damage.

Brigade Targeting (Overview/Example)

Ref: ATP 3-60, Army Targeting (Aug '23), chap. 3.

Targeting at the tactical level is largely defined by the capability and capacity of personnel and equipment. The sheer numbers of personnel at the brigade are far less than that of a corps or Theater Army. The staff and targeting team must assess capability and requirements to determine the feasibility and level of detail when applying the D3A methodology. The brigade is realistically the first echelon where a formal process will be conducted. The application of the methodology will be very similar at echelon; however, processes and procedures may be abbreviated. Targeting decisions at a higher HQ affect targeting decisions at subordinate HQ. The brigade staff may use the targeting products of the division or produce their own, depending on time and personnel available, to coordinate and integrate targeting actions. The brigade will focus on synchronizing assets to meet the commander's objectives within their assigned AO and submit requests for support or target nominations to division for effects they are unable to create with task organized capability or organic assets and enablers. Targeting at these echelons is typically more focused on specific threat systems, equipment, or individuals than on larger threat capabilities or functions such as integrated air defense systems, integrated fires commands, or logistics.

Operational Framework

The understanding of the operational framework allows commanders and staff the ability to geospatially, as well as temporally understand their assigned area of responsibility. Furthermore, it provides them the understanding to define the rear, close, and deep fight of the organization. Additionally, it provides an appreciation for the depth and breadth of their assigned AO across the multi-domain battlefield to include the electromagnetic spectrum, and information dimension. It ensures the correct orientation of forces while constantly assessing the adversary's ability to contest friendly forces across all domains. True understanding of the operational framework better allows the commander to make informed targeting decisions that quickly creates windows of opportunity and convergence of effects.

Although BNs rarely, independently conduct organizational targeting, they play a valuable role in the tactical success of the brigade. The brigade applies the organizational targeting process to synchronize effects and to strike the enemy rapidly at decisive points to set favorable force ratios and enable conditions for continuous consolidation of gains. To accomplish this, the brigade leverages the operational framework to focus fires and effects ahead of friendly forces probable line of contact.

The operational framework extends from the brigade to theater army level in a single avenue in depth. Once identified, the brigade delineates a rear, close, and deep area to describe the physical arrangement of friendly forces in time, space, and purpose. This provides focus to the brigade targeting team and requires the commander and staff to examine the capabilities and operational reach of its forces and assigned enablers. Once applied, the brigade develops, as well as applies measures that favor the employment, timeliness, and synchronization of effects. Organizations apply the capabilities and limitations of its assigned sensors and weapon systems to inform placement. This is useful when articulating requirements to higher HQ or requests for assistance.

Figure 3-1 *(facing page)* contains the brigade operational framework, which highlights the targeting effort at the brigade level and shades other echelon's AOs denoting they are outside of the brigade's responsibility. Despite being outside the brigade's responsibility, it is within their area of interest and brigade targeting teams should be conscious of activities in the areas that may impact their AO.

Brigades may build off the division's targeting products to ensure integration with higher HQ plan and to save time. This does not preclude the brigade staff from inherent targeting functions to meet its own commander's intent. These functions and products are explained in detail in the corresponding appendices.

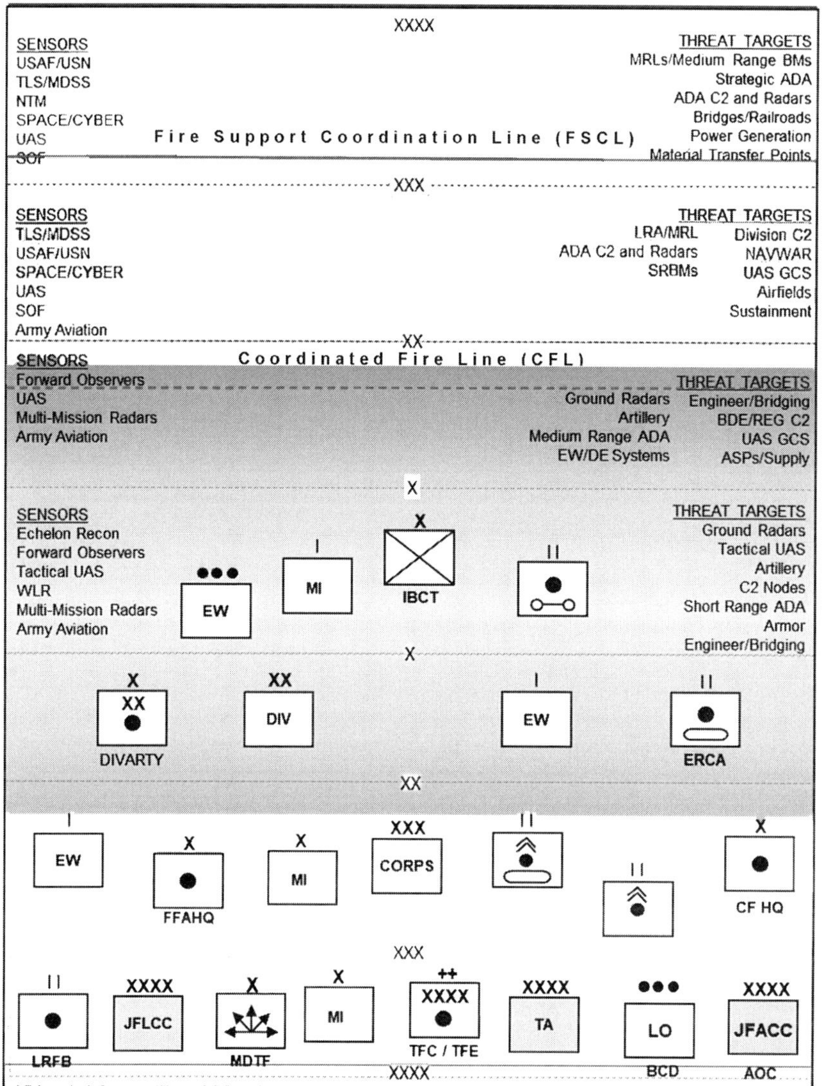

ADA – air defense artillery, AOC – air operations center, ASP – ammunition supply point, BCD – battlefield coordination detachment, BDE – brigade, BM – ballistic missiles, C2 – command and control, CF – counterfire, DE – directed energy, DIV – division, DIVARTY – division artillery, ERCA – extended range cannon artillery, EW – electromagnetic warfare, FFAHQ – force field artillery head quarters, GCS – ground control station, HQ – headquarters, IBCT – infantry brigade combat team, JFACC – joint force air component commander, JFLCC – joint force land component commander, LO – liaison officers, LRA – long-range artillery, LRFB – long range fires battalion, MDSS – multidomain sensing system, MDTF – multidomain task force, MI – military intelligence, MRL – multiple rocket launcher, NAVWAR – navigation warfare, NTM – national technical means, REG – regiment, SOF – special operations forces, SRBM – short range ballistic missile, TA – target acquisition, TFC – theater fires command, TFE – theater fires element, TLS – terrestrial layered system, UAS – unmanned aircraft system, USAF – United States Air Force, USN – United States Navy, WLR – Weapons Locating Radars

JOINT **ARMY**

Ref: ATP 3-60 (Aug '23), fig. 3-1. Brigade operational framework. Refer to ATP 3-60 for discussion of division targeting (chap. 4) and theater targeting (chap. 5).

Army Targeting Methodology (D3A Overview)

Ref: ADP 3-19, Fires (Jul '19), pp. 3-7 to 3-9.

The **Army targeting process** organizes the efforts of the commander and staff to accomplish key targeting requirements. This methodology is referred to as the D3A. D3A assists the commander and staff decide which targets must be acquired and engaged and to help develop options to engage those targets. Options may include lethal or nonlethal, organic or supporting assets at all levels, including maneuver, electronic attack, psychological operations, attack aircraft, surface-to-surface fires, air to surface fires, other information-related capabilities, or a combination of these options.

The D3A methodology is an integral part of the MDMP. As the MDMP is conducted, targeting becomes more focused based on the commander's guidance and intent. Certain targets may require special considerations or caution, because engaging them improperly could create unintended effects. Examples include targets that should be handled with sensitivity due to potential political and or diplomatic repercussions and targets located in areas with a high risks of collateral damage, to include weapons of mass destruction facilities. These measures are incorporated in the coordinating instructions and appropriate annexes of the operation plan or operation order.

I. Decide *(See pp. 8-12 to 8-18.)*

Decide is the first function in targeting and occurs during the planning portion of the operations process. It is the most important function, requiring close interaction between the commander, intelligence, plans, operations, the fires cell, and staff judge advocate. It begins during the mission analysis portion of the MDMP and continues throughout the operation.

II. Detect *(See pp. 8-18 to 8-20.)*

Detect is the second function in targeting and occurs initially during the prepare portion of the operations process, continuing throughout the operation. A key resource for fires planning and targeting is the intelligence generated through information collection to answer the targeting information requirements. Commanders express requirements for target detection and action as priority intelligence and information requirements. During large-scale combat operations, it might be challenging to prioritize the detection of targets and could require the opening of windows of opportunity for specific collection capabilities in support of fires. High-payoff targets must be integrated and support associated priority intelligence requirements. Their priority depends on the importance of the target to the friendly course of action and target acquisition requirements. Targets are prioritized through a quantitative and qualitative valuation methodology. An example of a valuation methodology is the target value analysis process that prioritizes targets based on the target's criticality, accessibility, recuperability, vulnerability, effect, and recognizability. Targeting working groups incorporate priority intelligence and information requirements that support acquisition of high-payoff targets into the overall information collection plan along with named areas of interest, target areas of interest, and engagement areas.

III. Deliver *(See pp. 8-20 to 8-22.)*

Deliver is the third function in targeting and occurs primarily during the execution portion of the operations process. The main objective is to engage targets in accordance with the commander's guidance or engagement authority's direction. The selection of a weapon system or a combination of weapons systems leads to the tactical decision of time of engagement and then the technical solution for the selected weapon.

IV. Assess *(See pp. 8-22 to 8-24.)*

Assess is the fourth function of targeting and occurs throughout the operations process. The commander and staff assess the results of mission execution. The assessment process is continuous and directly tied to the commander's decisions throughout planning, preparation, and execution of operations.

Methodology Tools

Ref: ATP 3-60, Army Targeting (Aug '23), p.

Understanding the D3A methodology itself is quite simple but the creative thinking and synchronization of resources required to be successful during operations takes repetition and experience. Tools are available to assist the targeting team with effective and efficient execution. A targeting checklist covering the D3A methodology can be found in Appendix A and the diagram below. Figure 2-3 provides a visual example of assets and desired effects that can be used during each function of the methodology.

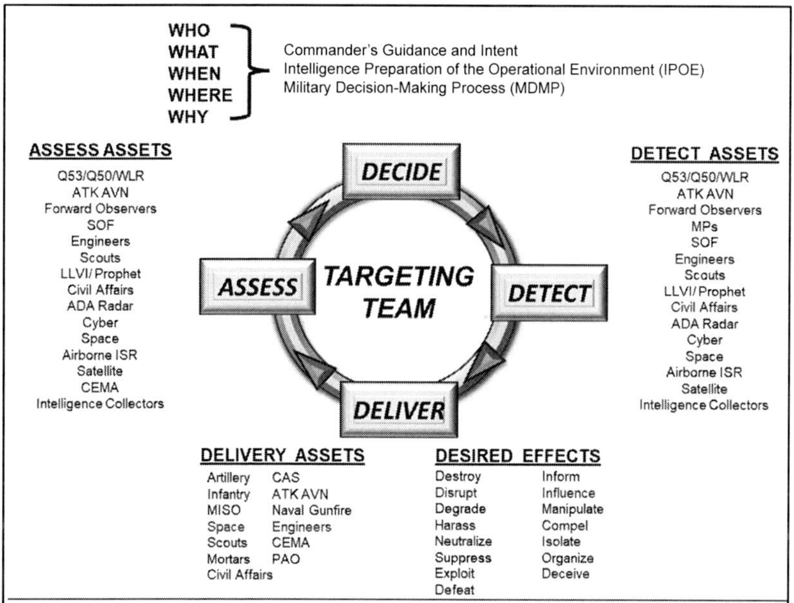

ADA- air defense artillery, ATK AVN- attack aviation, CAS- close air support, CEMA- cyber electromagnetic activities, IPB- intelligence preparation of the battlefield, ISR- intelligence surveillance reconnaissance, LLVI- low level voice intercept, MDMP – military decision-making process, MISO- military information support operations, MP- military police, PAO- public affairs officer, Q50 – lightweight counter mortar radar, Q53 – weapons locating radar SOF- special operations forces, WLR- weapons locating radar

Ref: ATP 3-60 (Aug '23), fig. 2-3. Decide, detect, deliver, and assess methodology and assets.

In large-scale combat operations, targets in all domains are arrayed across the depth and breadth of the OE. These targets far exceed the number of resources available to acquire and create desired effects. It is paramount that higher HQ provide adequate guidance and must anticipate requirements subordinates may have. Targeting is a top-down driven process with a substantial need for bottom-up refinement. This will be applicable in any circumstance requiring lead time or insertion into supporting planning or execution cycles. The importance of what targets to attack and with what available capabilities must be planned and prioritized.

Refer to AODS7-1: The Army Operations & Doctrine SMARTbook, 7th Ed. (with SMARTupdate 1) for complete discussion of the fires warfighting function from ADP 3-19 (Jul '19). Sections include fires as a warfighting function; fires in unified land operations, execution of fires across the domains; and integrating Army, joint & multinational fires (to include airspace planning/integration and air and missile defense planning/integration).

In large-scale combat operations, targets in all domains are arrayed across the depth and breadth of the OE. These targets far exceed the number of resources available to acquire and create desired effects. It is paramount that higher HQ provide adequate guidance and must anticipate requirements subordinates may have.

The targeting process provides an effective method for matching friendly force capabilities against enemy targets in order to achieve the commander's desired effects to achieve the commander's objectives. There are two general ways to frame the D3A methodology. During planning and the MDMP, the staff uses the methodology as a process to assist in product development and visualization. The methodology is applied to analyze commanders' guidance in order to determine the right targets, at the right place, at the right time. It is used through MDMP to make these decisions and apply the required assets to create desired effects. Further, the methodology is discussed through developing courses of action and wargaming them. It is also applied to assessment requirements and the analysis needed to facilitate future decisions. As an integrating process during execution, the second application enables the staff to apply the targeting products (HPTL, TSS, AGM, intelligence collection synchronization matrix (ICSM), and TSM) to facilitate operations. The staff is continuously making decisions, adjusting detection methods or location, and revising delivery options based on changes to threat courses of action. D3A is a flexible, simple, repeatable four-function process and is not designed to be time constrained or rigidly sequential.

I. Decide

The decide function begins the targeting cycle and is applied throughout the Operations Process (Plan, Prepare, Execute, Assess). Decisions made during planning will be evaluated and revisited during preparation and execution, ultimately leading to effective assessments. The targeting team must be conscious of the decisions made and adjust them during rehearsals, individual, and collective training. The endstate is to ensure the proper capabilities are integrated and synchronized in the proper geographic locations, at the right time to create desired effects. Following engagement, how and what assessment must be decided upon. The application of commander's guidance is implemented throughout the steps of MDMP and refined until mission completion.

The outcome of success results in efficient employment of capabilities, favorable force ratios, and the accomplishment of all assigned tasks and objectives. The following are the consequences of failure:

- Enemy prevents friendly forces from achieving tasks and objectives potentially leading to mission failure.
- The commander is presented with unachievable solutions (capabilities are out of range).
- Resources are misused (inaccurate weaponeering or firing solution, mismanagement of ammunition, employment of wrong capability).
- Force ratios are unfavorable (Failure to see the enemy, lack of anticipation to support subordinates).
- Subordinate echelons improperly resourced (Task organization, information collection, delivery capabilities).

With this information, the targeting team and appropriate staff officers can prepare their respective running estimates. From the standpoint of targeting: the FS, intelligence, information capabilities, influence capabilities, and operations estimates are interrelated and closely coordinated among each cell.

This function provides the overall focus and sets priorities and criteria for information collection and engagement planning. The decide function draws heavily on the staff's knowledge of the enemy (to include their tactics, culture, and ideology), a detailed IPB, and continuous assessment of the situation. Targeting priorities must be addressed for each phase or critical event of an operation. The targeting team works in concert to draft targeting guidance and priorities for the commander. Once approved, the decisions made are reflected in the following foundational products *(see following pages).*

Intelligence Preparation of the Operational Environment (IPOE) *

IPOE helps the commander to selectively apply and maximize combat power at critical points in time and space. IPB helps the commander and staff visualize the operating environment, how the natural environment affects friendly units and likely enemy COAs.

IPOE is a continuous staff planning activity undertaken by the entire staff. Intelligence preparation of the battlefield is the systematic process of analyzing the mission variables of enemy, terrain, weather, and civil considerations in an area of interest to determine their effect on operations (ATP 2-01.3). Although the G-2/S-2 leads the IPOE, various staff sections contribute to it. IPOE, sets the context for operational planning and targeting. The process is flexible and well suited to account for lethal and nonlethal considerations (across the physical, information, and human dimensions) during operations. IPOE:

- Describes the totality of the OE that may impact friendly, threat, and neutral forces at that echelon.
- Accounts for all relevant domains and dimensions of the OE that may impact friendly and threat operations.
- Identifies possible defeat mechanisms and relative advantages that friendly forces might be able to create and exploit.

Initial IPB, which occurs during mission analysis, is important as a means to focus the staff on various aspects of the OE specific to that unit, mission, time, and place, as well as a means to create products that are important to subsequent planning and targeting. As planning progresses, beyond mission analysis, IPB products are refined to support planning and targeting. Some of the more important products include the modified combined obstacles overlay, threat models (with generic HVTs), situation templates, event template, event matrix, and draft HVT list. (*Refer to ATP 3-60, appendix B and ATP 2-01.3 for more information.)*

** IAW Change 2 to ATP 2-01.3 (Jan '24) modifies all instances of the term "intelligence preparation of the battlefield (IPB)" to "intelligence preparation of the operational environment (IPOE)." (Refer to BSS7-1, The Battle Staff SMARTbook, chap. 3 for 50 pages of detail of the IPOE process.)*

A. High-Payoff Target List (HPTL)

The high-payoff target list is a prioritized list of high-payoff targets by phase of the operation. The HPTL is a key element commander's guidance and communicates priority to the staff. Other considerations include the following:

- The sequence or order of appearance of the system, function, or capability being targeted.
- The ability to detect, identify, classify, and track the target. (This decision must include sensor availability and processing timeline considerations).
- The degree of accuracy available from the acquisition system(s).
- The ability to engage the target.
- The ability to create the desired effects based on attack guidance.

Targets are prioritized according to the considerations above within specific time windows. The targeting working group recommends priorities for the targets according to its judgment and the advice of the targeting officer and the FAIO. Target list worksheets give a recommended priority and engagement sequence. If the target list worksheet or war gaming departs from the commander's guidance, it is noted on the proposed HPTL to inform the commander of the conflict. The target category

Example Targeting Product Formats (Decide)

Ref: ATP 3-60, Army Targeting (Aug '23), app. D.

There are no prescribed formats for the targeting tools. Each unit will develop tools that work best for them. Factors to consider in developing formats are as follows: type, level, or role of the command; operational environment; assets available; missions, & SOPs.

Regardless of the formats used, the targeting methodology applicable for the unit's role must be followed. Targeting products can be presented in many forms. It is important forms be sharable in oral presentations, hard copy publications, or electronic format.

A. High-Payoff Target List (HPTL) *(See p. 8-13.)*

High-Payoff Target List		
Phase or Critical Event		
Priority	***Category***	***Targets***
1	ADA	SA-15, SA-17, Tracking Radar, Engagement Radar
2	Fire Support	HVY MRL, Cannon ARTY, CTR BTY Radar
3	R&S	UAS GCS, Scouts
4	Lift	ATK helo, FARP, Utility helo, Airfields
5	Sustainment	Bulk fuel, logistics, ammunition, maintenance
ADA – air defense artillery, ARTY– artillery, ATK – attack, BTY – battery, CTR – counter, FARP – forward arming and refuel point, GCS – ground control station, helo – helicopter, HVY – heavy, MRL-multiple rocket launcher, R&S – reconnaissance and surveillance, SA – surface-to-air, UAS – unmanned aircraft systems		

B. Target Selection Standards (TSS) *(See p. 8-16.)*

Target Selection Standards			
High-Payoff Target	***Timeliness***	***Accuracy***	***Size***
Tor	15 minutes	200 meters	Battery
HVY MRL	30 minutes	100 meters	Battalion
Ammunition Holding Point	6 hours	200 meters	
UAS ground control station	1 hour	100 meters	Individual
Counter-battery radar	30 minutes	200 meters	Section
Print Facilities	2 hours	50 meters	Individual
Forward arming and refueling point	1 hour	200 meters	
Access Point (Internet Activities)	10 minutes	50 meters	Individual
HVY – heavy, MRL– multiple rocket launcher, Tor – surface-to-air missile system, UAS – unmanned aircraft systems			

C. Attack Guidance Matrix (AGM) *(See p. 8-17.)*

Attack Guidance Matrix				
HPT	***When***	***How***	***Effect***	***Remarks***
BM-30	A	Field Artillery	Neutralize	Coordinate and ensure CFFZ and CFZ are emplaced and active
Tor	A	UAS	Destroy	Maneuver engages target and calls in UAS strikes as necessary.
UAS Ground Control Station	P (1330Z)	Electromagnetic Attack	Neutralize	Jam Communications at H-1
WLR	I	SCAR	Destroy	Monitor for signals transmission to trigger engagement
Print Facilities	A	AI/Cyber	Neutralize	Coordination required to synchronize lethal engagement with cyber effects.
A – as acquired, ACM – airspace coordinating measures, AI – air interdiction, BM – boyevaya mashina, CFFZ – call for fire zone, CFZ – critical friendly zone, H – hour, HPT – high-payoff target, I – immediate, P-planned, SCAR – strike coordination and reconnaissance, Tor – surface-to-air missile system, UAS – unmanned aircraft system, WLR – weapons locating radar				

D. Information Collection Synchronization Matrix *(See p. 8-17.)*

UNCLASSIFIED

BCT Action		Phase I: Area defense and security operations along PL Python							Phase II: Prepare to support passage of lines							Phase III: Clear enemy in zone		
Enemy Action		SPF and RISTA elements conduct screening operations							Mechanized infantry battalions establish battle positions to block friendly forces									
Collection Focus		Identify SPF and RISTA disposition within the disruption zone							Identify enemy defensive positions along avenue of approach within the battle zone									
Local Time		0200	0300	0400	0500	0600	0700	0800	0900	1000	1100	1200	1300	1400	1500	1600	1700	1800
H-Hour		H-4	H-3	H-2	H-1	H	H+1	H+2	H+3	H+4	H+5	H+6	H+7	H+8	H+9	H+10	H+11	H+12
EAB	COMINT										NAI 3004							
	ELINT										NAI 3004							
	GEOINT						NAI 3001/TAI 1					NAI 3005/TAI 5						
	MASINT						NAI 3001/TAI 1											
BCT	A Troop						NAI 3001/TAI 1		Movement									
	B Troop								NAI 3003/TAI 3									
	C Troop								Movement									
	HCT 1						NAI 3001/TAI 1											
	HCT 2																	
	HCT 3						NAI 3002/TAI 2											
	CI																	
	Prophet 1	Maintenance			NAI 3002/TAI 2					NAI 3003/TAI 3								
	Prophet 2	Maintenance											NAI 3004					
	Shadow 1				NAI 3002/TAI 2								NAI 3004					
	Shadow 2									NAI 3003/TAI 3								
	Shadow 3													NAI 3005/TAI 5				

☆ decision point ⌗ requested asset

BCT brigade combat team
CI counterintelligence
COMINT communications intelligence
EAB echelons above brigade
ELINT electronic intelligence
GEOINT geospatial intelligence
HCT human intelligence collection team
H-hour specific hour at which a particular operation commences
MASINT measurement and signature intelligence
NAI named area of interest
PL phase line
RISTA reconnaissance, intelligence, surveillance, and target acquisition
SPF special purpose forces
TAI target area of interest

E. Target Synchronization Matrix (TSM) *(See p. 8-18.)*

DECIDE			DETECT		DELIVER		ASSESS	
P	**Category**	**HPTs**	**Agency**	**Asset**	**Agency**	**Asset**	**Agency**	**Asset**
1	Fire Support	BM-30 BM-27 1V18 1V19 Counter-battery Radar	FAB CAB G-2 FAB E-MIB G-2	Q-53 FMV JSTARS Q-53 NOS Rivet Joint	1-FAB 2-Avn DIVARTY FAB FAB FAB CAB	1-MLRS 2- AH-64E Artillery MLRS MLRS MLRS AH-64E	E-MIB FAB G-2 E-MIB E-MIB G-2 CAB	FMV Q-53 JSTARS Rivet Joint FMV Rivet Joint AH-64E FMV
2	ADA	Tor Buk Tracking Radar	G-2 CAB E-MIB	Rivet Joint FMV Rivet Joint	AI FAB AI	F35, F16 MLRS F22, F16	AOC CAB G-2	MSNREP FMV NOS
3	R&S	UAS ground control station Scouts	CEMA AMD G-2 BCT	NOS Sentinel FMV	FAB BCT FAB	MLRS Artillery Organic Artillery	AOC G-2 BCT	MSNREP EAD ELINT Prophet

of the HPT is shown, either by name or by number, on the list. The category name and number are shown on the target list worksheet. The number of target priorities should not be excessive. Too many priorities will dilute the intelligence collection acquisition and engagement efforts. The approved list is shared with the targeting team and staff. It is used as a planning tool to determine attack guidance and to refine the collection plan. This list may also indicate the commander's operational need for BDA of a specific target and the time window for collecting and reporting it.

Note. Any format for prioritizing HPTs serves the purpose of a HPTL for linking targets with phases of battle.

One way to organize the HPTL is to group all HPTs into target sets that reflect the capabilities and functions the commander has decided to engage. Target sets are identified and prioritized for each phase of the operation. Within the sets, individual targets are ordered by target value, sequence of appearance, importance, or other criteria that satisfy the commander's desired effects. Additionally, the HPTL may be broken down based on individual targets within a system. Rather than listing integrated fires commands or integrated air defense systems, a systems approach to analysis may highlight key components of those systems. Staffs must understand the criticality of the system or system components when constructing their HPTLs. In this way, the targeting working group reduces, modifies, and reprioritizes HVTs while ensuring that HPTs support the concept of operations.

The commander's guidance may require changes, which should be annotated on the HPTL. The target name or number and description are placed on the list for specific HPTs in each category. Once the commander approves or amends the HPTL, it goes back to the targeting working group to help them develop the AGM and collection plan.

B. Target Selection Standards (TSS)

TSS are criteria applied to enemy activity (acquisitions and battlefield information) and used in deciding whether the activity is a target. TSS put nominations into two categories: targets and suspected targets. Targets meet accuracy and timeliness requirements for engagement. Suspected targets must be confirmed before any engagement. Units may develop their own worksheet format.

TSS are based on the enemy activity under consideration and available weapon systems by using the following:

- Weapon system target location accuracy requirements such as target location error (TLE).
- Size of the enemy activity (point or area).
- Status of the activity (moving or stationary).
- Timeliness of the information.

Different TSS may exist for a given enemy activity based on different weapons system. For example, an enemy artillery battery may have a 150-meter TLE requirement for attack by cannon artillery and a 1¬kilometer requirement for attack aircraft. TSS are developed by the FSE in conjunction with the intelligence section. Intelligence analysis use TSS to quickly determine targets from battlefield information and pass the targets to the FSE. Weapon system managers such as FSEs, fire control elements, or fire direction centers use the TSS to identify targets for expeditious attack. Commands can develop standard TSS based on threat characteristics and doctrine matched with the standard available weapon systems.

TSS worksheets are given to the intelligence officer by the FSE. The FAIO uses TSS to confirm or verify targets that are forwarded to an FSE. Intelligence analysts evaluate the source of the information as to its reliability and accuracy, confirm that the size and status of the activity meet the TSS, and then compare the time of acquisition with the dwell time. Accurate information from a reliable source must be verified before declaring it a target if the elapsed time exceeds dwell time.

Note. Dwell time is the length of time a target is doctrinally expected to remain in one location.

Some situations require intelligence assets to confirm positive identification before approval to engage is given. HPT that meets all the criteria should be tracked until they are attacked in accordance with the AGM. Location of targets that do not meet TSS should be confirmed before they are attacked.

C. Attack Guidance Matrix (AGM)

Knowing target vulnerabilities and analyzing the probable effect an engagement will have on enemy operations allows a staff to propose the most efficient available engagement option. Emphasis is placed on the desired effect and commander's guidance. During war gaming, decision points linked to events, areas of interest, or points on the battlefield are developed.

The attack guidance matrix is a targeting product approved by the commander, which addresses the how and when targets are engaged and the desired effects (FM 3-09). The AGM is a compilation of engagement criteria in a format that can be easily understood. Attack guidance applies to both deliberate and dynamic targets and incorporates lethal and nonlethal engagement against both target types. One AGM may be produced for execution at any point in the operation, per targeting cycle, or by phase.

Based on commander's guidance, the targeting working group recommends how each target should be engaged in terms of the desired effects and engagement options. Desired effects are translated into automation system default values to engage targets more effectively.

The decision of what engagement means or system to use is made at the same time as the decision on when to acquire and engage the target. Coordination is required when deciding to engage with two or more means such as electromagnetic attack, information capabilities, or attack aviation. Joint Munitions Effectiveness Manual Weaponeering Software (referred to as JWS) is the digital weaponeering tool used to assist in attack guidance decisions.

The commander, with recommendations from the targeting working group, must approve the attack guidance. Attack guidance may address specific or general target descriptions and is provided to weapons systems managers via the AGM. This guidance should detail the following:

- A prioritized HPTL.
- When, how, and desired effects of engagement.
- Any special instructions.
- BDA requirements.

The AGM is a tool usually disseminated as a matrix. The AGM includes as a minimum:

- Phase or time period covered by the AGM.
- Specific HPT.
- Timing of engagement.
- How targets are engaged.
- Desired Effects.
- Remarks to include restrictions.

D. Information Collection Synchronization Matrix

The information collection plan focuses the collection effort to answering PIRs and identifying HPTs. To support the detection of HPTs on the battlefield, the targeting cell provides indicators and specific information requirements (SIRs) to the collection management team for each HPT. The information collection matrix accounts for PIRs and other intelligence requirements. The collection management team accounts for targeting intelligence requirements, if developed, in a targeting addendum to the information collection matrix. If the collection management team does not develop a targeting addendum, the team must graphically account for TAIs and HPTs through other means. For example, the team could add both TAIs and HPTs as columns to the information collection matrix or develop a supporting crosswalk of HPTs with supporting SIRs, NAIs, TAIs, an NAI, and collection assets.

E. Target Synchronization Matrix (TSM)

The TSM visually illustrates the HPTs and is designed to list specific targets with locations, in each category. The matrix has entries to identify if a target is covered by a NAI; the specific detect, deliver, and assess assets for each target; and engagement guidance. Once completed, the TSM serves as a basis for updating the information collection plan and issuing a fragmentary order once the targeting working group and targeting board concludes their sessions. In addition, the TSM facilitates the distribution of results produced by the targeting working group and targeting board. The TSM ensures planned activities are coordinated and execution is feasible in time and space.

II. Detect

Operations officers, at all levels, are responsible for directing the effort to detect HPTs identified in the decide function. To identify the specific who, what, when, and how for target acquisition, operations officers must work closely with the targeting team. The collection manager is a critical piece of this effort. The detect function is planned during the decide function and conducted during the execution of the OPORD.

The outcome of success achieves relative advantage in all dimensions, (human, information, and physical). It provides for positional, information, capability, and decision advantage. Successful detection enables timely and accurate employment of capabilities. The following are outcomes and consequences of failure to detect:

- Enemy prevents friendly forces from achieving tasks and objectives potentially leading to mission failure.
- PIR/information requirements remain unanswered and decision points are affected.
- The enemy retains critical capabilities to achieve mission success.
- HPTs are not affected, and friendly objectives remain unaccomplished.
- May require rapid decision making and synchronization process (RDSP).
- Increased risk to mission and risk to force.

The effective and efficient delivery of capabilities is achieved through a comprehensive collection plan. Detect and track targets by the maximum use of all available capabilities to create an effective layered collection plan. A collection plan is a systematic scheme to optimize the employment of all available collection capabilities and associated processing, exploitation, and dissemination resources to satisfy specific information requirements (JP 2-0). The collection manager must focus the collection efforts on approved PIRs, targeting intelligence requirements and other information requirements. The collection manager considers the availability and capability of all collection assets. The collection manager translates PIR and targeting intelligence requirements into indicators and specific information requirements. Dissemination of targeting information from collection will be through pre-determined channels and in accordance with unit SOPs.

Information Collection

The accurate and timely detection of targets is dependent upon a comprehensive collection plan and facilitates effective and efficient employment of capabilities. The collection manager must focus the collection efforts on approved PIRs, targeting intelligence requirements and other information requirements. The collection manager considers the availability and capability of all collection assets. The collection manager translates PIR and targeting intelligence requirements into indicators and specific information requirements. Dissemination of targeting information from collection will be through pre-determined channels and in accordance with unit SOPs.

HPTs must be detected in a timely and accurate manner, and reported in sufficient detail to enable the commander, through various lethal and nonlethal capabilities, the ability to produce the desired effects. Clear and concise tasks must be given to the reconnaissance units, information collection elements, weapons locating radars or any asset that can detect a given target within the information collection plan (Annex L). Information collection assets collect information and report back within operations and intelligence or targeting channels. Some collection assets provide actual targets, while other assets must have their information processed to produce valid targets. Processing, exploitation, and dissemination (PED) capabilities play an important role within targeting. The target priorities developed in the decide function are used to expedite the processing of targets. As the situation changes during operations the current operations or staff, led by the collection management team, assesses collection, and updates the collection management plan as needed to meet new requirements.

The information collection plan focuses the collection effort to answering PIRs and identifying HPTs. To support the detection of HPTs on the battlefield, the targeting cell provides indicators and SIRs to the collection management team for each HPT. The information collection matrix accounts for PIRs and other intelligence requirements. The collection management team accounts for targeting intelligence requirements, if developed, in a targeting addendum to the information collection matrix. If the collection management team does not develop a targeting addendum, the team must graphically account for TAIs and HPTs through other means. For example, the team could add both TAIs and HPTs as columns to the information collection matrix or develop a supporting crosswalk of HPTs with supporting SIRs, NAIs, TAIs, an active NAI, and collection assets.

The information collection matrix and information collection synchronization matrix are products used by the collection manager to ensure that collection tasks are tied to the scheme of maneuver in time and space, effectively linking reconnaissance and surveillance to maneuver and effects.

The target priorities developed in the decide function are used to expedite the processing of targets. As a part of information collection, successfully acquiring and identifying HPTs, tracking HPTs, and conducting BDA are often critical to the success of the mission. This effort is collection asset-, intelligence analysis-, and support-intensive. Depending on circumstances, this effort is often controlled out of the current operations cell in conjunction with the targeting team and intelligence section. In some situations, a unit may choose not to conduct BDA on an HPT, but then the unit assumes an operational risk because it will not know the effects of fire on and the status of the HPT.

Actions Within Detect

The first portion of the detect function is positive identification. This is confirming the threat entity acquired in collection is in fact an HPT. Depending on the collection capability of this initial acquisition it may require the cueing of another asset to confirm identification of the HPT.

Mobile HPTs present an additional requirement within the detection function. Tracking is often required to maintain a current target location. Tracking priorities are based on the commander's concept of the operation and targeting guidance. Target tracking requires a specific level of timeliness, accuracy, and detail necessary to support the execution of fires. The HPT is tracked as it either moves to the TAI as planned or as subsequent planning occurs to develop a new TAI to engage the target. In accordance with the TSS, the FSE tells the collection manager the degree of accuracy required and dwell time for a target to be eligible for engagement. The collection manager must match accuracy requirements to the TLE of the collection systems. If the target type and its associated signatures (for example, electronic, visual, and thermal) are known, the most capable collection asset can be directed against the target. The asset is positioned according to estimates of when and where the enemy target will be located.

When an HPT is detected, the information is quickly disseminated to the FAIO, or designated targeting officer, located in the intelligence cell for necessary action. These actions include validating if the target is an HPT, and determining the target's priority, and if the target complies with TSS. In those cases where identification of the HPT is too uncertain, the collection manager and current operations integration cell must cue other information collection assets to conduct further collection and the FAIO, or designated targeting officer, must notify the intelligence cell and targeting team of the need for subsequent analysis.

As the assets collect information, it is forwarded to the intelligence analysts of the ACE. The ACE uses the information in developing situational awareness (SA) and situational understanding and target refinement. Continuous updates to the collection plan and ICSM are required as targets are detected. Once an HPT is identified, it is passed from G-2 to operations.

Essential Target Information

Target recommendations are passed from the G-2 to the FSE by a number of means either digital or analog. It is important that the essential information be passed for proper analysis and engagement to take place. At a minimum, the target report (digital or analog) must include the following:

- Reporting agency.
- Sensor type.
- Report Date Time Group.
- Acquisition Date Time Group.
- Target Description.
- Posture.
- Activity.
- Size of the target.
- Target location.
- TLE.
- Dwell time.

Refer to ATP 3-60, appendix D for a sample target report.

III. Deliver

Deliver is the third function in the Army targeting methodology and occurs during the execution portion of the operations process. This function executes the target attack guidance and supports the commanders battle plan once the HPTs have been located and positively identified. The main objective is to create effects as planned in the decide phase and in accordance with the commander's guidance. Additional activities include the prosecution of targets that were not selected for action in sufficient time to be included in deliberate targeting. The selection of a capability or a combination of capabilities to meet the desired effect requires a synchronized time of engagement dependent on the tactical situation and the technical solution for the selected capability.

The outcome of success allows friendly forces to achieve relative advantage in the human, physical and information dimensions. Consequences of failure are as follows:

- Tasks, Objectives, and End-states not met resulting in mission failure.

Desired Effects (Deliver)

Ref: ATP 3-60, Army Targeting (Aug '23), app. C.

Desired effects should be created with the outcome in mind (see table C-1). The outcome must be perceivable providing the targeting team the ability to measure created effects. A desired effect represents a condition for achieving an associated objective, while an undesired effect could inhibit progress toward the same objective.

TASK	Effect/Outcome
Attrit	To wear down or weaken (an opponent or enemy).
Compel	1) To force, drive or constrain. 2) To make necessary.
Convince	1) To overcome by argument. 2) To bring to belief, consent, or a course of action (COA).
Damage	To reduce the soundness, effectiveness, or perfection of.
Deceive	To cause to believe what is not true.
Defeat	To render a force incapable of achieving its objectives.
Degrade	1) Damage done to the function is permanent, but only portions of the function were affected; that is, the function still operates, but not fully. 2) A function's operation is permanently impaired, but the damage does not extend to all facets of the function's operation.
Deny	1) To hinder the enemy the use of space, personnel, or facilities. It may include destruction, removal, contamination, or erection of obstructions. 2) Damage done to the function is only temporary, but all aspects of the function were affected. 3) A function's operation is impaired over the short term, but the damage extends to all facets of the function's operation.
Delay	1) To slow down the arrival of a unit on the "battlefield." 2) An operation in which a force under pressure trades space for time by slowing down the enemy's momentum and inflicting maximum damage on the enemy without, in principle, becoming decisively engaged.
Destroy	1) To damage the condition of the target so that it cannot function as intended nor be restored to a usable condition. 2) Damage done to the function is permanent, and all aspects of the function have been affected. 3) A function's operation is permanently impaired, and the damage extends to all facets of the function's operation.
Diminish	1) To make less or cause to appear less. 2) To reduce the effectiveness of an activity. This is similar to degrade without the kinetic overtones.
Disrupt	1) To break apart, disturb, or interrupt a function. 2) Damage done to the function is temporary, and only portions of the function were affected. 3) A function's operation is impaired over the short term and the damage does not extend to all facets of the function's operation.
Divert	To restrict the enemy's capabilities to pursue a particular COA.
Enhance	To increase or make greater the capabilities of a force or a people.
Exploit	To gather information that will enable opposition ability to conduct operations to induce other Effects.
Expose	1) To make known or cause to be visible to public view. 2) To make visible, to reveal something undesirable or injurious.
Harass	To disturb the rest of enemy troops, curtail their movement and lower morale by threat of loss.
Influence	1) To affect or change how someone or something develops. 2) To cause a change in the character, thought, or action of a particular entity.
Inform	To impart information or knowledge.
Manipulate	1) To influence or control someone to your advantage, often without that person knowing it. 2) Control or change information, information systems, and/or networks in gray or red cyberspace to create physical denial effects, using deception, decoying, conditioning, spoofing, falsification, and other similar techniques.
Negate/Neutralize	1) To render an enemy weapon system and maneuver units ineffective or unusable for a specific period of time. 2) To render ineffective, invalid or unable to perform a particular task or function. 3) To counteract the activity or effect of.
Prevent	1) To deprive of hope or power of acting or succeeding. 2) To keep from happening, to avert.
Protect/Safeguard	1) To cover or shield from exposure, damage, or destruction. 2) To keep from harm, attack, injury or exploitation. 3) To maintain the status or integrity of.
Suppress(ion)	1) Involves temporary or transient degradation of an actual or suspected enemy weapons system for the purpose of degrading its performance below the level needed to fulfill its mission objectives at a specific time for a specified duration. 2) Temporary or transient degradation by an opposing force of the performance of a weapons system below the level needed to fulfill its mission objectives.

Ref: ATP 3-60, Army Targeting (Aug '23), table C-1. Desired Effects

- Attrition of Friendly Forces and unfavorable force ratios.
- Increase in human and material cost.
- Loss of initiative, inability to shape/set conditions.

Targeting products developed during the decide phase of the targeting process simplify and expedite delivery decisions during execution. A dedicated effort by the targeting team during planning enables predictability and assists the staff in executing the commander's guidance in support of the decision support matrix (DSM). During the initial targeting product development, when more time is available, the staff may not have all necessary information. As operations progress, the staff will inherently gain additional information. This will force the transfer of considerations into execution under increased time constraints and potentially reduced asset availability. This ultimately causes the targeting team to react to operational changes. When deviating from planned guidance (AGM, TSS), the decision maker needs to weigh the operational risks of tactical patience balanced with the immediacy of directed action.

Unplanned or unanticipated targets always have the potential to be identified during operations. Unforeseen variables in the operation can cause a change in commander's or attack guidance.

Dynamic targets or targets of opportunity are processed the same as planned targets. Targets of opportunity are first evaluated to determine when, or if, they should be engaged. The decision to engage targets of opportunity follows the engagement guidance and is based on a number of factors such as the:

- Criticality of target compared to other targets being processed for engagement.
- Activity of the target (risk to friendly forces).
- Dwell time.

If the decision is made to engage immediately, the target is processed further. The availability and capabilities of assets to engage the target are assessed. If the target exceeds the capabilities or availability of the unit delivery asset, the target should be sent to a higher HQ for immediate engagement. If the decision is to defer the engagement, then continue tracking, determine decision point(s) for engagement, and modify collection tasking as appropriate.

Engagement Considerations

The engagement of targets must satisfy the targeting guidance developed in the decide phase and be integrated with the friendly scheme of maneuver. The plan for target engagement is typically developed via the targeting working group and presented for approval in the targeting board. Target engagement requires several decisions and actions. These considerations fall into two categories: tactical and technical.

Tactical considerations:

- Time of the engagement.
- **Desired effect**, degree of damage, or both *(see previous page).*
- Capability to use for engagement.
- Potential for collateral damage.

Technical considerations:

- Number, type, and duration of effect (munition/payload/action).
- Unit/agency/organization to conduct the engagement.

IV. Assess

Assessment occurs at all levels and across the range of military operations. Staffs help the commander by monitoring the numerous aspects that can influence the outcome of operations and provide the commander timely information needed to make decisions.

During the planning and preparation for operations, the targeting team must take all aspects of assessment into consideration. Assessments are difficult and time consuming. They require dedicated personnel and resources and must be included in the collection plan and synchronized through the collection and targeting working group. The staff must prioritize the use of collection assets and consider the risk of tasking assets for assessment versus other requirements.

The outcome of success is effective targeting, efficient use of resources, and lower risk to mission and force. The following are consequence of failure to assess:

- Tasks, Objectives, and End-states not met resulting in mission failure.
- PIR/information requirements remain unanswered and decision points are affected.
- The enemy retains critical capabilities to achieve mission success.
- Inaccurate understanding of the operational situation.
- Inability to capture Munitions effectiveness or make reattack recommendations.
- Inability to measure effectiveness of target engagement.

Combat assessment is the determination of the overall effectiveness of force employment during military operations (JP 3-60). Combat assessment is composed of three elements: **battle damage assessment (BDA), munitions effectiveness assessment (MEA), and reattack recommendation.**

In combination, BDA and munitions effectiveness assessment inform the commander of effects against targets and target sets. Based on this information, the enemies' ability to make and sustain war and centers of gravity are continuously estimated. During the review of the effectiveness of operations, reattack recommendations are proposed or executed.

A. Battle Damage Assessment (BDA)

BDA includes known or estimated enemy unit strengths, degraded, neutralized, or destroyed enemy weapon systems, and all known captured, wounded, or killed enemy personnel during the reporting period. BDA in targeting pertains to the results of lethal and nonlethal engagements on targets designated by the commander. Producing BDA is primarily an intelligence responsibility but requires coordination with operational elements to be effective. BDA requirements may be translated into PIR.

BDA has three components. They are:

1. Physical Damage Assessment

Physical damage assessment estimates the quantitative extent of physical damage through munitions blast, fragmentation, and fire damage effects to a target. This assessment is based on observed or interpreted damage.

2. Functional Damage Assessment

Functional damage assessment estimates the effect of engagement on the target to perform its intended mission compared to the mission objective established against the target. This assessment is inferred based on all-source intelligence and includes an estimate of the time needed to replace the target function. A functional damage assessment is a temporary assessment (compared to target system assessment) used for specific missions.

3. Target System Assessment

Target system assessment is a broad assessment of the overall impact and effectiveness of all types of engagement against an entire target system capability, for example, enemy air defense artillery systems. It may also be applied against enemy unit combat effectiveness. A target system assessment may also look at subdivisions of the system compared to the commander's stated mission objectives. The target system assessment is a relatively permanent assessment (compared to a functional damage assessment) that will be used for more than one mission.

BDA may take different forms besides the determination of the number of casualties, or the amount of equipment destroyed. Other information of use to the targeting working group includes, but not limited to, the following:

- Whether the targets are moving or hardening in response to the engagement.
- Changes in enemy behavior and techniques.
- Increased communication efforts as the result of jamming.
- Whether the damage resulting from an engagement is affecting the enemy's combat effectiveness as expected.

BDA results may change plans and earlier decisions. The targeting working group must periodically update the targeting decisions and products made during the decide function.

B. Munitions Effectiveness Assessment (MEA)

The operations officer, in coordination with the FSE and targeting working group, conducts munitions effectiveness assessment concurrently and interactively with BDA, as a function of combat assessment. Munitions effectiveness assessment is an assessment of the military force in terms of the weapon system and munitions effectiveness. Munitions effectiveness assessment is conducted using approved weaponeering software and provides the basis for recommendations to increase the effectiveness of:

- Methodology
- Tactics
- Weapon systems
- Munitions
- Weapon delivery parameters

The targeting working group may recommend modifying commander's guidance concerning:

- Unit Basic Load (UBL)
- Required Supply Rate (RSR)
- Controlled Supply Rate (CSR)

C. Reengagement Recommendation

Failure to achieve BDA, or failure to create necessary effects as a result of BDA, requires a decision from the commander. The targeting team and current operations must assess operational risks associated with the HPT and provide options to mitigate the risks. Based on the BDA and munitions effectiveness assessment, the intelligence officer in conjunction with the FSCOORD or deputy FSCOORD and operations officer consider the level to which objectives have been achieved and make recommendations to the commander. Reattack and other recommendations should address objectives relative to:

- Targets.
- Target critical elements.
- Target systems.
- Enemy combat force strengths.
- Friendly maneuver.

Battle Damage Assessment & Operational Assessment

Commanders continuously assess the OE and the progress of operations and compare them to their initial vision and intent. Commanders adjust operations based on their assessment to ensure objectives are met and the military end state is achieved. Operational Assessment is normally performed only at echelons above brigade. At the echelons above brigade this is typically achieved through the assessments working group.

Chap 9

I. Air Defense Artillery (ADA) Equipment

Ref: FM 3-01.44, Short-Range Air Defense Operations (Jul '22), app. A.

Understanding the integration of **air defense artillery (ADA) assets** is critical when planning and conducting counter–unmanned aircraft system (C-UAS) operations because ADA provides the structural foundation and doctrinal framework for layered, defense-in-depth protection against aerial threats. *(See chap. 3.)*

The Army Air and Missile Defense System

ADA is composed of composite systems (systems consisting of sensor, weapon, and C2 components) and singular systems (a sensor, weapon, or C2 system) that provide sense, engage, and interoperable capabilities to counter aerial threats to maneuver forces and other high priority assets. ADA systems are arrayed throughout a theater. Composite ADA systems are generically referred to as high-to medium-altitude air defense (HIMAD), and singular systems typically in the context of short-range air defense (SHORAD). Singular systems require sensors or weapons networked by and with C2 capabilities.

Short-Range Air Defense (SHORAD) *(See pp. 9-2 to 9-10.)*

Short-range air defense are those capabilities that provide air defense against low-altitude air threats. Short-range air defense (SHORAD) consists of dedicated ADA and non-dedicated air defense (known as combined arms for air defense) capabilities which deny an enemy the use of airspace and provide the supported commander freedom of action. SHORAD capabilities do this by deterring, defeating, or destroying low-altitude air threats to defended critical fixed and semi-fixed assets and maneuvering forces. SHORAD systems are M-SHORAD, Avenger, Stinger man-portable air defense (MANPAD), Counter-Small Unmanned Aircraft System (C-sUAS) variants, Land-Based Phalanx Weapon System (LPWS), Sentinel, and the forward area air defense (FAAD) C2 system.

High-to-Medium Altitude Air Defense (HIMAD) *(See pp. 9-11 to 9-12.)*

High-to-medium altitude air defense are those capabilities that provide defense against longer-range air and ballistic missile threats. HIMAD addresses capabilities that are primarily employed to counter ballistic missile threats, ranging from close-range to intermediate-range variants, in theater. Most HIMAD systems have diverse, system-specific C2 sub-systems that require conventional manual processes and methodology to perform integrated AMD operations, though, as noted above, IBCS is being fielded to some Patriot units. Patriot and THAAD are HIMAD systems.

See chap. 3, Air and Missile Defense.

A. Maneuver Short-Range Air Defense (M-SHORAD) *(See pp. 1-42 to 1-43.)*

Ref: FM 3-01.44, Short-Range Air Defense Operations (Jul '22), pp. A-1 to A-2.

M-SHORAD provides air defense protection of Stryker and Armor BCT maneuvering forces against fixed-and rotary-wing aircraft and UASs throughout the range of military operations. It employs a mix of sensors and weapons. A mix of different sensors provides the capability to detect, identify and track a variety of aerial objects, including fast moving targets and small signature objects. A mix of different, selectable weapons enables the M-SHORAD gunner to precisely match munitions to targets, increasing efficiency and effectiveness against targets of all types.

M-SHORAD (Inc 3) is the Army's Next Generation interceptor to replace the legacy Stinger Missile, with production starting in Fiscal Year 2027. The Army will also field new 30 mm Proximity Airburst Ammunition to increase lethality against Unmanned Aerial Systems. (U.S. Army Acquisition Portfolio 2023-2024, p. 60.)

The M-SHORAD system consists of a digitized Stinger Vehicle Universal Launcher, four Multi-Mission Hemispheric Radars, maneuver and Forward Area Air Defense C2 systems, and associated fire control, power generation, and related components on a Stryker vehicle. The launcher mounts four Stinger missiles, a 30-milimeter cannon, a 7.62-milimeter coaxial machine gun, and sensor system with electro-optical and infrared cameras for target acquisition and general situational awareness. The radars detect, track, and identify low-altitude air targets up to 360 degrees using onboard acquisition and tracking sensor capability, including under obscured, day and night conditions. M-SHORAD is integrated with Forward Area Air Defense C2 and is interoperable with the Sentinel radar. The maneuver C2 system allows M-SHORAD to integrate into maneuver forces.

Directed Energy Maneuver-Short Range Air Defense (DE M-SHORAD) will be a complementary capability to M-SHORAD Inc 1. It is a three-soldier crew, Stryker-mounted, high-energy laser experimental prototype weapon system. DE M-SHORAD will support Multi-Domain Operations as part of the M-SHORAD Battery. (U.S. Army Acquisition Portfolio 2023-2024, p. 51.)

B. Avenger

Ref: FM 3-01.44, Short-Range Air Defense Operations (Jul '22), pp. A-2 to A-3.

The Avenger weapon system is a mobile lightweight, day or night, limited adverse weather fire unit employed to counter enemy reconnaissance, surveillance, and target acquisition efforts and low-level fixed and rotary-wing threats. The fire unit consists of two turret-mounted standard vehicle mounted launchers, carrying Stinger infrared homing, fire-and-forget missiles; a M3P .50 caliber machinegun; and a sensor package. The Avenger turret provides the gunner with unobstructed fields of view rotating through 360 degrees in azimuth.

Forward Area Air Defense System, Line-of-Sight, Rear (Pedestal Mounted Stinger) – Avenger is a lightweight, ground-to-air missile and gun weapon system that protects against unmanned aircraft systems, cruise missiles, and fixed- and rotarywing threats. An Identification Friend or Foe system aids in the identification of friendly aircraft to minimize the potential for fratricide. (U.S. Army Acquisition Portfolio 2023-2024, p. 53.)

The Avenger sensor package aids in the acquisition and identification of air tracks. The package includes an optical sight, forward-looking infrared radar, laser range finder, identification friend or foe, and a fire control computer. The optical sight is a driven reticle heads-up display which allows the gunner to manually acquire targets through the turret canopy and to aim the missiles. The forward-looking infrared radar provides enhanced acquisition capability in various environments: night, smoke, rain, background clutter, and haze. The laser range finder is used to determine if the target is within the normal performance range of the missile. The identification friend or foe aids the gunner in the identification of targets; it distinguishes aircraft as positive friend, possible friend, or unknown.

The Avenger mounts eight Stinger missiles, four in each launcher. Should the Avenger system or turret malfunction, the missile can be converted to a man-portable configuration by removing it from one of the Avenger missile pods. A grip stock and battery coolant unit is attached to the launch tube, and the Stinger becomes a ready-weapon.

Avenger links to the Forward Area Air Defense C2 for air battle management, early warning and cueing, and aids for track identification.

C. Man-Portable Stinger

Ref: FM 3-01.44, Short-Range Air Defense Operations (Jul '22), pp. A-3 to A-4.

Stinger is a shoulder-fired, infrared radiation homing, heat-seeking, guided missile system that tracks to the target through proportional navigation. It is designed to counter low-level fixed and rotary-wing aircraft and UASs. The Stinger components are a missile housed in a fiberglass launch tube; a permanently attached, hinged, open sight assembly located atop the launch tube, a grip stock with an identification friend or foe antenna; and a battery coolant unit.

The Stinger Block I missile is an advanced, fire-and-forget, short-range, man-portable, shoulder-fired, platform-mounted, guided missile system that provides the maneuver force and point defense assets air defense protection. Stinger's mission is to provide the force with low-altitude air defense against various threats. The Stinger missile is deployable from a variety of platforms. The missile is delivered as a certified round and requires no field testing or maintenance.(U.S. Army Acquisition Portfolio 2023-2024, p. 63.)

To assist the gunner in aircraft identification, the identification friend or foe antenna is unfolded and the identification friend or foe cable is connected. Interrogation of aircraft is accomplished by pressing the interrogate switch located at the rear of the grip stock. Stinger is a "fire-and-forget" system. This means that, immediately after firing, the gunner removes and discards the coolant unit from its receptacle, folds and stows the identification friend or foe antenna, unlatches the latch mechanism located at the forward end of the grip stock, and removes the grip stock from the launch tube. The gunner can now ready another missile round for engagement or opt to seek cover after launch. The grip stock is reused continually until failure. After firing, the gunner has no control over the missile and is required only to observe the missile's flight trajectory to the target.

D. Sentinel

Ref: FM 3-01.44, Short-Range Air Defense Operations (Jul '22), pp. A-4 to A-5.

The Sentinel radar is an X-Band, pulse-Doppler phased array radar that provides 360-degree curing and target identification. It's extremely low noise design and Doppler signal processing result in exceptional ability to detect small, low-flying targets in clutter. It can acquire, track, and classify cruise missiles, UASs, and fixed-and rotary-wing aircraft. The Sentinel can detect and classify rotary-wing aircraft, including hovering rotary-wings, at ranges beyond firing unit's maximum weapon detection and engagement range in a severe electronic countermeasures environment. Targets can be detected and tracked in all weather, day or night, under the most extreme battlefield conditions. Sentinel's air defense interrogator provides improved identification friend or foe to aid in positive identification and preventing fratricide. Due to its three-dimensional pencil beam and frequency-agile mode of operation, the Sentinel is difficult to detect and locate accurately, improving its survivability.

Sentinel consists of a trailer-mounted radar with prime mover, generator, Identification Friend or Foe system, and Forward Area Air Defense Command, Control, and Intelligence systems interface. Sentinel provides persistent surveillance and fire control quality data on Army and Joint networks enabling protection of U.S. and coalition forces as well as critical geopolitical assets against various threats. The Sentinel A4 is a modification to the current Sentinel A3 radar that will be a ground-based sensor system designed to locate, identify, and track threats. The simultaneous multi-mission Sentinel A4 system will provide 360-degree surveillance and tracking with the ability to handle concurrent threats arriving from different azimuths.(U.S. Army Acquisition Portfolio 2023-2024, p. 62.)

The Sentinel radar is towed by a heavy high mobility multipurpose wheeled vehicle with a mounted 10-kilowatt generator in the rear of the vehicle. Sentinel is capable of 24-hour operations and can be located within 10 kilometers of the forward line of own troops if the mission and situation so dictate. It should have a clear field of view (360 degrees), be positioned to ensure overlapping coverage with adjacent radars such that their search areas do not leave any seam in the defense that might be used by incoming threats, and should be placed away from power sources of similar radiating frequency bands to avoid interference.

Sentinel is air transportable by medium-lift rotary-wing aircraft for the high mobility multipurpose wheeled vehicle variant, and by transport fixed-wing aircraft for the medium tactical vehicle variant.

E. Land-Based Phalanx Weapon System

Ref: FM 3-01.44, Short-Range Air Defense Operations (Jul '22), pp. A-5 to A-6.

The Land-based Phalanx Weapon System (LWPS) is a fast reacting, short-range system used to detect incoming rockets and mortars, provide focused early warning, and destroy inbound rockets and artillery and mortar projectiles in the air before they hit their ground targets. The intercept system has an integral sensor for targeting and a multi-barrel gun system to destroy RAM threats.

Ref: Raytheon has won a $205m contract to provide the Phalanx land-based weapon system to the US Army. (Credit: Raytheon)

The Land-based Phalanx Weapon System consists of a trailer-mounted close-in weapon system. The close-in weapon system is a 20-millimeter Gatling gun, with six gun barrels, separate search-and-track radars, and forward-looking infrared radar. The gun system is capable of firing at a selectable rate of 3,000 to 4,500 rounds per minute, with an advanced Ku-band radar to provide autonomous target detection and engagement. Using the Ku-band radar, the Land-based Phalanx Weapon System detects threats early in flight and hands them over to the track mode only when those targets are determined to threaten the area protected by the system. Two 60-kilowatt generators mounted on the trailer supply power to the entire system.

The Land-based Phalanx Weapon System leverages ADA and field artillery sensors to provide alerting and cueing of incoming threats. The Sentinel radar and AN/TSQ-50 Lightweight Counter Mortar Radars are organic to the Land-based Phalanx Weapon System formation. The Lightweight Counter Mortar Radar is operated as either a counterfire sensor (Special Operations Command mode) or in the counter-RAM mode. It provides 360-degree surveillance and about 7.5-kilometer range. In either mode, the Lightweight Counter Mortar Radar detects RAM fire, and predicts the suspected points of origin and impact. In the counter-RAM mode, the Lightweight Counter Mortar Radar utilizes either the local area network or high-speed radios to send this data to the Forward Area Air Defense C2 to be used for track confirmation/correlation, cueing of other sensors, and countermeasures (by relaying data to friendly artillery and air assets). The Land-based Phalanx Weapon System also pulls data from the AN/TSQ-53 Firefinder radar while Firefinder is executing is primary counterfire mission.

F. RAM Warn System

Ref: FM 3-01.44, Short-Range Air Defense Operations (Jul '22), pp. A-6 to A-7.

The RAM Warn system capability provides localized RAM early warning and impact point prediction and **can be used within Land-based Phalanx Weapon System units or as an independent warning capability**. RAM Warn system uses local sensors and a network of broad area sensors, as available, to sense attacks and disseminates warnings using audio and visual methods, both outdoors and indoors. Warnings are limited to the zone of danger, not to an entire base or AO unless directed by the supported asset commander. Emplacement considerations in positioning RAM Warn system components include the size of the defended area, heavy traffic areas, designated logistical arrival and departure points, and dead space areas.

Ref: FM 3-01.44 (Jul '22), fig. A-6. RAM Warn system.

A Forward Area Air Defense C2 variant, with specific counter-RAM modifications, provides the requisite battle management. It links the weapons, sensors, and warning systems for Land-based Phalanx Weapon System intercepts. It provides situational awareness to the air and missile defense workstation and engagement commands to the Land-based Phalanx Weapon System.

G. Forward Area Air Defense C2

Ref: FM 3-01.44, Short-Range Air Defense Operations (Jul '22), pp. A-7 to A-8.

The Forward Area Air Defense Command and Control (C2) system-of-systems consists of common hardware, software, communications equipment, and shelters to meet the C2 and targeting needs of SHORAD battalions. Forward Area Air Defense C2 collects, processes, and disseminates real-time target tracking and cueing information to all SHORAD weapons and provides C2 for the Land-based Phalanx Weapon System.

Forward Area Air Defense Command and Control (FAAD C2) provides critical C2, situational awareness, and automated air track information by integrating engagement operations software to support multiple missions. FAAD C2 supports air defense and Counter-Rocket, Artillery, Mortar (C-RAM) weapon systems engagement operations. The system also performs C2 engagement operations. (U.S. Army Acquisition Portfolio 2023-2024, p. 52.)

Forward Area Air Defense C2 hardware consists of shelters (Army standard rigid wall and command post platform shelters) and fire unit components, such as the forward area computer terminal. The terminal provides Avenger fire units with airspace situational data, engagement commands, and weapon coordination, status, and control. C2 software provides engagement and force operations functions, air battle management and situational awareness, low level air picture, and automated integration with Army C2 systems.

Forward Area Air Defense C2 supports the AMD mission by providing real-time, correlated air tracks and C2 information, and intelligence assessments to higher, adjacent, and lower units. Computer displays allow commanders to access the air picture, situation reports, enemy assessments, and friendly forces. The Forward Area Air Defense C2 system has the capability to interface with joint and NATO C2 systems, as well as Army C2 systems. The engagement operations subsystem provides a joint air picture via tactical data links B and J, transmitted using various joint range extension applications protocols.

The C2 software assists with digitization of the battlefield by providing air situational awareness to the supported force and alerting and cueing to SHORAD systems. The Forward Area Air Defense C2 systems support Land-based Phalanx Weapon System units by receiving and correlating sensor inputs and then alerting the intercept system and the sense and warn elements of an impending RAM attack.

H. Indirect Fire Protection Capability (IFPC)

Ref: U.S. Army Acquisition Portfolio 2023-2024, p. 55.

Indirect Fire Protection Capability Increment 2 (IFPC Inc 2) is a mobile, ground-based weapon system designed to defeat unmanned aircraft systems (UAS), Cruise Missiles (CM), and Rocket, Artillery, and Mortars (RAM) projectiles. The IFPC Inc 2 capability will be developed in three blocks, each a separate acquisition program. IFPC Inc 2 will provide a ground-based weapon system designed to acquire, track, engage, and defeat subsonic CMs, UAS, RAM, and other aerial threats. The IFPC Inc 2 system supports the Threshold CM and UAS defeat mission. The objective counter-RAM mission employs both alternative kinetic and non-kinetic defeat solutions.

(U.S. Army Acquisition Portfolio 2023-2024, p. 55.)

The Army's plan to add four new IFPC battalions reflects the critical need for integrated air and missile defense against evolving aerial threats. IFPC units bridge the gap between short- and long-range air defense systems, providing a mobile, layered defense against rockets, artillery, mortars, cruise missiles, and unmanned aerial systems (UAS). These formations will likely complement the Army's forthcoming Long-Range Hypersonic Weapon (LRHW) batteries, offering protection for maneuver and support elements in high-threat theaters.

See pp. 1-38 to 1-39 for an overview and further discussion.

I. Iron Dome Defense System - Army (IADS-A)

Ref: U.S. Army Acquisition Portfolio 2023-2024, p. 56.

The U.S. Army uses Iron Dome as an interim capability to protect forces from short-range threats like cruise missiles, drones, and rockets, artillery, and mortars (C-RAM). Two batteries were procured and delivered, with one deployed to Guam and the other at Joint Base Lewis-McChord, and the U.S. is also incorporating the technology into its future Indirect Fires Protection Capability (IFPC) program.

The Iron Dome Defense System – Army (IDDS-A) is an Israeli-developed system that provides an initial cruise missile defense capability. IDDS-A engages many simultaneous threats arriving from different azimuths.(U.S. Army Acquisition Portfolio 2023-2024, p. 56.)

Interim capability: The Army procured two Iron Dome batteries to immediately address the need for a defense against short-range threats that could not be met by existing systems.

Deployment: One battery was sent to Guam for testing and training, while the other is located at Joint Base Lewis-McChord in Washington state.

Purpose: The system is intended to protect U.S. forces and assets from a variety of aerial threats in contested areas.

Future integration: The Army plans to incorporate parts of the Iron Dome technology into its longer-term IFPC program, which will provide a more comprehensive defense against a wider range of threats including cruise missiles, drones, rockets, artillery, and mortars.

J. PATRIOT (PAC3)

Ref: U.S. Army Acquisition Portfolio 2023-2024, p. 61.

Phased Array Tracking Radar to Intercept of Target (PATRIOT) Advanced Capability-3 Missile Segment Enhancement (PAC-3 MSE) is a high-velocity, hit-to-kill, surface-to-air missile capable of intercepting and destroying threats. The PAC-3 MSE is the follow-on variant of the PAC-3 missile. The PAC-3 MSE has improved capability and employs kinetic energy to destroy targets. It provides expanded battlespace performance against complex threats.

(U.S. Army Acquisition Portfolio 2023-2024, p. 61.)

Lower Tier Air and Missile Defense Sensor (LTAMDS)

Lower Tier Air and Missile Defense Sensor (LTAMDS) expands the battlespace for the Phased Array Tracking Radar to Intercept of Target (PATRIOT) Advanced Capability-3 Missile Segment Enhancement interceptor and is a lower-tier sensor component of the Army's Integrated Air and Missile Defense architecture.

(U.S. Army Acquisition Portfolio 2023-2024, p. 58.)

K. Terminal High Altitude Area Defense (THAAD) System

The Terminal High Altitude Area Defense (THAAD) system is a U.S. Army mobile, ground-based air and missile defense system designed to intercept and destroy short-, medium-, and intermediate-range ballistic missiles during their terminal phase of flight using hit-to-kill technology. Operating at high altitudes both inside and outside the atmosphere, THAAD provides a critical upper-tier layer in the U.S. and allied ballistic missile defense architecture, complementing lower-tier systems such as Patriot and higher-level systems like Aegis. It integrates advanced radar (AN/TPY-2), interceptor launchers, and a fire control and communications network to detect, track, and engage threats with precision. THAAD enhances protection for deployed forces, population centers, and critical infrastructure by providing a rapidly deployable, globally transportable capability that contributes to joint and multinational air and missile defense operations.

The first of two Terminal High Altitude Area Defense (THAAD) interceptors is launched during a successful intercept test. (Ralph Scott/Missile Defense Agency.)

Originally a United States Army program, THAAD has come under the umbrella of the Missile Defense Agency.[7] The Navy has a similar program, the sea-based Aegis Ballistic Missile Defense System, which also has a land component ("Aegis Ashore").

Sometimes called Kinetic Kill technology, the THAAD missile destroys missiles by colliding with them, using hit-to-kill technology, like the MIM-104 Patriot PAC-3, although the PAC-3 also contains a small explosive warhead. This is unlike the Patriot PAC-2, which carried only an explosive warhead detonated using a proximity fuze. Although the actual figures are classified, THAAD missiles have an estimated range of 125 miles (200 km), and can reach an altitude of 93 miles (150 km). A THAAD battery consists of at least six launcher vehicles, each equipped with eight missiles, with two mobile tactical operations centers (TOCs) and the AN/TPY-2 ground-based radar (GBR).

II. Counter-Unmanned Aircraft (C-UAS) Systems

Ref: CRS Report R48477, Department of Defense Counter Unmanned Aircraft Systems, Daniel M. Gettinger (Mar '25).

DOD Systems for Countering UAS (Overview)

Following its establishment in 2020, the JCO endorsed eight interim counter-UAS systems for potential acquisition and use by the military services, some of which have since progressed to programs of record. The selected systems included kinetic and nonkinetic effectors, command-and-control systems, and sensors. In the years since 2020, the JCO has expanded its list of systems it recommends to the military services based on product evaluations it has conducted with the Rapid Capabilities and Critical Technologies Office and the Army's Program Executive Office (PEO) Missiles and Space. Army officials have said that the service expects to continually evaluate and field new counter-UAS systems in order to maintain pace with technological developments. Collectively, with these systems and others, DOD seeks to provide what Army doctrine describes as a "layered" defensive capability against UAS threats.

Repurposing Older-Generation Missiles as Counter-UAS Interceptors (the Stinger and Its Successors)

The conventional example of man-portable and handheld equipment for countering UAS is the Stinger surface-to-air missile, which the Army fielded both as a component of Cold War-era short-range air defense (SHORAD) systems like the Avenger and Linebacker. Today, the Stinger is a key element of the Army's LIDS and Marine Corps's MADIS counter-UAS systems. In the mid-2010s, the launched the Stinger Service-Life Extension Program (SLEP) to improve the missile's ability to intercept drones,89 which the Army described as designed to provide an organic air defense capability to soldiers outside of the Air and Missile Defense (AMD) community. The United States has also provided Stingers to Ukraine in an effort to boost that country's air defenses against attacks by Russian aircraft, cruise missiles, and drones.

The Army has launched at least two projects—the Next-Generation Short-Range Interceptor (NGSRI) and the eXtended Range Counter-sUAS Missile (XRC)—that aim to replace the Stinger missile. The NGSRI is a conceptual surface-to-air missile for the Maneuver Short Range Air Defense (M-SHORAD) Increment 3 platform. In 2023, the Army reportedly selected Lockheed Martin and Raytheon Technologies to develop prototypes of the NGSRI.

The military services have also sought to repurpose older-generation air-to-air and air-to-ground missiles as counter-UAS interceptors, an option that some experts have said could provide a lower-cost cost option than conventional surface-to-air air defense missiles. For example, counter-UAS systems like the Navy's VAMPIRE and the Army's EAGLS and Containerized Weapon System (CWS) each feature the Advanced Precision Kill Weapon System (APKWS), a modified, precision-guided version of the Cold War-era Hydra-70 rocket. The Army has also reportedly explored using the Hellfire Longbow air-to-ground anti-tank missile in a counter-UAS role,100 and the Navy has reportedly begun fielding repurposed Hellfire missiles as counter-UAS interceptors on its Littoral Combat Ships.

Electronic Warfare Countermeasures

In addition to kinetic countermeasures, each of the services have fielded nonkinetic counter-UAS equipment, some of which have been adapted from technology developed to counter improvised explosive devices (IEDs) in Iraq and Afghanistan. For example, the Navy derived its Drone Restricted Access using Known Electromagnetic Warfare (DRAKE) system from the Joint Counter Radio-Controlled Improvised Explosive Device Electronic Warfare (JCREW) system, one of the projects of the Joint IED Defeat Organization (JIEDDO) in the 2000s. The DRAKE is designed to be carried as a backpack and to jam the communication links between the operator of the adversary drone and the aircraft. Meanwhile, the Air Force Research Laboratory's Negation of Improvised Non-State Joint Aerial Threats (NINJA), an Air Force program of record, is designed to detect, track, locate, identify, and defeat UAS threats using the radio frequency communications between the drone and the operator. In congressional testimony in 2023, then-Secretary of the Air Force Frank Kendall said that 99 Air Force locations had installed the NINJA system.

Handheld and Soldier-Worn Equipment

Other examples of counter-UAS equipment fielded or considered by the military services include handheld and soldier-worn devices (sometimes referred to as "dismounted" solutions). This equipment is designed to be ultra-portable and for use against small drones at short ranges, and in some cases as a last resort in the event that other solutions have failed or are unavailable. Possible examples of such equipment include the "Dronebuster," a handheld device first acquired by the Army and Air Force in 2017, that uses jamming to disrupt the communication signals of drones. The Army and Marine Corps have also acquired equipment that is designed to improve the accuracy of rifle fire—known as "SmartShooter"—potentially enhancing the ability of ground-based troops to shoot down small drones. In 2024, the Marine Corps issued solicitations for a "buckshot-like" counter-drone solution that could be fielded with small units, such as squads and platoons

II. New Units (with C-UAS Capabilities) (Army Force Structure Transformation)

The *Army Force Structure Transformation White Paper (2024)* outlines the U.S. Army's ongoing effort to reshape its organization and capabilities to meet emerging threats across multiple domains of warfare—land, air, maritime, space, and cyberspace. Central to this transformation are four new types of units that represent the Army's drive to integrate advanced technologies, increase mobility and survivability, and enhance joint interoperability.

Collectively, these new formations reflect a pivot toward multi-domain operations (pp. 1-21 to 1-32) and layered defense integration (see chap. 5.)

A. Multi-Domain Task Forces (MDTFs) *(See pp. 1-36 to 1-37.)*

At the core of transformation are the Multi-Domain Task Forces, which embody the Army's shift toward integrated operations across domains. The Army is building five MDTFs—tailored, theater-level units capable of combining intelligence, cyber, space, and long-range fires effects. Each MDTF includes specialized elements: a headquarters, a multi-domain effects battalion, a long-range fires battalion, an Indirect Fire Protection Capability (IFPC) battalion, and a brigade support battalion. These task forces are designed to sense, target, and strike across great distances, protecting joint and coalition forces while shaping operational environments through synchronized non-kinetic and kinetic means. Full fielding is expected in the near term, establishing a scalable blueprint for future formations.

B. Indirect Fire Protection Capability (IFPC) Battalions *(See pp. 1-38 to 1-39.)*

The Army's plan to add four new IFPC battalions reflects the critical need for integrated air and missile defense against evolving aerial threats. IFPC units bridge the gap between short- and long-range air defense systems, providing a mobile, layered defense against rockets, artillery, mortars, cruise missiles, and unmanned aerial systems (UAS). These formations will likely complement the Army's forthcoming Long-Range Hypersonic Weapon (LRHW) batteries, offering protection for maneuver and support elements in high-threat theaters.

C. Counter-Small Unmanned Aerial Systems (C-sUAS) Batteries *(See pp. 1-40 to 1-41.)*

Recognizing the proliferation of small drones on modern battlefields, the Army is standing up nine C-sUAS batteries to bolster defense against this rapidly growing threat. These units will integrate into both IFPC and divisional air defense battalions, employing a variety of detection, jamming, directed-energy, and kinetic systems to defeat Group 1–3 UAS. The creation of dedicated C-sUAS units marks a doctrinal acknowledgment that drone defense is now a core requirement for every echelon of combat operations.

D. Maneuver Short-Range Air Defense (M-SHORAD) Battalions *(See pp. 1-42 to 1-43.)*

To restore organic air defense capabilities lost after decades of counterinsurgency focus, the Army plans to form four additional M-SHORAD battalions. Mounted on Stryker platforms and future variants, these mobile formations provide protection against low-altitude threats such as rotary-wing, fixed-wing, and unmanned aircraft. M-SHORAD enhances survivability and freedom of maneuver for maneuver brigades, particularly Infantry and Armored Brigade Combat Teams (IBCTs and ABCTs).

Low-Slow-Small Unmanned Aircraft System (LIDS) Family of Systems (Overview)

KEY

A: AN/TPQ-50 Radar

B: Counter unmanned aircraft electronic warfare system-direction finding

C: Forward Area Air Defense Command and Control

D: Electro-optical infrared camera

E: Wireless encrypted communications

F: Fixed-site Coyote Block 2 launcher

G: Fixed-site Ku Band Multifunction Radio Frequency System Radar

H: M-LIDS electronic warfare vehicle

I: M-LIDS kinetic defeat vehicle

J: Coyote Block 2 Interceptor

(Graphic by IF/RCO, U.S. Army)

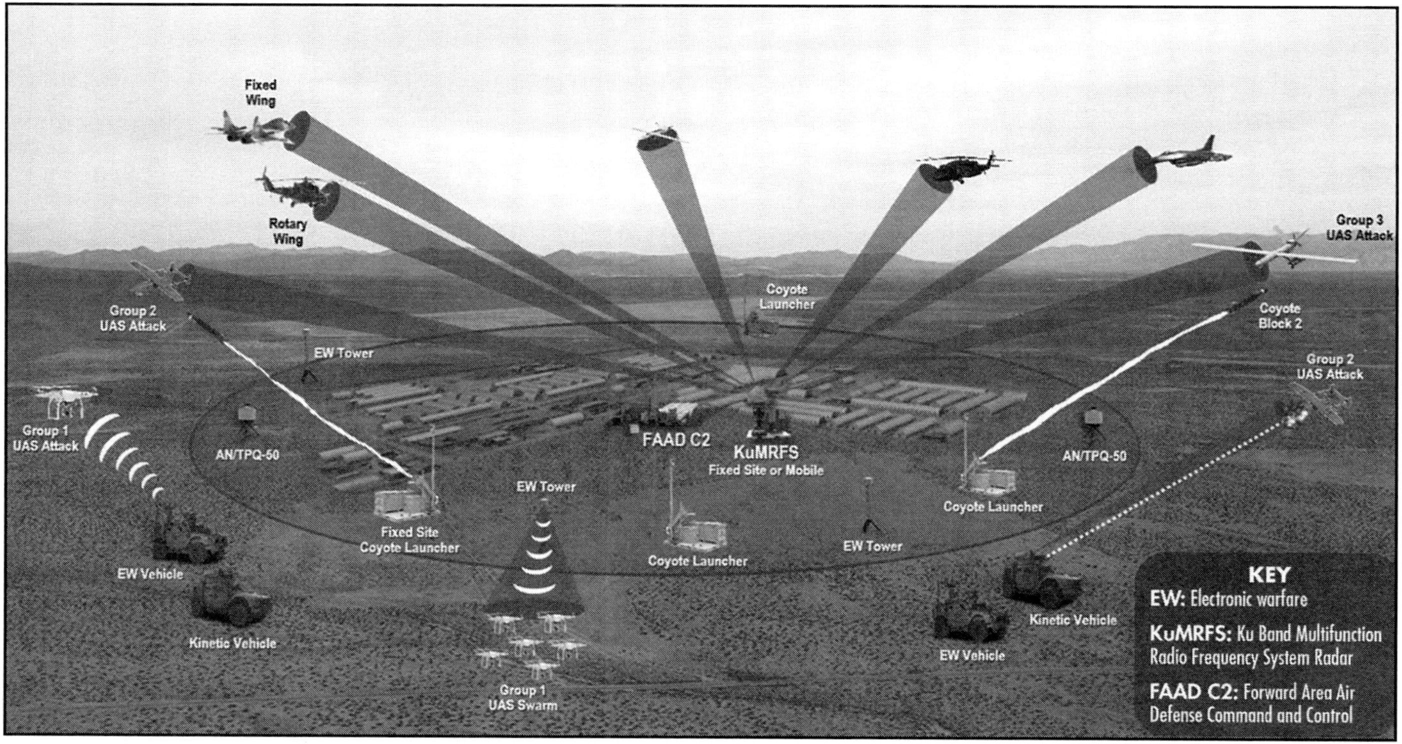

LIDS CONCEPT: The LIDS family of systems uses a range of passive and active sensors to detect, track and identify UAS and non-hostile aircraft. LIDS uses a comprehensive network to communicate, allowing the operator to task appropriate mitigation techniques, ranging from electronic warfare to full physical intercepts. (Graphic by IF/RCO, U.S. Army)

A. Low-Slow-Small Integrated Defense System (LIDS) Platforms

Ref: ATP 3-01.81, Counter-Unmanned Aircraft System (C-UAS) Operations (Mar '25), pp. 67 to 69 and 75 to 76.

Fixed-Site LIDS (FS-LIDS)

The Fixed Site-Low, Slow, Small, Unmanned Integrated Defeat System (FS-LIDS) is an integrated CUAS system-of-systems that can detect, identify, track and defeat UAS threats. The FS-LIDS employs an EW capability, via the Duke and direction finding antennas, to provide improved detection and defeat probability. It uses an electro-optic/infrared camera with automatic slewing capability working with radar to capture and track the target for positive identification.

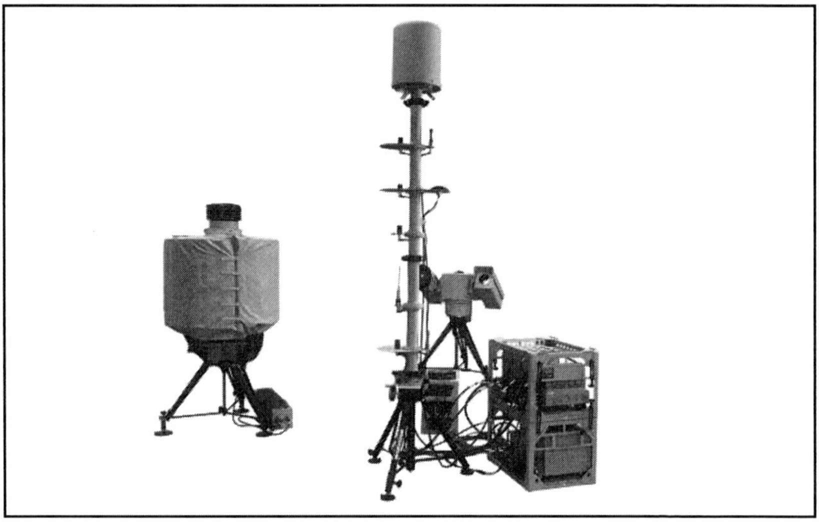

Ref: ATP 3-01.81, fig. A-2. FS-LIDS.

AN/TPQ-50 Multi-Mission Radar

An AN/TPQ-50 multi-mission radar actively detects and tracks UAS while the Nighthawk electro-optical infrared camera and the Counter-Unmanned Aircraft Electromagnetic Warfare System provides secondary location, verification, identification, and classification of detected UAS. The Counter-Unmanned Aircraft Electromagnetic Warfare System suite defeats UAS threats through its EA capability to disrupt or jam threat UAS position, navigation, and timing, command link, and video downlink signals. The FS-LIDS has lethal capability when coupled with the Coyote family of UAS interceptors.

FAAD C2 *(See p. 9-31.)*

The Forward Area Air Defense Command and Control system provides engagement operations, aircraft avoidance, and fratricide prevention. Forward Area Air Defense Command and Control system enables air defense C2 system interoperability with Link-16.

Mobile LIDS (M-LIDS)

The Mobile-Low, Slow, Small Unmanned Aircraft Integrated Defeat System (known as M-LIDS) is a C-UAS system that can detect, identify, track, and defeat UAS threats. It is mounted on two mine-resistant ambush protected vehicles for employment at the short halt. M-LIDS uses Forward Area Air Defense Command and Control system software and a mission data recorder connected via secure communications links. It includes one EW vehicle and one lethal defeat vehicle.

(U.S. Army Photo by Spc. Damian Mioduszewski.)

Electromagnetic Warfare (EW) Vehicle

- Incorporates LIDS based configuration on a mobile platform.
- Provides electronic detect, identify, track, and defeat capabilities.
- Incorporates a .50 caliber machine gun for protection with an electro-optical infrared camera gun system (target acquisition range 2 kilometers).

Kinetic Defeat Vehicle

- Provides detect, identify, track, and lethal defeat capabilities.
- Incorporates the Coyote Block 2 interceptor; XM914, 30-caliber machine gun; and M240, 7.62-millimeter medium machine gun.
- Mounts an electro-optical infrared camera gun system (target acquisition range 2 kilometers).
- Includes KuRFS mobile radar that provides air surveillance and fire control at the short halt.
- Vehicle should remain within 50 meters of EW vehicle with an active CREW system
- Vehicle must be cleared of personnel and material to 150m at +/- 40 degrees at the boresight of the launcher to account for sabot ejection at time of launch.

FAAD C2 *(See p. 9-31.)*

The Forward Area Air Defense Command and Control (FAAD C2) system provides engagement operations, aircraft avoidance, and fratricide prevention. Command and Control system enables air defense C2 system interoperability with Link-16

B. Ku-Band Multi-Function Radio Frequency System (KuRFS)

Ref: ATP 3-01.81, Counter-Unmanned Aircraft System (C-UAS) Operations (Mar '25), pp. 69 to 70.

The Ku-Band Multi-Function Radio Frequency System (known as KuRFS) is a proven, true multi-mission radar capable of providing simultaneous C-UAS, counter-rocket, artillery, and mortar (C-RAM), air surveillance and effector support missions that can be either mobile or palatized (see figure A-3 on page 70). The KuRFS radar provides target tracks for all categories of UAS and can discern air targets from ground targets. KuRFS provides sensing and fire control capability, delivering a clear air picture and improved probability of kill across an array of effectors, including the high energy laser (HEL) and EW effectors, and lethal-defeat missiles and guns for C-UAS such as the Coyote Weapon System, 50-caliber guns, 30-millimeter cannons, the Land-based Phalanx Weapon System (LPWS).

Mobile Ku-Band Radio Frequency System (KuRFS) (Photo credit: Raytheon.)

The KuRFS radar family consists of two options:

- The **palletized radar** is intended for fixed sites and can be packaged on a medium tactical vehicle class for transportable capability to remote and temporary locations. The palletized radar enables SHORAD/C-UAS protection of high-value assets and military facilities.
- The **mobile configuration** of the KuRFS, which fits within the size, weight and power limits of a mine resistant ambush protection all-terrain vehicle, joint light tactical vehicle or other light tactical vehicle. The mobile radar provides the range capability to support light expeditionary and small mobile unit operations.

C. Coyote Block 2C (Missile)

Ref: ATP 3-01.81, Counter-Unmanned Aircraft System (C-UAS) Operations (Mar '25), pp. 70 to 71.

The Coyote Block 2C is an expendable, rail/tube launched, turbine engine-powered missile for highspeed C-UAS missions. It is a ground-to-air effector proven against small maneuvering UAS threats with rapid time-to-target. It is command and control guided to target using KuRFS family of radars until the onboard seeker takes over for terminal engagement. Destruction of the UAS is completed with detonation of the low collateral damage, forward-firing blast fragmentation warhead. The KuRFS enables surveillance of the airspace near the Coyote Block 2C launcher to allow operators to verify that airspace is clear of friendly platforms.

FS-LIDS equipped with a lethal defeat solution includes a palletized KuRFS radar and a ***palletized Coyote missile pod*** *Ref: ATP 3-01.81 (Mar '25), fig. A-2. FS-LIDS kinetic defeat solution.*

Coyote Block 2C is integrated into both fixed site and mobile platforms for flexibility of mission employment. It can be ground launched, or vehicle mounted. The ground launch configuration is one 4-pack launcher; the mounted version is one two-pack launcher. Integration with Forward Area Air Defense Command and Control allows engagement of multiple targets, with multiple Coyote Block 2C warheads simultaneously.

D. Counter-Remote Control Model Aircraft Integrated Air Defense Network (CORIAN)

Ref: ATP 3-01.81, Counter-Unmanned Aircraft System (C-UAS) Operations (Mar '25), pp. 71 to 72.

The Counter-Remote Control Model Aircraft Integrated Air Defense Network (known as CORIAN) is a C-UAS system that can detect, identify, track and defeat threat UAS. The system is configured with multiple EW capabilities enabling non-lethal engagements. The system consists of passive and active sonar nodes for detection, geolocation and disruption of UAS in its target library.

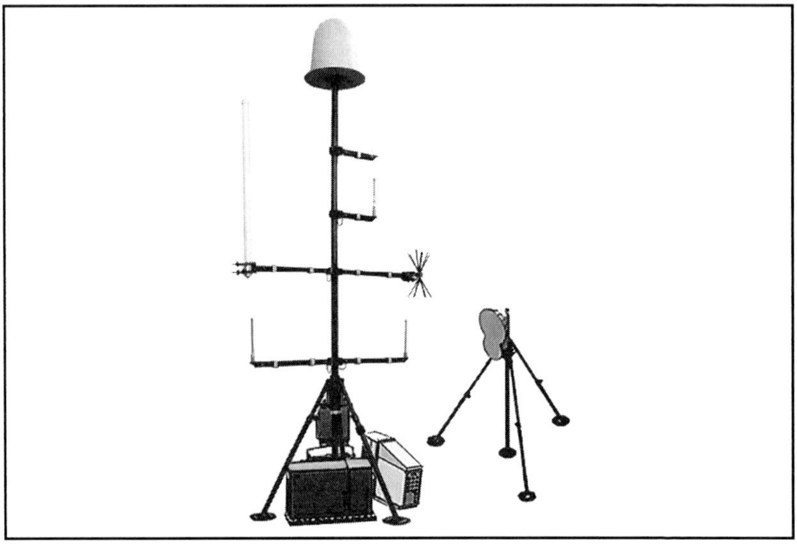

Ref: ATP 3-01.81 (Mar '25), fig. A-5. Counter-Remote Control Model Aircraft Integrated Air Defense Network (CORIAN).

The CORIAN system-

- Provides continuous, automated monitoring in all weather conditions.
- Recognizes and tracks UAS radio links to identify and locate UAS operating in restricted or protected airspace.
- Can locate UAS ground control stations.

Passive sensors-

- Detect and geolocate targets in a 360-degree field of view.
- Create an electronic perimeter detection boundary.
- Provide time-difference of arrival geolocation information based on sensor data collected by three or more sensors and displays probability ellipses on the graphic user interface map.

Active Sensors-

- Disrupt non-Wi-Fi targets by attacking C2 uplinks and targeting RF bands.
- Can attack a single target or jam a wide area depending on type of attack +-initiated..

E. Negation of Improvised Non-State Aerial-Threats (Ninja)

Ref: ATP 3-01.81, Counter-Unmanned Aircraft System (C-UAS) Operations (Mar '25), pp. 72 to 73.

The Ninja is a C-UAS system that can detect, identify, track and defeat threat UAS using a library based, RF capability. The Ninja disrupts threat UAS communications through RF detection, tracking and EA.

Ref: ATP 3-01.81 (Mar '25), fig. A-6. Negation of Improvised Non-State Aerial-Threats (NINJA).

> The Air Force Research Laboratory's Negation of Improvised Non-State Joint Aerial Threats (NINJA), an Air Force program of record, is designed to detect, track, locate, identify, and defeat UAS threats using the radio frequency communications between the drone and the operator. In congressional testimony in 2023, then-Secretary of the Air Force Frank Kendall said that 99 Air Force locations had installed the NINJA system.
>
> *- CRS Report R48477*

The Ninja-

- Passively detects, identifies, and tracks UAS.
- Actively defeats hostile UAS without indiscriminate interference to friendly systems.
- Enables multiple defeat options depending on threat signal protocols.
- Can operate as a stand-alone system or networked with other Ninja nodes, compatible electronic jammers (EGON, DRAKE, Duke), or autonomous interceptors.
- Ninja also has a mobile variant (Ninja Transmit and Receive) and a dismounted variant (Ninja Manpack).

F. Windtalker System

Ref: ATP 3-01.81, Counter-Unmanned Aircraft System (C-UAS) Operations (Mar '25), pp. 74 to 75.

The Windtalker system provides warfighters the ability to detect, identify and track DJI devices. It provides long-range detection of DJI aircraft and associated ground control stations and can identify whether the UAS is transmitting a valid identification, friend or foe IFF code.

It is easy to operate and low maintenance. The system user interface can provide situational awareness across multiple sensors through a cloud-based infrastructure. The system is internet based and requires secure login credentials to operate, enabling secure, unclassified usability by the warfighter.

Ref: ATP 3-01.81 (Mar '25), fig. A-8. Windtalker.

G. Anvil (Quadcopter)

Ref: ATP 3-01.81, Counter-Unmanned Aircraft System (C-UAS) Operations (Mar '25), pp. 73 to 74.

Anvil is a quadcopter that when cued by the system, navigates autonomously to intercept potential threat UAS. It provides visual feedback to the operator, employing artificial intelligence (AI) and computer vision technology, to obtain positive identification. Once authorized, the Anvil uses physical speed and onboard terminal guidance to deliver lethal effects to the threat with minimal collateral damage

Ref: ATP 3-01.81 (Mar '25), fig. Figure A-7. Anvil.

H. High Energy Laser (HEL)

Ref: ATP 3-01.81, Counter-Unmanned Aircraft System (C-UAS) Operations (Mar '25), pp. 76 to 77.

The high energy Laser (known as HEL) uses pure energy to detect, identify and defeat UAS threats . HEL employs a multi-spectral targeting system that emits beams of light to defeat hostile UAS. The system is mounted on a Polaris MRZR all-terrain vehicle.

Ref: High energy laser (HEL)

The HEL has-

- Capability of targeting a single UAS with precision.
- Intuitive remote single user interface.
- Recharge capability for continuous operation.
- Ability to accept target cueing from existing surveillance, tracking radars.
- Modular architecture that can be integrated on multiple ground or airborne vehicles in palletized or podded configuration.

I. FIM-92 Shoulder-Fired, Man-Portable Air Defense System (MANPADS) Stinger

Ref: ATP 3-01.81, Counter-Unmanned Aircraft System (C-UAS) Operations (Mar '25), pp. 77 to 78.

The FIM-92 man-portable air defense system (MANPADS) Stinger is an infrared homing shoulder-fired, portable surface-to-air missile. The missile can be deployed for defense against fixed- and rotary-wing aircraft, and UAS. The handheld launcher consists of a launch tube, battery coolant unit, separable gripstock assembly, sight assembly and separate IFF interrogator. The entire system weighs 35 pounds. The Stinger is operated by two operators and can be employed in all weather conditions.

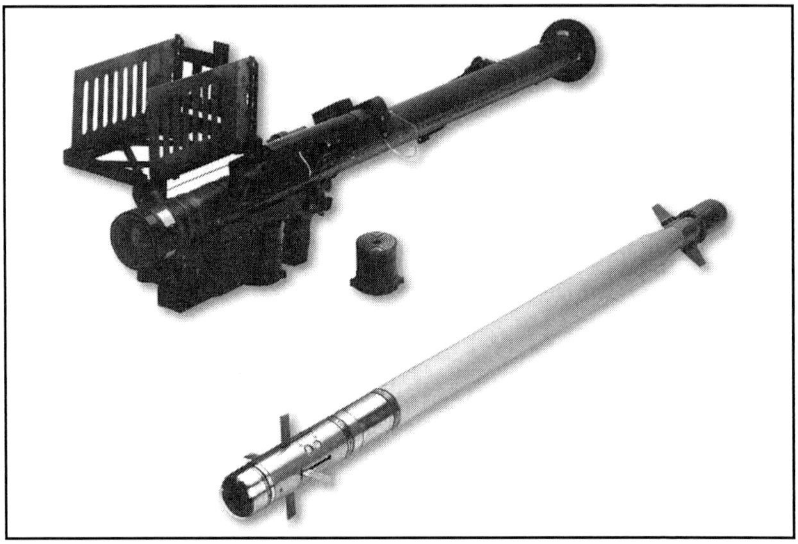

Ref: ATP 3-01.81 (Mar '25), fig. A-12. Stinger FIM-92 man-portable air defense system.

The MANPADS capabilities include-

- Fires basic and reprogrammable micro-processor block 1 missiles with no modification required.

Gripstock:

- Contains all pre-launch electronics and IFF antenna for MANPADS use.
- Reusable for multiple firings.

Battery coolant unit:

- Contains thermal battery and argon gas coolant.
- Expendable; supports one missile launch.

IFF Interrogator:

- Contains IFF system electronics and removable power source.
- IFF compatible with U.S./NATO equipment.
- Integrated Forward Area Air Defense Command and Control system Software.

J. Bal Chatri

Ref: ATP 3-01.81, Counter-Unmanned Aircraft System (C-UAS) Operations (Mar '25), pp. 78 to 79.

The Bal Chatri is a passive RF detection capability used to detect and identify threat UAS. The system employs a software defined radio RF detection system used to detect and identify enemy UAS. The radio can be configured for body-worn detection capability or for a small, fixed site location. Bal Chatri can also be integrated with a counter radio-controlled improvised explosive device electronic warfare system such as Modi. Limitations, system excludes automatic target detection and tracking source. Also, the operator must determine general directions of the target prior to system transmit.

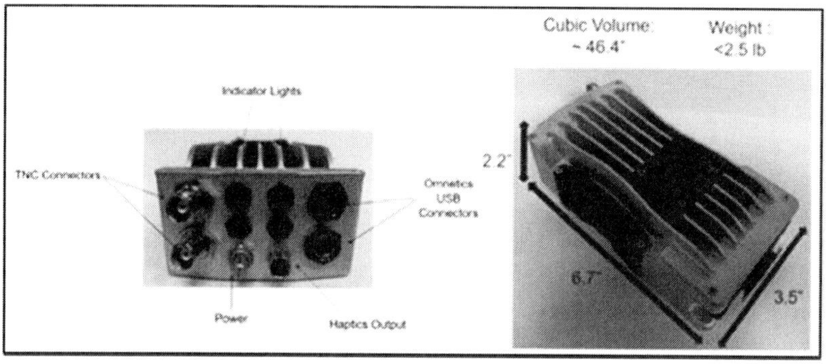

Ref: ATP 3-01.81 (Mar '25), fig. A-13. Bal Chatri.

System requires continual battery charging intervals to maintain system readiness. Bal Chatri Version 2 is a passive RF detection capability used to detect and identify enemy UAS. This system provides—

- Multi-band inter/intra team radio battery (up to 4 hours of battery life).
- Alternate power adapter provided.
- Universal serial bus and wireless networking capabilities.
- Built-in alerts.
- Improved light emitting diode visibility; night vision compatible.
- Ruggedized construction: meets defined test specifications for dust, water immersion, sand, vibration, and shock/drop.
- Flexibility to support multiple configurations and interface with multiple external systems for extended detect and defeat capabilities.
- Custom hardware includes dual simultaneous receive channels, high-speed detection scans (<5 seconds), and increased detection ranges.
- Spectral analysis and archiving tools.
- Also be integrated with counter radio-controlled improvised explosive device electronic warfare system such as MODI.

K. Dronebuster

Ref: ATP 3-01.81, Counter-Unmanned Aircraft System (C-UAS) Operations (Mar '25), pp. 79 to 80.

The Dronebuster provides detect and defeat capability against threat UAS. It is a lightweight, handheld, battery operated, dismounted, EA device specifically designed to defeat UAS. The Dronebuster requires one person and is a standalone system. User interface is controlled via switches on the unit and a trigger. Light emitting diodes provide visual cues of the reporting status of the unit, its battery status, and the presence of a target enemy UAS downrange.

A U.S. Army paratrooper assigned to the 173rd Airborne Brigade uses a Dronebuster 3B to disrupt enemy drones as part of Exercise Shield 23, April 20, 2023 in Pula, Croatia. (U.S. Army photograph by Sgt. Mariah Y. Gonzalez.)

The unit exploits weaknesses in UAS communication protocols enabling the operator to jam the control signal forcing the unmanned aircraft to execute pre-programmed "lost signal" protocols. The system—

- Defends against groups 1 and 2 UAS.
- Operates strictly LOS.
- Requires the operator to keep "eyes on target" throughout engagement.
- If LOS to an engaged target is broken, control may still be regained.
- Designed to disrupt remote control (915 MHz, 2.4 GHz, and 5.8 GHz) and PNT guided UAS from approximately 400 meters.
- Jams UAS command and control and PNT links simultaneously, depending on system settings.
- May or may not have effects on target depending on UAS flight mode/capability.

L. Smart Shooter

Ref: ATP 3-01.81, Counter-Unmanned Aircraft System (C-UAS) Operations (Mar '25), p. 80.

The Smart Shooter enables the operator to defeat UAS threats. It is a rail mounted, smart electro-optic developed to enhance lethality of units engaged in C-UAS operations. Smart Shooter is compatible with existing military rifles and can be mounted on any rail on a weapon system.

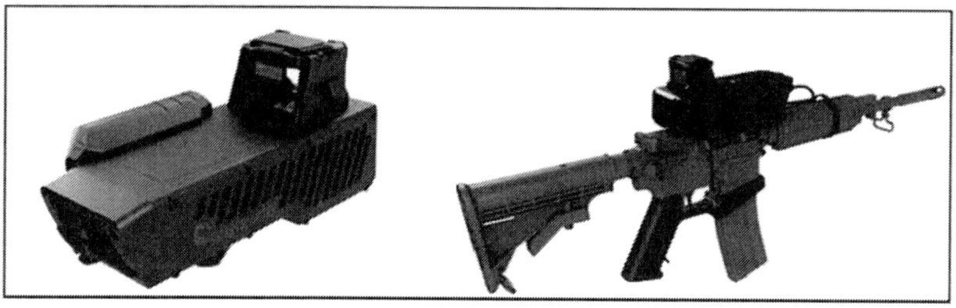

Ref: ATP 3-01.81 (Mar '25), fig. A-15. Smart shooter.

When the Smart Shooter is employed it—

- Performs target lock, track and fire control capabilities.
- Provides a reliable push button targeting capability using an internal computer and image processor to deliver real-time firing solutions and a trigger interrupt feature (assisted firing).
- Can track and hit group 1 UAS up to 120 meters.
- Uses image processing to automatically acquire a target from the sight's field of view.
- Displays a box around the target in the shooter's reflex sight.
- Enables the shooter to select and lock on to the target.
- Will fire only when the sight is aligned to hit the target (includes required "lead" on a moving target).

M. Modi

Ref: ATP 3-01.81, Counter-Unmanned Aircraft System (C-UAS) Operations (Mar '25), p. 81.

The Modi is a dismounted, man portable EW system that enables detect and defeat capability. Modi systems provide integrated tactical EW capabilities to defeat many UAS and other threats. The unit may be employed solely as a dismounted system or, using a mounted power amplifier, may be used in a fixed site or mounted configuration and dismounted when required.

(Photo credit: Sierra Nevada Corporation.)

The Modi—

- Dismounted weighs 40.25 pounds (with pack).
- Operates at a wide temperature range (-20 to 60 degrees Celsius/-4 to 140 Fahrenheit).
- Can be integrated with multiple C-UAS and EW systems.

N. Forward Area Air Defense (FAAD) C2

Ref: ATP 3-01.81, Counter-Unmanned Aircraft System (C-UAS) Operations (Mar '25), pp. 81 to 82.

FAAD C2 is a real-time safety critical C2 software application providing a single integrated air picture, airspace coordination and deconfliction, and fire control to support. This can be interconnected with the sheltered and handheld subsystems, the ADAM cell, Medusa (USAF), Avenger and various other FAAD C2 subsystems. Additionally, FAAD C2 software provides a three-dimensional (3D) view of objects displayed on the Battlefield Situation Display (BSD) of a visual representation of the flying aircraft and airspace control measures in the operator's area of responsibility.

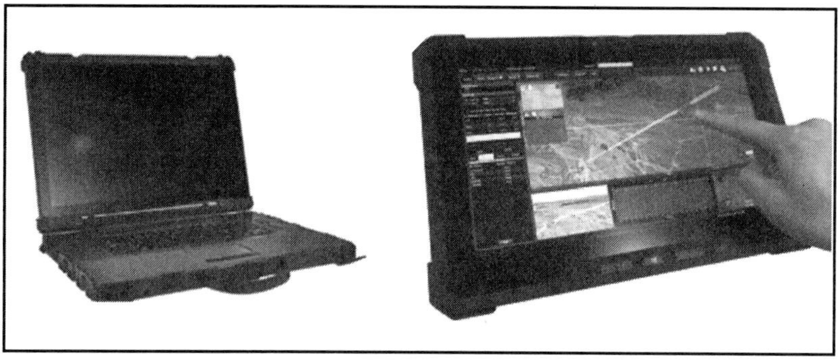

Forward Area Air Defense Command and Control (FAAD C2) provides critical C2, situational awareness, and automated air track information by integrating engagement operations software to support multiple missions. FAAD C2 supports air defense and Counter-Rocket, Artillery, Mortar (C-RAM) weapon systems engagement operations. The system also performs C2 engagement operations. (U.S. Army Acquisition Portfolio 2023-2024, p. 52.)

Missions that FAAD C2 support including:

- Counter-Rocket, Artillery, Mortar Intercept Land-Phalanx Weapon System
- Maneuver-Short Range Air Defense (M-SHORAD)
- Low, Slow, Small, UAS, Integrated Defeat System (LIDS)
- Avenger

FAAD C2 provides an air picture to engagement and force operations via government-owned interfaces and based on data gathered from various sensors. The FAAD C2 is integrated with over 25 sensors, including:

- AN/MPG-64 Sentinel
- AN/TPQ-50 Lightweight Counter Mortar Radar
- AN/TPQ-53 Counterfire Target Acquisition Radar
- Indirect fire protection capability
- High energy laser
- High power microwave
- Low, Slow, Small, UAS, Integrated Defeat System (LIDS)
- Ku-band Radio Frequency System Radar
- Lightweight Surveillance Tracking and Acquisition Radar
- Various EO/IR Systems.

O. RAJANT

Ref: ATP 3-01.81, Counter-Unmanned Aircraft System (C-UAS) Operations (Mar '25), pp. 82 to 83.

RAJANT utilizes wireless networking to form a mesh network. The network is mobile, self-integrating, self-meshing, self-healing, and secure. The RAJANT provides flexibility, adaptability, and simplicity ensuring rapid deployment of a broadband wireless network into any situation. The network can be deployed as a stand-alone wireless network or bridged to another network (such as the Internet) utilizing available communication links such as a digital subscriber line, cable, or satellite modem. The RAJANT provides high bandwidth for applications to stream video, audio as well as data over large distances. The network traffic can be secured by using a selection of available security features. This makes the network optimal for tactical deployments and mission critical activities. Example of the RAJANT and components shown in figure A-18.

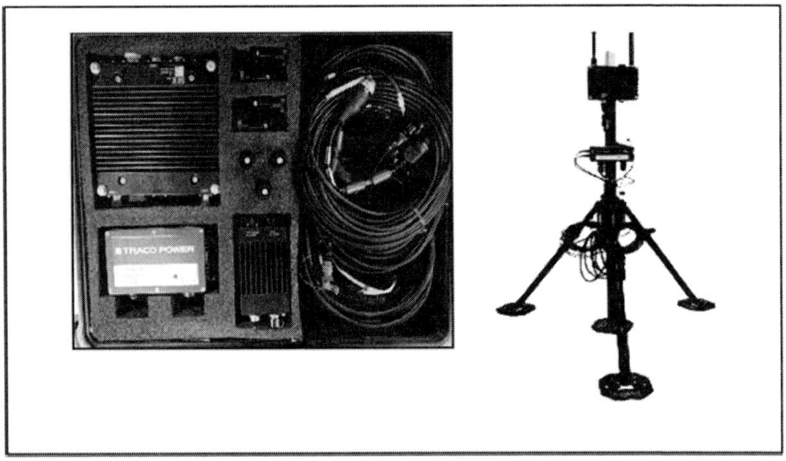

Ref: ATP 3-01.81 (Mar '25), fig. A-18. Rajant.

C-UAS/SUAS(1) Index

Index

D

E

F

G

H

I

R

S

T

U

V

W

Purchase/Order

www.TheLightningPress.com

SMARTsavings on SMARTbooks! Save big when you order our titles together in a SMARTset bundle. It's the most popular & least expensive way to buy, and a great way to build your professional library. If you need a quote or have special requests, please contact us by one of the methods below!

View, download FREE samples and purchase online:

www.TheLightningPress.com

Order SECURE Online
Web: www.TheLightningPress.com
Email: SMARTbooks@TheLightningPress.com

24-hour Order & Customer Service Line
Place your order (or leave a voicemail)
at 1-800-997-8827

Phone Orders, Customer Service & Quotes
Live customer service and phone orders available
Mon - Fri 0900-1800 EST at (863) 409-8084

Mail, Check & Money Order
12144 Stonelake Ranch Blvd.
Thonotosassa, FL 33592

Government/Unit/Bulk Sales

The Lightning Press is a **service-disabled, veteran-owned small business**, DOD-approved vendor and federally registered—to include the SAM, WAWF, FBO, and FEDPAY.

We accept and process both **Government Purchase Cards** (GCPC/GPC) and **Purchase Orders** (PO/PR&Cs).

Keep your SMARTbook up-to-date with the latest doctrine! In addition to revisions, we publish incremental "**SMARTupdates**" when feasible to update changes in doctrine or new publications. These SMARTupdates are printed/produced in a format that allow the reader to insert the change pages into the original GBC-bound book by simply opening the comb-binding and replacing affected pages. Learn more and sign-up at: **www.thelightningpress.com/smartupdates/**

VISA MasterCard PayPal